IMS: A Development and Deployment Perspective

IMS: A Development and Deployment Perspective

Khalid Al-Begain
University of Glamorgan, UK

Chitra Balakrishna
University of Glamorgan, UK

Luis Angel Galindo
Telefonica, Spain

David Moro
Telefonica, Spain

A John Wiley and Sons, Ltd., Publication

This edition first published 2009
© 2009 John Wiley & Sons Ltd.

Registered office
John Wiley & Sons Ltd, The Atrium, Southern Gate, Chichester, West Sussex, PO19 8SQ, United Kingdom

For details of our global editorial offices, for customer services and for information about how to apply for permission to reuse the copyright material in this book please see our website at www.wiley.com.

Library of Congress Cataloging-in-Publication Data

IMS : a development and deployment perspective / Khalid Al-Begain . . . [et al.].
        p. cm.
  Includes bibliographical references and index.
  ISBN 978-0-470-74034-7 (cloth)
 1.   Multimedia communications.   I. Al-Begain, Khalid.
  TK5105.15.I56 2009
  006.7–dc22

                                                      2009020178

A catalogue record for this book is available from the British Library.

ISBN 978-0-470-74034-7 (H/B)

Set in 10/12 Times New Roman by Laserwords Private Ltd, Chennai, India.
Printed and Bound in Great Britain by CPI Antony Rowe, Chippenham, Wiltshire

Contents

Preface

There are many books published to date covering the technological aspects of next-generation networks in mobile and fixed domains. These books explain in depth the architectures, protocols, interfaces and procedures as defined by the standardization bodies such as 3GPP, OMA, IETF and others. These standards prescribe a new generation of network architecture by deploying a signalling layer over the legacy infrastructures. They have enabled the evolution of the horizontally layered, all-IP, network architecture from the traditional vertically oriented stove-pipe networks.

The obvious question then is 'Why another book?' From the time the writing of the book began, the goal was clear not to produce just another book but to create a book that differed sufficiently from all the currently published work by way of providing a new understanding of the next-generation networks. The book aims to present the technology prescribed by the standards from service-centric and user-centric perspectives through innovative services examples and new business models. Finally, a new way of thinking needed to redefine the relationships between the various actors has been put forward.

The landscape of convergence can be viewed from two different perspectives. The first one is from the telecom perspective. The telecom industry players, such as operators, service providers, product manufacturers, vendors and application developers, have adopted a network-centric approach. The focus is on creating a new network and a new service platform rather than on providing innovative services catering to end-user requirements and demands. By taking such a stand, the telecom players are clearly retracing the path followed by the GPRS and the UMTS launches, where the operators marketed the technology instead of services. The only difference now is that the new network platform, called the IP multimedia subsystem (IMS), is deployed but is not being commercialized.

The other view of convergence is from the Internet perspective. In contrast to the approach of the telecom industry, the Internet has always followed a service-centric approach. The second version of the web called Web 2.0 has brought about a new paradigm, where the user takes the central role and service-oriented approach is now evolving into a user-oriented approach. With this new user-centric dimension, Internet service players are developing and deploying new services and applications best-suited to user's needs.

Influenced by the success of the Internet, both in terms of the service model as well as the economic model, telecom operators intended to gain control over the Internet by deploying IMS. Consequently, they created a parallel 'walled-garden' Internet that is controlled and governed by the operators. However, the social changes brought about by the new Internet created a confrontational atmosphere in the Telecom–Internet arena. This led to both players fighting to gain the central position by providing end-to-end services regardless of their previous expertise. This in turn raised questions on the usefulness and the future of the IMS.

Traditionally, some experts on IMS have defended the technology as being crucial for the convergence between fixed and mobile networks. However, what is imminent and of significance now is the convergence of the telecom and Internet worlds. In order to view IMS as more than just a platform for convergence, the pioneering initiative such as WIMS 2.0 is redefining the path to follow for survival in the convergence era.

To cover the above-mentioned aspects, the book has been divided into three parts. The first two parts of the book introduce those concepts that form pillars of the new network. They provide an understanding of IMS from service-centric and user-centric perspectives as well as the path ahead for IMS and web convergence. Evolution to an all-IP architecture is discussed along with the presentation of IMS as a service delivery framework and as a service integration enabler. The service architecture of the future network along with the service possibilities is discussed. IMS deployment concerns and the interoperability efforts are presented, all of which could be benefit engineers working on IMS technology, application developers, operators and the service providers who are on their road to IMS deployment. The distinctive core features of the book then follow, which explain new innovative services merging telecom and the Internet, define how and why a new and fair ecosystem is essential and define taxonomy of existing business models in the new merged world. All this is presented from the perspective of the WIMS 2.0 initiative, a world pioneering initiative.

The last part of the book reviews the new trends that may generate new business revenues in the new convergent environment; some of them promote the WIMS initiative. New devices like iPhone or software like Android are currently modifying the market. The openness of telecom capabilities to developers, the trends of digital marketplaces or the strategies on user generated services are some of the examples that validate the strategy presented in this book. The book will provide significant help to those operators who are defining the strategy of the companies in the convergent sector, to those service providers who are deciding the services to be delivered to the end user as well as to the subset of researchers and developers whose focus areas are on the converging telecom and Internet worlds.

Author Biographies

Prof. Khalid Al-Begain is the Director of the Integrated Communications Research Centre (ICRC) at the University of Glamorgan, UK. He received his PhD in Communications Engineering in 1989 and has been working in different universities and research centres in Jordan, Hungary, Germany and the UK.

He is the Chairman of the European Council for Modelling and Simulation, UNESCO Expert in networking, Senior Member of the IEEE, British Computer Society Fellow and Chartered IT Professional, member of the UK EPSRC College (2006–2009), Wales Representative to the IEEE UK&RI Computer Chapter MC, and UK representative to the COST290 Action management committee. He has led and is leading several projects in mobile Computing, multimedia multicasting wireless networking and performance evaluation funded by UK EPSRC, EU FP6 and FP7 projects.

Al-Begain has been the General Chair of twelve international conferences including two in 2009. In particular, he is the founder and general chair of the IEEE International Conference on Next Generation Mobile Applications, Services and Technologies with the third to be held in Cardiff in September 2009.

He has edited 10 books, co-authored one book and has more than 150 papers in refereed journals and conferences to his credit.

He is currently leading the project on the IMS based next generation service and applications creation environment with a bid of £8m supported by major industries.

Dr. Chitra Balakrishna received her Bachelor's degree in Telecommunications Engineering at Bangalore University, India. She has worked as a Network Engineer at the prestigious Education and Research Network Project at the Indian Institute of Science in Bangalore, India. She was part of a very talented team that pioneered in planning and establishing a Network Operation Centre [NOC] that connected all the Research Institutes of India. She has also worked as a freelance network consultant in IP Data Networking projects involving open source technologies. Following which, she was sponsored by the South-Asia scholarship to complete her Master's degree in Data Telecommunication and Networking at the University of Salford, Manchester, UK in 2004. She joined the University of Glamorgan as a Member of Research at the Integrated Communications Research Centre (ICRC) in 2004. At Glamorgan, she has spearheaded the setting up of a Next-Generation Service Creation Test Facility with the support from the industrial partners such as Orange, France Telecom R&D, Dialogic and Avaya. In 2008, she defended her Doctoral Thesis in the area of Service Enabler-based Quality Adaptation for Streaming Applications on the IP Multimedia Subsystem. She has been part many Next-Generation Service Creation projects, and has had many international publications to her credit. She has co-founded and chaired an IEEE-sponsored workshop on Open source Technologies and Next-Generation Networking.

Mr. David Moro received his Master Engineer Degree in Telecommunication Engineering in 2001 from Valladolid University, Spain. Previously, in 2000 he joined the Mobile Multimedia Systems department of Telefonica I+D performing tasks related to short-range radio wireless technologies and their applications for mobile networks. From 2001, he has been working for the Next Generation Mobile Services department in Telefonica I+D, involved in areas such as 3G packet network evolution, IMS and the service enabler architecture. As project manager, he has lead several consultancy activities for Telefonica Moviles related to the selection and roll-out of IMS platforms and solutions, as well as providing assistance in Telefonica relationships and participation in standard development organizations such as 3GPP, OMA and ITU. He also co-operated in pan-European R&D in the context of the European Union Framework Programmes for Scientific Research, in major Information Society projects such as Daidalos and WWI Ambient Networks. From 2007 onwards, he assumed a new role as technology expert in the NGN Service Platform area of Telefonica I+D, driving technical specifications for Service Delivery Platforms for Telefonica group, and researching in fields related to telco convergence with Web 2.0, open service capabilities and the Future Internet of Services. He is co-founder of the WIMS 2.0 initiative.

Mr. Luis Angel Galindo has an Executive MBA from IESE Business School, a Masters in Services and Security in IP networks from Technical University of Madrid, a Masters in Production Management and an M.Sc. degree in telecommunications from the UPM.

From the beginning he has been involved in wireless networks, firstly in Motorola, and from July 1996 in Telefónica Móviles Spain. He was the IMS Project Director in Telefónica.

He is the founder of the WIMS 2.0 initiative (www.wims20.org), where telco and internet worlds are merging to create innovative services and business models, that he is personally leading.

Luis is a socio-technologic, analyzing how technologies are modifying behaviours and trends in society. He is responsible for bringing innovation to the user.

In parallel, he is leading a new way of working with third parties, based on trust relationships with the aim of reducing time to market and capitalizing innovation.

With high innovative and creativity skills, Luis is a promoter of change inside his organization. Luis is a consultant in 2.0 strategy issues, where he has led several projects migrating companies from traditional ways of organizing to turn them into 2.0 organizations.

He represents Telefónica in different national and international congresses and fora. Luis is the author of many articles for technical magazines such as IEEE, GlobalComm, etc. He teaches in Spanish universities and also makes in-company courses for external enterprises. He is reachable at luisangel.galindo@gmail.com

Part I

Introduction

1

IMS Context

This chapter aims to lay a foundation to viewing the IMS from a service and deployment perspective. It is done by understanding the current IMS context and by discussing the past, present status and the future directions of IMS. The 'IMS-Past' represents the drivers of the convergence era, the origination of IMS architecture and the IMS-hype that followed. The 'IMS-Present' reveals that all the players of the communication industry are treading with caution with respect to IMS development and deployment. The chapter discusses the cause for the current status by presenting the misconceptions about IMS that exist among the telecom operators, service providers and Internet players. A reality check on the status of the IMS standards and commercial deployment is presented. The 'IMS-Future' represents the road ahead for IMS technology and its deployment. This is presented in the context of the possible convergence of IMS with web technologies.

1.1   Drivers of Convergence

Technological Drivers

The past decade has witnessed a rapid technological evolution in terms of network technologies, end-device technologies and multimedia related technologies. Figure 1.1 depicts the recent enhancements in data rates in the mobile networks and fixed networks. The evolution path indicates that the latest cellular access technologies such as UMTS and HSDPA would provide a connection bandwidth of 300 kbps to 10 Mbps and the future cellular technology based on orthogonal frequency division multiplexing (OFDM) provide data rates up to 50 Mbps. Similarly fixed access technologies such as ADSL and xDSL also provide data rates up to 10 Mbps. This implies that the data rate experienced by end-users connected either to mobile or fixed access networks is converging at 10 Mbps and beyond. In addition, multimedia technologies have evolved in terms of video compression technologies in the series of MPEG-1, MPEG-2, MPEG-4 and H.264 (Furht, 1995; Richardson, 2003). Most recent codecs, such as MPEG-4, can transmit DVD/VCR quality video. Lastly, the end-user

IMS: A Development and Deployment Perspective   Khalid Al-Begain, Chitra Balakrishna, Luis Angel Galindo and David Moro
© 2009 John Wiley & Sons, Ltd

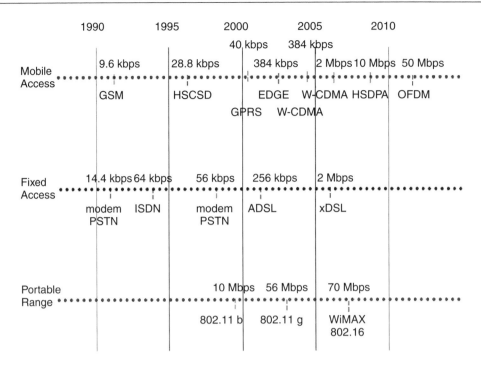

Figure 1.1 Evolution in mobile, wireless and fixed access data rates.

devices have evolved into high specification mobile devices with high processing speeds, memory and screen resolutions that include multimedia capabilities and support for media capture and playback features. They also come with a built-in java-enabled web browser such as ThunderHawk (2008). The above technological evolution has popularized multimedia services, Internet and Internet-like services among end-users. Furthermore, flat-rate pricing for unlimited access to data on 3G mobile networks as well as on the fixed broadband access networks has now become common practice among operators. This has only hastened the demand for multimedia applications. The result is the rapid shift in the communication content, from text-based to multimedia content. The nature of the applications demanded by end-users has metamorphasized, from being purely text-based applications on the Internet and purely voice-based applications on the cellular-telecom network to seamless multimedia-based applications on cellular networks and the Internet.

This communication revolution is further driven by the end-users and their changing usage pattern:

- Use of PC and Internet to make voice calls, and the use of mobile devices for accessing multimedia streaming and other Internet services.
- The use of a single device to access multiple services (e.g. mobile devices to access emails, multimedia streaming, voice, etc.) and the access of a single service from multiple devices (voice call continuity from a mobile phone to a fixed telephone) has driven 'device convergence'.

The changing user preferences have led to a whole new expectation. The end-users expect applications that run on a desktop computer to be ubiquitously available, and have the same quality and presentation (look and feel) regardless of the user being a fixed, mobile or nomadic user. This is also termed as the anywhere, anytime, anyhow communication paradigm. The effect of this changing expectation resulted in the evolution of the underlying networks to keep pace with the end-user demands. This therefore triggered the need for networks to transform from old paradigms to newer more cost-effective ones.

Economic Drivers – The Bit-Pipe Threat

The roll-out of wireline and wireless broadband access networks has lowered the entry barrier for VoIP service providers, allowing them to offer lower priced telephony services to consumer and enterprise customers. The average voice revenue per user for the telecom and cellular service providers is on the decline; hence many service providers are looking at alternative IP-based multimedia communication applications to replace the lost revenue. This has caused a strong shift in focus within the communication industry from digital telephony to more media rich applications (O'Connell, 2007).

Some important inferences can be made from tracing technological revolution, changing user preferences and economic trends:

1. The Internet is the largest and most popular public data network based on packet-switched technology (Cerf, 2002), while the telecom network is the largest voice network based on circuit-switched technology.
2. With the worldwide deployment of circuit-switched cellular networks (ITU, 2006) and with the growing subscriber base for mobile operators (CTIA, 2005), coupled with the availability of mobile devices with high processing speed and memory, the services offered are not limited to voice alone.
3. There is a proliferation of multimedia communications and the demand for Internet services on mobile devices.
4. There is a clear shift from fixed communication to mobile communication leading to the demand for applications and services to be delivered to end-users regardless of the type of network they are connected to and the type of access device they use.

From the above inferences, it is only natural to expect the future network architecture to bridge the divide between the circuit-switched cellular network and the packet-switched Internet and consolidate them into a single network providing a common set of services to users anywhere, anytime. On the one hand, the above demand is driving the convergence of the telecom and the Internet world, while on the other hand it threatens to turn the telecom network operator's infrastructure to mere bits-carrying pipes (Cuevas *et al.*, 2006). In order to avoid this, the standard bodies such as 3GPP (1999d) designed a converged networking architecture called the IP Multimedia Subsystem (3GPP, 2007g). The IMS architecture is designed to deliver applications to subscribers in an access-agnostic manner through a common IMS core network. The IMS architecture is service-agnostic and location-agnostic as well, allowing network operators to realize a truly converged core that removes service silos and reuses common functions. However, the ultimate success of IMS is dependent upon the services and not the networking core. The service logic resides in the *Service Layer*, which provides service enablers that can be blended for accelerated service creation and service roll-out. Thus, IMS enables the network operators to deliver innovative, personal, value-added services acting as more than a bit-pipe carrier, thereby enabling the telecom operators to play a significant role in service provision as well as in service creation.

1.1.1 Origin of IMS

The conceptualization of IMS was motivated by the success of the Internet as presented in the section above. Multimedia services on the packet-switched technology of the Internet experienced tremendous success. This instigated the cellular operators to introduce multimedia services on their largely voice-centric cellular networks. Cellular access networks had evolved from being a purely circuit-switched technology to third generation (3G) wireless networks that could support high speed data, voice and multimedia services on packet-switched technology (IP). 3G wireless networks were defined by the Third Generation Partnership Project (3GPP). The core network was divided into a circuit-switched domain based on the global system for mobile communication (GSM) and a packet-switched domain based on the general packet radio services (GPRS). The GPRS domain enabled users to access multimedia services and Internet applications using the IP protocol. 3G wireless access networks provided data rates up to 144 kbits/s for users moving at vehicular speeds, while users at pedestrian speeds had access to a maximum data speed of 384 kbits/s and stationary users could avail up to 2 Mbits/s (Chen and Zhang, 2004). However, bandwidth of the 3G cellular access networks (viz. GPRS, 3G) was still too scarce to support real-time multimedia services. Along with the network-related demands of multimedia services, cellular operators also had to deal with challenges posed by mobility where users roam across different access networks and administrative domains. Last but not the least, supporting end-to-end QoS across 3G wireless networks and the Internet was a complex task. IMS was introduced by the mobile operators in order to provide ubiquitous access to multimedia services on mobile and wireless access networks with guaranteed QoS, with service personalization, with a single billing feature, and with other services that the Internet failed to provide. IMS is the first and the only service control architecture that was standardized by the Third Generation Partnership Project in Release 5 (3GPP, 2007g). Right from its conception, IMS was not designed to compete with the Internet, but to complement the Internet with features from the telecom world. IMS was conceptualized to merge the cellular world and the Internet world in the true sense. It aimed to use the cellular technologies to provide seamless/ubiquitous access and Internet technology to provide innovative multimedia services, while the IMS's service control architecture was designed to provide the required QoS, charging and customization that were not possible on the Internet.

Convergence does enable the new personalized service model and offers many technological advantages to the operators, product vendors and end-users. However, the operator's main concern is the need to minimize operating costs and to provide cost-effective migration to an all-IP network. From an operator's perspective, the network convergence goal is to migrate the existing isolated circuit and packet-switch core networks to a unified core network that supports existing access technologies in both the fixed and mobile domains. This evolution, although it may take some time to complete, will be key to an operator's ability to reduce OPEX in the long term and increase competitiveness and profitability.

Operators need a strategy that defines their business objectives and solutions that map the various short-, medium- and long-term goals. In other words, convergence is expected to evolve along a well-defined migration roadmap. The solution must support the introduction of IP-centric multimedia services that can be delivered to a variety of terminals. Delivery must be cost-effective and employ complementary access technologies. It must also bring operating costs down and allow traditional services and applications to be retained. The first step in the evolution path is to optimize the circuit-switched core network in order to improve the delivery of regular voice services. The second is to enhance the packet core network in order to enable it to deploy new value-added IP multimedia services rapidly in the most cost-effective manner. A key feature of this strategy is the use of common components and service-specific extensions that reduce the cost of service development and implementation. This is best achieved using IMS as the service delivery engine within the unified core network. This unified network is optimized for the efficient delivery of services to users and facilitates the interworking of business partners and other networks while effectively managing the operator's day-to-day operations.

1.1.1.1 Initial Hype and Reality

The IMS architecture was one of the most hyped developments within the networking industry in the years 2006 and 2007. However, the technology became a 'reality' in 2008. Early, conflicting information from some vendors and service providers, both in favour of and against, distorted the understanding of IMS. A detailed analysis on the misconceptions of IMS is presented in Section 1.2. Proprietary IMS architectures created the illusion of inherent interoperability, reliability or availability problems in IMS-based services, while numerous companies claimed IMS capabilities for products that delivered little or nothing new (Lightreading, 2008).

Proponents currently see IMS as an open and practical standards-based blueprint for IP-enabling the carriers. For example, carriers providing video over IP could easily use IMS to deliver branded IPTV or related services regardless of the customer's location and access device. IMS's ability to measure IP streams, provide QoS and utilize service intelligence allows the network to know what services to supply and how much bandwidth to use based on who is requesting them and whether the customer is watching IPTV in their living room or opening a video chat on a smart phone far from home. Despite the lack of consensus, most service providers have begun either exploring or deploying IMS architectures of various stripes in staging as well as working networks. RFPs are being generated, roll-outs are being scoped and IMS is delivering on its promise of a cost-effective common platform for delivering converged IP services over wireline, cable, DSL, GSM, UMTS, WiFi and WiMAX networks (IMSForum and NGNForum, 2008).

Objections to the IMS architecture on the grounds of its lack of interoperability, its immaturity or its complexity are no longer tenable (refer to Chapter 4 for a reality check on the current IMS deployment status). Work is ongoing to put additional protocols and architectures such as IPTV on to IMS. Operators are making slow but steady progress towards simplifying their network architecture to all IP-based routers and soft switches. Meanwhile, the true benefits and the business case for IMS are now beginning to become clear.

1.1.2   IMS – Convergence Enabler

Convergence is a generic term used to represent the communication revolution that is taking place. As already presented, the technological revolution has facilitated the convergence of devices with the emergence of multimode terminals (devices with multiple network interfaces) and mobile devices with an in-built camera, music player and wide LCD display screens for video, etc. Further, there is convergence at the content level where the services are a mixture of text, audio and video content generated by the Internet space (youtube), broadcast space (BBC i-player) or telecom space (ring-tone downloads). In this context, this section investigates how IMS as a converged network architecture acts as the convergence enabler. At the outset IMS-based convergence is technological, i.e. it is a known fact that IMS facilitates the technological convergence of various access technologies. From the network architecture perspective, IMS is the common standard for next-generation fixed networks adopted by TISPAN (2000), cable network standardized by PacketCable (2007a) and mobile networks standardized by the 3GPP and the 3GPP2 standards. This phenomenon is currently being explored in the communication industry as the fixed mobile convergence, also known as FMC (2004). It is possible to achieve FMC through the deployment of a unique IMS core network for fixed and mobile access. Moving from two separate fixed and mobile networks to a single shared one clearly has cost-related advantages, both from a CAPEX and an OPEX perspective. This architectural convergence would lead to a certain homogeneity between fixed and mobile networks and facilitate both the network providers to support very similar services and may eventually find synergies between the fixed and mobile network organizations of the operator and vendors. It is also safe to infer that the user experience would benefit from this homogeneous core network support. Such an understanding of IMS as only a fixed mobile convergence enabler is limiting. FMC unifies the separate fixed and mobile core networks alone; it still views the fixed and mobile network subscribers as different. In order for the FMC to impact end-users, it is important to achieve convergence at the service level and at the user level.

It can be contended that IMS, which facilitates convergence at the access technology level and at the core network level, also facilitates convergence at the service level and at the user level. Through IMS, it is possible finally to unify the mobile and fixed dimensions of end-users.

1.1.2.1   Service-Level or User-Level Convergence

The user-level convergence is enabled by the intrinsic capability of SIP (Rosenberg *et al.*, 2002), the signaling protocol for IMS. SIP provides public user identities and is prescribed in the IMS standards, enabling a user-level convergence. Refer to Section 2.2.3 for detailed information on IMS user identities.

In order to understand the concept of user-level convergence, a brief description of IMS public user identity (IMPU) is essential. Typically, it is the home network operator that is responsible for the assignment of the private user identities and public user identities. Public user identities may be shared across multiple private user identities within the same IMS subscription. Hence, a particular public user identity may be simultaneously registered from multiple user equipments (UEs) that use different private user identities and different contact addresses. An IMPU can be either an SIP URI or a Tel URI, such as Sip: first.last@operator.com or Tel:+1-555-123-4567. The relationship for a public user

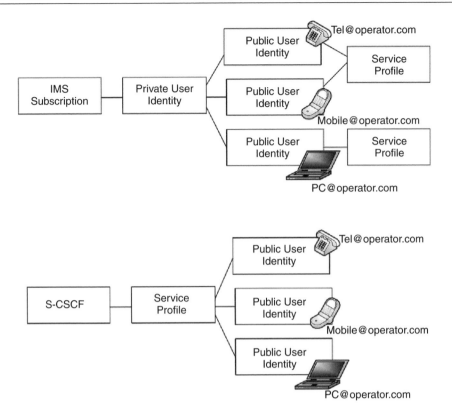

Figure 1.2 User identities and service profile relationship.

identity with private user identities, and the resulting relationship with service profiles and IMS subscription, is depicted in the Figure 1.2. An IMS subscription may support multiple IMS users.

The public user identities associated with the users are access independent. This implies that the IMPUs can be simultaneously registered from multiple user equipments that are connected to different access technologies.

For instance, a user who owns multiple personal communication devices such as a mobile phone, a fixed telephone and a PC can concurrently register the same identity from each one of these devices. When an SIP request is addressed to this identity, the network implements standard SIP mechanisms called forking to alert the user on different devices or to look for the best device to reach the user. This is performed by a mechanism termed a parallel forking or serial forking. The IETF RFCs relating to callee capabilities (the ability to register the services and features supported by a device) and caller preferences (the ability to instruct how forking should be performed according to callee capabilities) facilitate a more intelligent way to route the requests. IMS enables the IMPUs to be associated with so-called 'service profiles' stored in the network user database, the HSS. This service profile determines how SIP requests generated from or addressed to a particular identity would be routed to IMS applications. The service profile is the key for the door to IMS services. The fact that the service profile is associated with a user identity and that the user identity could be shared across multiple user devices that may be connected to multiple access technologies makes all services deployed on the IMS uniformly accessible from all user devices and all access technologies employed by the user. Therefore IMS ensures that all services hosted on IMS are inherently converged.

Service implementation logic can then ensure that the service is adapted to the specific characteristics of a device or the access technology currently employed by the user. It is important to note that such a user-centric fixed mobile convergence is totally network-based. It does not require the end devices to be able to connect to multiple access technologies. This means that IMS-based user-level convergence does not add any requirement on the devices other than their ability to register with and access IMS over IP.

Therefore, it can be inferred that the deployment of the IMS core network enables convergence at the transport-level (access-agnostic), at the core network-level (FMC) as well as at the user-level. All IMS-enabled devices making use of IMS-compatible access technology such as WiFi, WiMax, xDSL, GSM, UMTS, Cable can be converged at the service-level.

1.2 IMS Misconceptions

Convergence, fixed mobile convergence (FMC), all-IP architecture and next-generation networking (NGN) have been the buzzwords in the telecom industry for the past few years. The IP multimedia subsystem is a technology that has been common to the above-mentioned buzzwords. It offers open standards architecture, rapid service deployment features, network and service independence, reduction in operational expenses and many other benefits to operators, service providers and consumers. Despite the above technological and business advantages, it is a surprising fact that IMS is yet to witness large-scale commercial deployment. The lack of innovative multimedia services and the lack of bandwidth on the mobile access links to support real-time multimedia could be considered as the deterrents to the commercial deployment of IMS. Lack of support from the mobile device manufacturers has led to scepticism among mobile operators about the future of IMS. However, the deployment of high speed downlink packet access (HSDPA) allowing 3G wireless networks to have higher data transfer speeds and capacity up to a theoretical maximum of 14.4 Mbits/s have brightened the chances for IMS. This increased availability of data rate is expected to encourage mobile operators to provide multimedia-rich applications and eventually deploy IMS commercially. One critical reason that could be attributed to IMS still being in an explorative stage is the contradictory views about IMS that exist among the various players of the communication industry. The 3GPP specifications that could be interpreted in so many different ways have only added to this contradiction. One of the most prevailing misconceptions of IMS is that it re-creates the traditional telecom world over IP.

1.2.1 IMS – A Replacement or New or Complementary Domain

IMS was introduced in the telecom standards panorama as an enabling architecture for new multimedia person-to-person services that either could not be offered via the previously existing circuit-switching infrastructure or, by migrating it to the IP switching domain, could offer a capacity increase or cost reduction. The rationale for introducing this new technology in the telecom infrastructure in any operator network has to be driven by one or more business factors among the following:

- cost reduction, which will correspond to the IP-boosted increase of capacity for classical services;
- the new possibilities for enhanced or new services in interpersonal communication portfolio of services; or
- the enrichment of other service types, like content-oriented services and information access, by the synergetic application of the former communication services to the latter.

As reasoned in the next section, IMS was not targeted at all at replacing the services provided by the circuit-switching domain or as the IP flavour of the IN architecture of the legacy telephone networks, although design principles are of course influenced by the telco way of doing things, which comes from the era of circuits.

Apart from this, there are two related prominent misconceptions around and caused by IMS. Firstly, there is the belief that IMS will be utilized to provide all kinds of services within a telco portfolio. Secondly, IMS constitutes a loosely coupled or even independent domain within the telecom service infrastructure, barely related to the other pieces affecting service delivery. It is depicted in Figure 1.3.

To clarify the first issue, services in a telco portfolio can be simply classified as follows:

- Broadband access to the Internet and intranets. In general, these are the connectivity services that are usually provided with broadband technologies for fixed and mobile: FTTH, xDSL, 3G family of evolved technologies, etc.
- TV, media content and infotainment services. This includes the delivery of IPTV services as well as media/content services for mobile devices like music stores, video

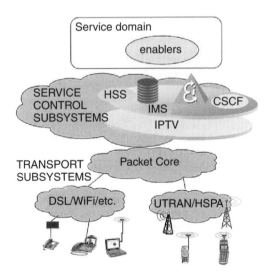

Figure 1.3 A convergent vision of a telco network including IMS.

streaming, content download including gaming, etc. In this category also falls any
kind of leisure information access services like mobile portals, mobile web browsing,
IPTV equivalents and value-added services for PCs offered on top of broadband
connectivity.
- Interpersonal communication services. This is the classical domain of expertise for a
  telcom operator, and includes all types of services that address the need for users to
  communicate and stay connected to other users, e.g. voice calls, video telephony, all
  flavours of messaging (SMS, MMS, etc.), sharing and so on.

In its original conceptual design, IMS is geared at the enhancement of the latter category. Precisely, IMS is an essential technology for a telco since it addresses the services
where a telco has historic strength and a profitable and solid business. IMS therefore cares
for making this communication business sustainable and improved in the future. Beyond the
classical voice call telephony and SMS messaging, IMS brings a new set of new services
such as the power of multimedia telephony, which enriches the indispensable conversational speech with other media; the instant sharing of multimedia contents and live streams
between communicating peers; the simple-to-use and ad-hoc conferencing; new ways of
communication, like the push-to-talk or conversational messaging, the unified experience of
multimedia exchange of information; collaboration tools and multipeer gaming; the presence
in everything, and all of its combinations. In a few words, IMS brings multimedia, instan-
taneity and spontaneity to characteristics of the core personal communications services of
a telco in order to modernize them and keep the match between the offered portfolio, the
eager ever-demanding user and never-fully-satisfied-with-operator customer.

After building the foundations of a technology to fulfil all these necessities and requirements, i.e. the IMS and associated technologies, the industry has been analysing and paving
the way to extend the IMS scope to the other two categories of services described before.
Obviously, IMS has nothing to do with the first category of services, connectivity. However, there is indeed a synergetic potential with the second category: the TV, media content
and infotainment services. The combination of personal communications with content and
media consumption can be obvious and immediate. Users enjoying content (video clips,
music, cinema, TV channels, information, etc.) usually feel the necessity to disseminate,
give opinions and share experiences and feelings caused by the content with other peers.
The first step towards achieving such a synergy with standardized solutions would be the
consideration of the different convergence options between IPTV and IMS, which is ongoing. Besides this, operators and vendors are taking their own strategies in cross-platform
products and combinations between the media world and the communications world.

This leads to the second issue of misconception, related to the belief that IMS constitutes
a new service architecture independent of existing platforms in the telecom service delivery
infrastructure. Obviously, this cannot be realizable, and furthermore, for the sake of the
success of any potential roll-out, the decision maker needs to understand that IMS is to be
fully integrated with existing networks, e.g. the circuit domain, and with existing service
platforms and enablers. From an architectural point of view, IMS provides pieces to different
layers of the service delivery domain of an operator. Chapters 2 and 3 will provide a detailed
vision on this. In a very simplistic manner, one could say that IMS provides a core network
piece or component at the basics, and a set of pieces or components on top, thus enabling
new services, i.e. the service enablers.

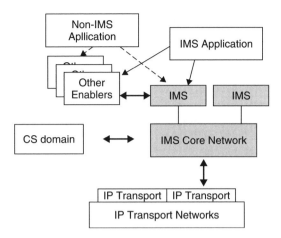

Figure 1.4 IMS integration in the telco service delivery domain.

Following this simplistic approach, different reasoning can be given to illustrate why a full integration of the IMS pieces and components with the rest of the service delivery domain infrastructure is required (see Figure 1.4):

- The IMS core network component needs to be integrated with the transport networks providing IP connectivity and to be capable of meeting transport requirements in terms of capacity and performance.
- Additionally, the IMS core network component needs to be integrated with the circuit-switching (CS) domain of the telco network to prevent the isolation of IMS users from other non-IMS users. Any subscriber to a telephony service needs to be able to establish a call to another party, whether this is attached to IMS or to a legacy network.
- The IMS enablers may need integration to other enablers or service capability platforms in the service delivery domain of the telecom operator. For instance, a presence enabler may integrate and utilize location information from a non-IMS location platform or a push-to-talk IMS enabler may require access to a user directory for a user identity conversion.
- The service logic for a pure IMS application may require invoking not only IMS enablers but also other numerous platforms in the network. For instance, an IMS application may require to be accessed by users via the mobile web, and thus the service platform will need to integrate with a browsing enabler or to access a platform providing identity single sign-on, etc.
- The service logic for a non-IMS application that is not utilizing IMS for media exchange might be enhanced by accessing an IMS enabler. For instance, an alert content delivery application via MMS might adapt the time of delivery by checking the recipient availability in the presence of an enabler of IMS.

In conclusion, IMS is, and shall be understood to be, an integral part of the infrastructure constituting the service delivery domain of an operator.

1.2.2   IMS – Telecom Network over IP?

From the cellular operator's perspective, IMS was never seen as a tool merely to transform the existing circuit-switched technology to packet-switched technology. 3GPP had introduced soft-switch architecture in Release 4 (3GPP, 2006a) to support voice services over IP prior to the standardization of IMS Release 5. This validates that IMS was not designed to be a voice-centric platform either. From the very start, IMS was designed to be an enabler to provide 'new multimedia services and Internet or Internet-like services' on a largely voice-centric cellular network. With increasing demand for multimedia services, the operators envisioned that IMS would eventually replace the circuit-switched core network and the telecom networks would evolve to being an all-IP core network. However, 3GPP specifications that define the IMS architecture (3GPP, 2007g) project IMS as a complex, centralized core network architecture. The specifications describe the functioning of the IMS service architecture and represent the application layer only as complementary to the core network. This is similar to the telecommunication IN architecture, where IN application servers complement the circuit-switched core network. The IMS service control (ISC) interface (3GPP, 2007b), which connects the application layer with the IMS core, is defined as part of the IMS core network architecture and there are no specifications defining its role in service provisioning. For readers not familiar with the IN architecture, it is a service framework that resides on top of the traditional telecom network. It provides network and service independence by separating the switching and the service control network elements as well as by providing service decomposition. The intelligent node (IN) is a nonswitching node that contains the service logic and the data accessible for switching nodes. Switching nodes use hooks to access the remote intelligent node and the SS7 network enables real-time signalling interconnections of nodes, as shown in Figure 1.5.

In an IN network, the switch interfaces with the application server through a dedicated control interface (e.g. INAP, CAP and an IS.41 subset). The application servers are invoked through triggers stored in the home location register (HLR). It permits the application server to control the switch by issuing instructions via the call control protocol (e.g. ISUP). These instructions essentially serve the purpose of controlling voice calls, as the circuit-switched network is voice centric. Here, the basic voice services are supported by the core network, which is complemented by application servers for the delivery of 'supplementary' or 'value-added' services. With this architecture as the background, it is very easy to misinterpret the IMS architecture as an IP replica of the intelligent network (IN) architecture.

Figure 1.6 depicts the similarity between the IMS service architecture and the IN architecture. The ISC interface that interfaces the IMS application servers with the IMS core can be perceived as the IMS equivalent to INAP/CAP. SIP could be viewed as the call control protocol similar to ISUP. The serving call/session control function (S-CSCF) could be perceived as the soft-switch of the IN architecture. IMS application servers being invoked by triggers called initial filter criteria (iFC) are stored in the HSS, which is the HLR equivalent and IMS application server, being similar to the service control points in the IN architecture. The two gateways, the IM-SSF, which is the gateway to CAMEL service environment, and the OSA/PARLAY gateway, explicitly link the IMS architecture with the IN architecture. The above comparison reinforces the idea that IMS is an attempt to recreate

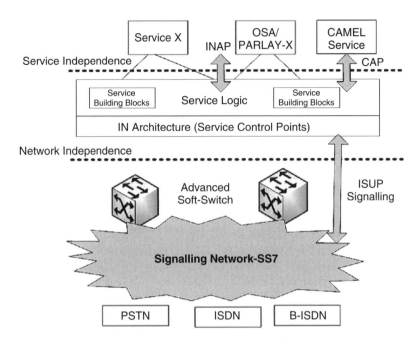

Figure 1.5 Telecommunications intelligent network (IN) architecture.

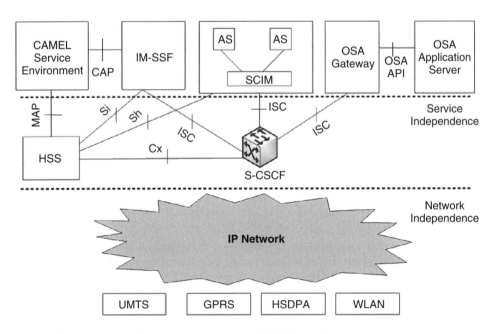

Figure 1.6 Author's representation of 3GPP's IMS service architecture.

the intelligent network (IN) over IP and the IMS application servers are only an extension to the IMS core network. This has strengthened the popular opinion that IMS creates another walled garden.

Although there is an architectural similarity between the IMS and the telco's IN, there are functional differences between the two. These differences prove that the IMS service architecture is not a mere replica of the telecom network over IP but is a generic multimedia service delivery framework. Some crucial differences are discussed in the following paragraph.

Firstly, ISC is not a control protocol that permits an application server to instruct the S-CSCF. ISC is an SIP interface, and its role is to extend SIP reachability to IMS application servers. Secondly, SIP is very different from ISUP, as it is not limited to the voice control function alone; rather it is a powerful and a generic session control protocol. SIP is capable of much more than session control and the potential of SIP's service control is discussed in Section 3.1.2. Thirdly, the initial filter criteria (iFC) that form service profiles in IMS service architecture are different from IN triggers. iFC are generic rules and can apply to any SIP messages. They are not merely used to invoke the application server but are capable of analysing the SIP messages. The iFCs can analyse the direction of the message if the subscribed user is the originator or the recipient of the message. The iFC checks for the SIP method if it is an INVITE, a SUBSCRIBE, a MESSAGE, etc., based on which each SIP request could be treated differently. iFC also analyses the existence of headers and the value of the headers with the possibility of using wildcards for preferential treatment of the SIP requests.

When the S-CSCF receives an SIP message and applies iFCs to it, the potential consequence is that this SIP message may be forwarded to one or more application servers. Similarly, when an application server generates an SIP message, the role of the S-CSCF will be to route it to another SIP entity (application server or device). In this interaction, the S-CSCF only acts as a proxy between two SIP entities. The real service control is what may happen between these two SIP entities.

Consequently, in the IMS service architecture, when you combine normal SIP routing and iFC-based routing, the S-CSCF simply acts as an SIP service router, which can support SIP-based interactions between IMS clients and IMS application servers. These interactions may relate to session control, but not necessarily.

1.2.3   Intelligent Core versus Dumb Edge

Among all the benefits the IMS core network can provide to its application layer, one of the most important is its transparency. An application should only care about what the IMS architecture can provide to it, not about the sophisticated details of how this is done. The application should only know that by issuing this SIP message to this user identity, it will be able to magically access this enabler (e.g. the presence information of this user).

The IETF initially defined SIP as a protocol permitting intelligent endpoints (devices or servers) to be implemented and making use of a relatively simple network. Over time, SIP extensions related to, for example, presence, resource list management and usage, or instant messaging, tended to add more and more intelligence in the SIP network in order to support the endpoints better. IMS defines a service architecture that permits it to add intelligence in the network optimally, without removing it from the endpoints.

Consequently, IMS services may either be implemented in IMS and the Internet endpoints or in the core network or have their logic more or less evenly distributed between endpoints and the network. IMS is therefore an intelligent network for intelligent endpoints.

1.2.4 IMS – Walled Garden?

IMS was from the beginning designed to permit end-to-end SIP signalling between IMS and non-IMS endpoints, and if the non-IMS endpoint does not support SIP, the IMS service architecture permits an easy integration of protocol gateways. 3GPP always intended to keep IMS open to non-IMS networks, and more especially the Internet. Creating new walled gardens is not a strategy that will be sustainable for operators in the years to come. Given the proliferation of Internet and Internet services, the eventual success of IMS would be proportional to the traffic generated between IMS and the Internet. An IMS with very low traffic to and from the Internet would be an IMS that has failed to deliver any added value to end-users and the users may prefer to bypass IMS to access services directly on the Internet. One proof that IMS is open to integration with non-IMS networks, especially the Internet, is the dependency between the two standards organization, the 3GPP and the IETF.

1.2.4.1 3GPP-IETF Dependency List

3GPP maintains a long list of dependencies on IETF specifications. Some may interpret it as a sign that IMS is full of telco-specific extensions to the IETF protocols it uses, which are mainly SIP and Diameter protocols. This view could be limiting as 3GPP essentially reuses specifications elaborated within the IETF space. More often than not, IMS has been an essential driver to improve and extend SIP-related specifications in the IETF. Without the push from the IMS and the telco industry, the IETF SIP community would not have diversified and be as dynamic as it is today. The reason for this list is that it is essential for 3GPP specifications to be based on stable RFCs, and not drafts that have an expiration time. The list permits 3GPP to identify clearly the IETF drafts that are of importance for IMS and to speed up their progress towards RFC status.

In the course of standardizing the IMS core network, 3GPP created new SIP headers and submitted them for endorsement to the IETF. Some extensions were accepted, as they were deemed to be useful to all types of SIP implementations. Others were related to user authentication, QoS or charging that emphasized the fact that, unlike the Internet, IMS is a network of privately owned networks, whose usage would involve money exchange and commercial implications.

Some compromises were made and a consensus reached between the 3GPP and the IETF hardliners. They agreed to have the questionable IMS SIP extensions to be specified in IETF RFCs, but they would have a specific 'proprietary' status, meaning that their usage is not recommended for general SIP usage. These headers are easy to recognize, as their name starts with 'P'. These P headers do not threaten end-to-end interworking with non-IMS clients:

- Many of them are used between IMS network entities and are invisible to IMS or SIP endpoints.

- Some might be visible to non-IMS endpoints, but the endpoint does not absolutely need to understand them.
- Gateways between IMS and the Internet may perform relevant SIP adaptations, if needed.

IMS specifications explicitly support session setup with non-IMS endpoints In order to offer a user an experience that matches the one in a circuit-switched network for a voice or video session, IMS adopted a session setup model that is more complex than the basic IETF SIP one. This extended session setup is also possible in the Internet, as it does not make use of any 3GPP-specific extension, but is only optional and may not be supported by all SIP clients.

3GPP then started a study to address the issue of interworking between IMS-compliant clients and clients that would not support the optional features required by the IMS session setup. This resulted in the requirement specified in Chapter 5.4.2 of TS 23.228 (3GPP, 2007g), named 'Interworking with Internet', that an IMS client or a user agent in the network acting on its behalf must be able to fall back to basic SIP session setup procedures if the peer cannot support the IMS-required extensions. All IMS enablers rely on IETF specifications and can be implemented on the Internet. There is not such a thing as an IMS-specific presence, IMS-specific conferencing or IMS-specific messaging. These enablers and services are implemented in IMS with a specific service architecture, but the protocol(s) and data model(s) used are standard IETF ones.

1.2.4.2 Gateways to IMS

The IMS service architecture makes provision to deploy protocol converters and gateways, which removes the need for any specific standardization. Any SIP message originated from or addressed to a user can be routed to such a protocol converter/gateway through the user's service profile provisioned in the HSS. The message can then be routed to any domain using any protocol. Conversely, these converters/gateways can generate IMS-compliant signalling from anything they receive. It is even possible for an IMS client to use non-SIP user addresses in its SIP messages. For instance, John may issue an SIP/IMS instant message towards Mary, whose IM service is not based on SIP and whose IM address is not an SIP or TEL URI. The IMS client could address the IM to, for example, foo:Alpha. The address is not routable in an IMS network, but this is not an issue. John's request is first routed to an IMS S-CSCF serving John. The S-CSCF then checks John's service profile. John's service profile may define that a request addressed to a user with an addressing scheme 'foo:' should be routed to an IM gateway, which will perform the necessary conversion. Only if no gateway was accessed through John's service profile would the S-CSCF look into the recipient's address, realize that there is a problem with the addressing scheme and therefore reject the request.

As depicted in the Figure 1.7, the distinct routing features of IMS enables multiple application servers to provide services to IMS users. As shall be discussed in Section 3.2.1, the usage of filter criteria by the S-CSCF enables end-user requests to be routed to different application servers, including third party application servers. However, it is assumed that the application servers are SIP based. As explained in the section above, non-SIP-based application servers can also be supported via appropriate gateway components or service brokers.

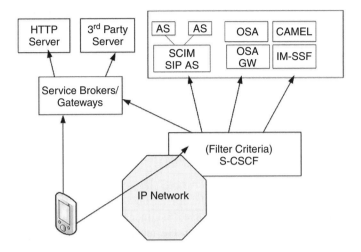

Figure 1.7 IMS – integration with multiple application servers.

1.2.5 IMS – An Integral Part of the Telecom Service Delivery Infrastructure

As discussed previously, IMS shall not be regarded as an independent piece or domain within the telecom operator infrastructure, but as a set of pieces forming part of the service delivery domain. Following a comprehensive approach, the infrastructure of a telco operator is divided into two conceptual domains (see Figure 1.8):

- The service delivery domain, which comprises all kinds of networks and service platforms utilized for the actual delivery of services to the end-user, from the application level to the access and terminal level. Formally, this domain is divided into the

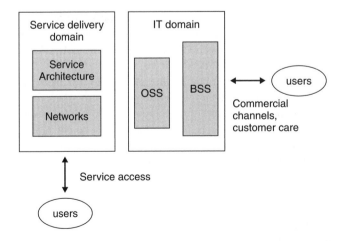

Figure 1.8 IT and service delivery domains of an operator infrastructure.

underlying network, which includes access and transport aspects, providing the basic connectivity, and the service architecture, on top of the networks, which comprises all service elements, service intelligence and reusable functionalities for service creation and execution.

- The IT domain, which comprises all the systems supporting the business and commercial activity of the operator as well as systems and platforms supporting the operational aspects of the business. Those are, respectively, the BSS (business support systems) and OSS (operation support systems). By the combined or chained action of these systems, the telco operator puts customer care and commercial channels in contact with the operating aspects and ultimately with the delivery platforms and networks.

There is a strong relationship between both the service delivery domain and the IT domain for the complete lifecycle of telecom products and services, for processes such as contracting and activating a service for a customer, the charging and billing of a user for the delivery of a service, the supervision and managing aspect of the services, and for the technical assistance to customers in the usage or misfunctioning of a given service or product.

This book focuses on the service delivery domain, and specifically how it can be enhanced and evolved with IMS technology and service innovation. For achieving a holistic and comprehensive view of this domain, and how the different IMS pieces fits into the complete service delivery domain of a telco, the service-centric layered model depicted in Figure 1.9 is utilized thoroughly through the subsequent chapters. This model, elaborated by the WIMS 2.0 initiative, responds to the fragmented view of the service delivery domain infrastructure of the different standard development organizations, like the 3GPP, ETSI and OMA, which are devoted to a specific part or layer in the model, while neglecting or simplifying the rest, and thus risking a global understanding of a multilayer technology like IMS.

The representation of the model depicted here does not differentiate the circuit-switching subdomain from the packet-switching subdomain. The following layers are represented in this model:

- Transport layer, which provides the basic access and transport for the service traffic, enabling a user via its user equipment to enter the service delivery domain. This layer is constituted by the access networks and the core networks providing a bearer services, either in circuit-switching mode.

- Service control layer, which embodies the intelligence for basic tele-services in the network and gathers all elements implementing the control plane of telecommunication services. In combination with the transport layer, it provides a set of basic telecom services to the user. This layer is formed by the packet-switching service control, i.e. the IMS core network, and the circuit-switching control, e.g. the MSC servers. In this layer also reside elements common to both circuit and packet modes, like subscriber databases and policy and charging elements. This layer interfaces and provides mechanisms for applications to control the service rendering to the user within a given value-added service. In the case of the IMS, that would be the dynamics of the ISC interface.

- Service brokering layer. This layer implements brokering functions to control the way elements in the upper layers of the service architecture can interact with the

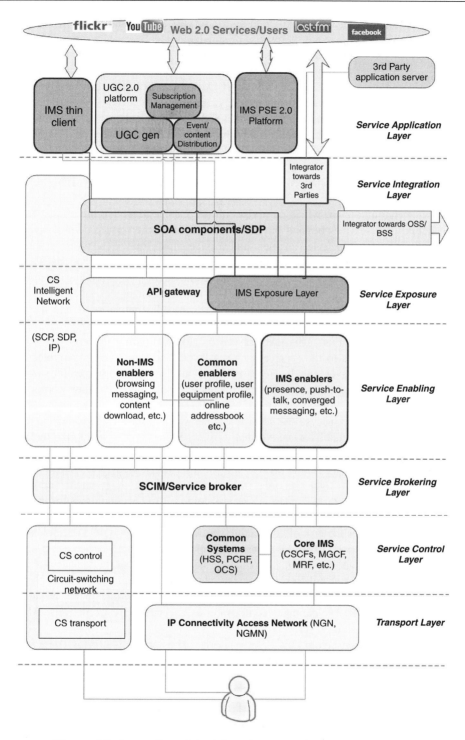

Figure 1.9 Layered model of the telecom service delivery domain.

telecommunication services provided by the service control and transport layers in an ordered manner to provide the user with an adequate delivery of a given value-added service or application, as well as a coherent and orchestrated execution of a portfolio comprising several applications or services that might be invoked under the same user-initiated event. Normally an SCIM or service broker can be found in this layer. These elements are, for instance, utilized to run legacy IN services on top of IMS basic tele-services, e.g. a VPN valid for CS telephony or IMS multimedia telephony, or to make compatible the chained execution services that naturally are not, e.g. a multi-SIM service and call forwarding and a missed call alerting service. Elements in this layer are required to be IMS-enabled as a minimum, and normally implement enhanced functionality beyond the basic ISC dynamics described in Chapter 3.

- Service enabling layer, which comprises all service elements that provide an enabling capability for services beyond the basic capabilities of the service control layer. This usually refers to all service enablers and service capabilities, as well as base services that can be re-utilized for composing value-added services and applications. This layer comprises IMS enablers like presence, instant messaging, conferencing, push-to-talk, converged messaging, xDM, etc., as well as non-IMS enablers like location, browsing, user directory, user profiling, content download, content push and streaming, converged address book, device synchronization, etc.

- Service exposure layer. This layer is utilized for facilitating the access for service container platforms to the service enabling layer. The functionality provided by the service enabling layer can be accessed by a myriad of communicating protocols, which in many cases are telco-specific. This exposure layer presents homogeneous service interfaces in the form of remote APIs that can be utilized from IT-based service platforms to access the service enabler without an awareness of the specific protocols that the enabler runs. Normally, web services and other IT-like technology for service interfaces and distributed computing are utilized in this layer. An example of elements pertaining to this layer would be a Parlay X web services gateway.

- Service integration layer. In order to accelerate service deployment and time-to-market, service platforms containing service logic runs IT technologies for distributed computing, like the ones following the service oriented architecture (SOA). An integration layer eases the task of connecting service containers with the reusable services offered by the service exposure layer and implemented by the service enabling layer. This layer offers additional functionality for the governance of the service architecture and the application of policies and security mechanisms at service runtime. Normally, this layer permits a secure and controlled access for external third parties to the operator service assets. The elements conforming to a service delivery platform (SDP) are mostly located in this layer.

- Service application layer. This layer comprises the service platforms containing the service logic that implements the network side of value-added services and applications. The service containers might be in-house owned by the operator or run by third parties. The access from these platforms to service assets of the service delivery domain of the operator can be via the service integration layer or due to the nature of services, performance or cost reasons, and service platforms may access directly to the service exposure layer or to the enabling layer. This layered model does not prevent an element in any layer accessing another element residing in a different nonadjacent layer without passing through the adjacent layers, due to performance or other reasons. Furthermore, in the case of a service container, a service logic might access a service feature A utilizing the integration layer, while for service feature B it might only utilize the service enabling layer and for service feature C it might directly use the enabler providing that feature, or directly interfacing the brokering or control layers.

As seen in this model, different functionalities and entities of the IMS technology are present at different layers:

- The IMS core network, located in the service control layer, will be described in Chapter 2. This chapter provides an overview of the access and core network supporting IMS.
- The IMS service control interface and IMS service architecture, which can be located at the interface between the service control layer and the upper layers, will be described in Chapter 3.
- IMS brokering aspects, positioned in the service brokering layer of the model, will be covered in Chapter 3.
- IMS enablers and service capabilities (refer to Chapter 3).
- Strategies for opening capabilities of IMS and other service features, located in the service exposure layer, will be tackled in Chapter 5.
- The SDP and SOA, part of the service integration layer although not strictly related to the IMS technology, will be tacked in Chapter 6, due to its importance to leverage IMS service features for application development.
- IMS services and applications, part of the service application layer, will be described in Chapter 3 from a pure telco perspective and in Chapter 5 from a convergence viewpoint.

1.2.6   Architectural Concepts Related to Service Infrastructures

Service architecture, service delivery platform, service creation environment, service development architecture, service features, service oriented architecture, etc., are just a few words tangling in the ICT landscape. There are a number of concepts under the umbrella of service infrastructures in the telco industry, forming a blur of definitions of functionalities, technology and architectures that usually require clarification. Accompanying the WIMS 2.0 model of the telecom service delivery domain, a collection of architectural concepts are defined, which might assist the reader in not losing the global view.

1.2.6.1   Service Architecture

This is the set of functional entities in the service delivery domain above the service control layer (IP multimedia core network subsystem, in the case of the IMS architecture) and below the application servers or service logic containers. The service architecture mostly presents re-utilizable functions to provide complex services, and some supporting and controlling functional entities for a normal service delivery. When talking about IMS, an IMS-enabled service architecture is the one that enables IMS service features in the services and applications built on top of the service architecture. The previously presented layered model provides a representation of an IMS-enabled service architecture. In certain contexts, an IMS service architecture can refer to the part of the service architecture that is composed by strictly speaking IMS functional entities, e.g. IMS service enabler, IMS service broker, IMS Parlay X service capability functions, and so on. In other contexts, IMS service architecture might refer to the IP multimedia core network subsystem (a.k.a. IMS core network or, simply, IMS core) specific architecture, i.e. interfaces and entities opening the control of network capabilities to service-level logic, i.e. the ISC interface dynamics, including iFC trigger philosophy. The IMS service architecture following the latter meaning will be thoroughly described in Chapter 3.

1.2.6.2 Service Delivery Platform

This is a set of functions, tools and infrastructure that can help an operator to create and deliver services with a reduced time-to-market by providing easy integration to the OSS/BSS in the IT domain and integration to the service architecture (network service capabilities and service enablers, mostly), as well as integration to third party platforms, or vice versa. Thus, the SDP prime principle is integration, to achieve the reduction of time-to-market. An SDP is de facto realized today by applying the SOA paradigm to the telecom operator, service delivery domain.

1.2.6.3 Service Oriented Architecture (SOA)

This is a paradigm for creating complex services by invoking and orchestrating autonomous, independent services with a clear remote service interface (a remote API). Normally it is assumed that services expose their functionality with an HTTP-based interface. This means that SOA might be using the HTTP fabric for connecting services and enabling service invocation. A de facto assumption is Web Services/SOAP as a specific technology used to materialize the SOA paradigm, while others think that the REST philosophy is also an SOA type of paradigm. SOA and web services are proven technologies to achieve good integration in the service sphere and thus good candidates for an SDP implementation.

1.2.6.4 Service Creation Environment

This might be the set of tools (e.g. Eclipse) that can be utilized to create services that later might be deployed in the service delivery domain on top of the service architecture, by utilizing an SDP or directly utilizing functions from the service architecture. Also, an SCE might include a sandbox environment, i.e. a miniature of the service delivery domain with testing tools and dummies, just to simulate the end-to-end behaviour of a service in a similar environment to the service delivery domain.

An IMS flavoured example of this would be Ericsson SDS (Ericsson, 2009). It includes plug-ins and tools for testing and some simulation for network IMS servers.

1.3 IMS Standards Status

As already discussed, IMS was originally designed to provision multimedia services on 3G wireless networks. International Mobile Telecommunications-2000 (IMT-2000) (ETSI,

2002) is a global standard for 3G networks. The Third Generation Partnership Project (3GPP, 1999d) and Generation Partnership Project 2 (3GPP2, 2000) are its two standard bodies. IMS was originally defined by an industry forum called 3G.IP led by the AT&T Wireless. However, it was standardized by 3GPP and forms the key standardization body involved in the development of IMS architecture along with 3GPP2. While 3GPP standardizes UMTS (W-CDMA radio interface with GSM network technology), 3GPP2 bases its standards on the cdma2000 radio technology and IS-45 network. A similarity between the 3GPP IMS and the 3GPP2 IMS is that both use the Internet protocols standardized by the Internet Engineering Task Force (IETF, 1986). The other key standard bodies involved in the IMS standardization are the Next Generation Networks Focus Group (NGN-FG) (NGN, 2006) created by the International Telecommunications Union (ITU-T) (ITU, 1989) and Telecoms and Internet Converged Services and Protocols for Advanced Networks (TISPAN) (2000) created by the European Telecommunications Standards Institute (ETSI). Just as 3GPP standardized IMS to provide multimedia services to a converged wireless and mobile access network, ETSI TISPAN has standardized the NGN architecture to provide similar services for converged next-generation fixed networks. The following section provides detailed functioning of each of the standardization bodies involved in defining the IMS architecture and their collaboration.

1.3.1   3GPP

3GPP was created in 1998 by the signing of 'The Third Generation Partnership Project Agreement'. It mainly consists of mobile operators and major network product vendors. The original scope and purpose of 3GPP was to produce, approve and maintain technical reports and technical specifications for the 3G mobile system based on evolved GSM core networks and the radio access technologies that they support. The scope was subsequently amended to include the maintenance and development of the global system for mobile communication (GSM) technical specifications and technical reports, including evolved radio access technologies such as the general packet radio service (GPRS) and enhanced data rates for GSM evolution (EDGE). The latest 'Scope and Objectives' document of the 3GPP includes four other important objectives, such as to prepare, approve and maintain technical specifications (Specifications, 1989):

1. The 3GPP core network and evolutions includes the 3G networking capabilities originally evolved from GSM. These capabilities include mobility management, global roaming and utilization of relevant Internet protocols.
2. Terminals to access the above-mentioned access technologies. This includes specifications for a user identity module (UIM).
3. An evolved IMS developed in an access independent manner.

1.3.1.1   Organization

3GPP consists of five Technical Specification Groups (TSGs) as depicted in Figure 1.10. Each of these TSGs consists of various working groups as depicted in the figure. The detailed information about each of the TSGs can be obtained from the 3GPP website: http://www. 3gpp.org/Specification-Groups. However, in the context of IMS architecture the following

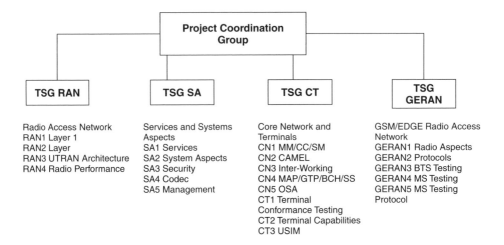

Figure 1.10 3GPP standards organization.

groups and the specifications published by them are of relevance to the operators, telecommunication engineers and the product vendors:

- TSG SA WG1 works on setting high level requirements for the overall IMS system and provides this in a stage 1 description in the form of specifications and reports. This work includes: definition of service and feature requirements, framework for services, specification of services (stage 1), specification of service capabilities (stage 1), identification of technical and operational issues to meet market requirements, and charging and accounting requirements. This group mainly belongs to the operators, and the specifications published by this group are numbered as TS 22.XX and are accessible via the specifications array on the 3GPP website (http://www.3gpp.org/ftp/Specs/html-info/SpecReleaseMatrix.htm).
- TSG SA WG2 develops the stage 2 of the 3GPP network based on the services requirement elaborated by TSG SA1 WG1. It identifies the main functions and entities of the network and how these entities are linked to each other along with the information they exchange. The output of SA WG2 is used as input by the groups in charge for the definition of the precise format of messages in stage 3. This is one of the most strategic groups in the 3GPP and mainly consists of network product vendors and a few operators. Stage 2 specifications are numbered as TS 23.XX.
- TSG CT WG3 specifies the bearer capabilities for circuit- and packet-switched data services, and the necessary interworking functions towards both the user equipment in the UMTS PLMN and the terminal equipment in the external network. In addition CT is responsible for end-to-end QoS for the UMTS core network in Release 5 and beyond. The relevant specifications are numbered as TS 24.XX and TS 29.XX.

In the context of multimedia service provisioning and multimedia service creation, the following groups are relevant to service providers and application developers:

- TSG SA1 WG4 deals with the specifications for speech, audio, video and multimedia codec, in both circuit-switched and packet-switched environments. The group also works on quality evaluation, end-to-end performance and interoperability aspects with existing mobile and fixed networks from a codec point of view.

- TSG CT WG6 is responsible for development of test specifications for the 3GPP smart card applications and the interface with the mobile terminal. It includes
  - Subscriber identity module (SIM);
  - USIM (universal subscriber identity module), which is used by 3GPP systems;
  - ISIM (IMS services identity module).
- TSG CT WG1 and TSG CT WG4 deal with the IMS protocol related specifications and would be particularly useful for application developers. CT WG1 is responsible for the SIP call control protocol and the SDP protocol for the IM subsystem.

1.3.1.2 Specifications Status

3GPP works through a series of Releases where each Release generally adds further features and capabilities, and addresses the gaps identified in previous Releases. With the introduction of IMS in Release 5, 3GPP has currently frozen Release 7 and the work on Release 8 is in progress at the time of writing this book.

In Release 5, the core network is evolved from its previous releases and an end-to-end all-IP network architecture called IMS is defined.The core of the IMS comprises a number of nodes that form the call session control function (CSCF). The CSCF is basically an SIP (session initiation protocol) architecture which was initially developed for the fixedline world and is one of the core protocols for most voice-over IP telephony services available on the market. While the CSCF is responsible for the call setup and call control, the user data packets, which, for example, include voice or video conversations, are directly exchanged between the end-user devices. A media gateway control function (MGCF) is only necessary if one of the users still uses a circuit-switched phone. With the UMTS radio access network it is possible for the first time to implement an IP-based mobile voice and video telephony architecture. With GPRS in the GSM access network, roaming from one cell to another (mobility management) for packet-switched connections is controlled by the mobile station. As the IMS has been designed to serve as a universal communication platform, the architecture offers a far greater variety of services than just voice and video calls, which are undoubtedly the most important applications for the IMS in the long term. By using the IMS as a platform for a standardized push-to-talk (PTT) application, it is possible to include people in talk groups who have subscriptions with different operators.

In Release 6, the IMS core architecture was enhanced with additional interworking entities and new reference points. UMTS Release 6 introduces an uplink transmission speed enhancement called high speed uplink packet access (HSUPA). In theory HSUPA allows data rates of several Mbit/s for a single user under ideal conditions. HSUPA also increases the maximum number of users that can simultaneously send data via the same cell and thus further reduces the overall cost of the network. Whereas HSDPA optimizes downlink performance, high speed uplink packet access (HSUPA) constitutes a set of improvements that optimizes uplink performance. These improvements include higher throughputs, reduced latency and increased spectral efficiency.

In Release 7, the multiple-input multiple-output (MIMO) technology is introduced. MIMO is a very promising technology for empowering UMTS networks by providing more throughput than HSDPA/HSUPA. IMS features were further enhanced based on 3GPP extensions developed in the IETF. Combinational services over IMS and CS networks and VCC emergency calls were introduced. It also focused on the requirements for fixed broadband networks to access IMS via TISPAN and cable access networks to access IMS via Cable 2.0. Release 6 and Release 7, which is currently frozen, form the basis for deployment within the telecommunication industry.

In Release 8, the network architecture for LTE (long-term evolution) is introduced along with the new radio technology called E-UTRAN. New core network architecture to support the high-throughput low-latency LTE access system is defined in the form of SAE (system architecture evolution). Release 8 continues to evolve IMS further as per the IETF enhancements and takes the TISPAN and Cable 2.0 requirements further. Release 8 is also adding the capability for IMS centralized services to allow the delivery of enhanced telephony by controlling legacy mobiles from the IMS core. Although the 3GPP standards and specifications may appear complex to a novice, they provide detailed and complete information about the IMS architecture, operation and services. They also provide an insight into how the telecommunication industry functions.

1.3.2 IETF-3GPP

The Internet Engineering Task Force (IETF) is a large open international community of network designers, operators, vendors and researchers concerned with the evolution of the Internet architecture and the smooth operation of the Internet. It is open to any interested individual. Technical work is done in working groups (WGs); the current WG areas are: Applications, General, Internet, Operations and Management, Routing, Security, Sub-IP, Transport Area. Results from working groups are usually a set of RFCs. Rosenbrock *et al.* (2001) present a set of principles and guidelines that serves as the basis for establishing the collaboration between 3GPP and IETF, with the objective of securing timely development of technical specifications that facilitate maximum interoperability with existing (fixed and mobile) Internet systems, devices and protocols. Refer to Section 1.2.3 for further details on the 3GPP-IETF dependency and collaboration.

The following working groups are of relevance to 3GPP and next-generation networking:

- IPv6 – 6man (6MAN, 2008), mext (MEXT, 2008) and v6ops (v6ops, 2008);
- Mobile IP (MIPSHOP, 2008);
- AAA (authorization, authentication and accounting);
- DIME – diameter protocol maintenance and extensions (DIME, 2008);
- SIP (session initiation protocol) (SIP, 2008);
- SIPPING (session initiation protocol project investigation) (SIPPING, 2008);
- Mediactrl – media control framework and media server control (MEDIACTRL, 2008);
- AVT – audio-video transport (AVT, 2008).

Apart from the 3GPP, IETF, ETSI and ITU, there are a few other organizations that have become involved in recent years in the IMS standards related activities, particularly in defining standards profiles, ensuring interoperability and dealing with the actual commercial deployment of IMS. The following subsection presents details of a few such organizations and their contribution to IMS technology and deployment.

1.3.3 TISPAN

Although the intention was to make IMS access-agnostic, initial releases of IMS standards by the 3GPP were oriented towards the characteristics and requirements of a mobile environment. Therefore, the assumptions based on which the 3GPP has defined the IMS architecture do not entirely match the requirements of the wireline environment. Some of the assumptions are as follows:

- 3GPP defines wireless access via the GGSN, while there is no provision for the fixed broadband networks to access the IMS infrastructure. There is no account of B-RAS for xDSL networks or CMTS for cable networks.
- 3GPP has concentrated only on SIP-signalled applications, while the large majority of Internet applications and broadband applications are not accounted for.
- Assumptions made from a security and authentication view do not hold good for wireline networks. For instance, security in wireless networks is based on secure IDs or SIM cards; hence the security mechanisms built in such networks are based on the above assumption. In fixed networks, however, hosts are identified by MAC addresses or hostline IDs that are not accounted for.

TISPAN works closely with the 3GPP to bring about the fixed mobile convergence to the NGNs. It proposes to use the 3GPP IMS for SIP-based applications but adds new components and functional blocks to handle non-SIP applications and other requirements of fixed broadband networks that are not addressed by the 3GPP IMS.

The TISPAN Release 1 architecture which was published in December 2005 is based on the 3GPP IMS Release 6 architecture but is intended to address the wireline provider's requirements. The architecture uses a 3GPP IMS core where subsystems share common components, while allowing the addition of new components over time in order to cover new demands and service classes. It ensures that the network resources, applications and user equipment that are inherited from IMS are common, thereby ensuring user, terminal and service mobility. Release 1 concentrated on the architecture and functional components and not on the NGN applications. Release 2 is expected to address this gap by defining real-life NGN services such as media streaming, IPTV and others.

1.3.4 PacketCable

Similar to TISPAN, Cable Labs created the PacketCable (2007a) architecture to address the requirements of the cable access networks to connect to the IMS infrastructure. It intends to extend cable IP services beyond today's VoIP to embrace wireless, fixed mobile convergence, video communications and other cross-platform features. PacketCable 2.0 architecture is based on 3GPP Release 7 and is currently complete.

1.3.5 Multiservice Forum (MSF)

The Multiservice Forum is an industry body whose goal is 'to promote multivendor interoperability as part of a drive to accelerate the deployment of next generation networks'. It does this by 'developing Implementation Agreements, Product Specifications or Architectural Frameworks, which can be used by developers and network operators to ensure

interoperability between components from different vendors'. Part of this involves producing a reference architecture, now on Release 3.0, issued in mid-2006, which is closely tied to
IMS. The MSF is running an NGN certification and interoperability program (MSF, 2008).

1.3.6 IMS Forum

The IMS Forum is the industry association for promoting IMS applications generally and for
their interoperability. It organizes the series of IMS Plugfest events for IMS services interoperability verification and certification, and, among other activities, is developing a certification program for IMS interfaces, including pre-certification in qualified labs (IMSForum,
2008). Another industry body, the SIP Forum, concentrates on SIP-related matters. It runs
an ongoing programme of 'SIPit' interoperability tests (SIPForum 2008).

1.3.7 Open Mobile Alliance (OMA)

The Open Mobile Alliance (OMA) is the main standardization body for mobile service
enablers in the NGN environment. It has an extensive programme of service enablers (e.g.
charging, client provisioning, digital rights management, email notification and so on – there
are about 50 of them) that provide architectures and interfaces that are independent of
the underlying wireless networks and platforms and are intended 'to work across devices,
service providers, operators, networks, and geographies'. There is also a programme of
interoperability testing. One current focus of development is Push-to-X (more formally
OMA POC Re.l2.0), which is the IMS-based multimedia evolution of today's push-to-talk
for cellular networks (POC). IMS itself is included as IMS in OMA Enabler Release
1.0, which contains general requirements and guidelines but does 'not specify detailed
requirements that should be tested or that by themselves can be implemented in products'
(OMA, 2008).

1.4 IMS Deployment Status

IMS network operators are implementing IMS on a per-service basis. For the major carriers,
the time for watching and waiting appears to be over, with many carriers implementing IMS
components in particular areas in which a favourable return on investment is clear. In other
words, the marketplace is segmenting various versions and pieces of IMS and carriers
are deploying them because they represent the best solution for a particular application
in that network at that point in time. Overall, it is believed that IMS offers the means
of a profitable migration path to IP. Service providers and carriers have for years been
attempting to reduce the cost of delivering existing services while providing a platform for
new ones, but the urgency has perhaps never been as keenly felt as it is today, given the
renewed thrust of the smartphone market and new choices for IP communications such as
Google's Android, launched in October 2008. That said, service providers have little choice
but to IP-enable their networks to meet these challenges with an eye towards the future.
IMS gives them a standard set of blueprints to do just that. The convergence of IP and
legacy communications technology is not only finally becoming real but is now occurring
at an accelerated rate. Given their different goals, business models and sometimes even
radically different infrastructures, the service providers will each respond differently. In the

end, many will select various features and specifications for numerous application-delivery scenarios, seizing upon what serves their needs and discarding the rest. IMS will continue to evolve over time relative to each service provider's specific needs. As new aspects of the business case become clear, additional devices and protocols will be brought up to speed. For example, the IMS standard has been shown to interoperate correctly with at least one software set-top box, but no hardware set-top boxes have been available for testing yet because the standard for interoperating with hardware set-top boxes has yet to be finalized. In a sense, then, the IMS infrastructure is evolving from the inside out, starting with individual applications at the core service layer and spreading from there. The immediate need served is service deployment and the long-term benefit is the beginning of a common foundation for existing and new mobile and converged services as they emerge. What is certain is that it is expensive for providers to continue to deliver new services using outdated infrastructure or proprietary solutions that are not interoperable. It remains an open question whether we will ever see a proliferation of all-IMS networks, i.e. the complete architecture as described by the various standards bodies. However, we will most certainly see networks built using those elements of IMS that are fundamental to building IP-enabled infrastructures. While service providers may exploit IMS in a piecemeal fashion to suit their own unique needs, new killer applications may yet emerge once the infrastructure has been put in place. However, that killer application may in fact not be a single application but the sum of evolutionary changes in data, communication and media services and how people use them. We may be seeing the first stage of this in the advent of the Apple iPhone and other new network appliances designed to give users ubiquitous access to multiple services whenever they want no matter where they may be. Of course, additional pieces must fall into place to support the continued evolution of IMS networks. This we address in the next section.

1.5 Future

The immediate future of the IMS as a core network and service enabling technology is determined by the standardization plans, which in turn are normally market-driven requirements posed by those stakeholders from the industry participating in the standard development bodies. In order to understand what the future holds for the IMS and related technologies, it is important to understand the future trends in the telecommunication and adjacent industries. In other words, IMS represents the keystone for telecoms evolution, and thus every single event affecting the telecom world would directly impact the IMS adoption or, furthermore, it will be leveraged by an IMS-enabled telecom network. This section analyses the effect IMS or IMS adoption may have on the ICT perspectives in the near future.

1.5.1 The 2.0 Era: A Consolidated Trend

For the past few years, the ICT industry has been experiencing the so-called '2.0 revolution', motivated by the eruption in the Internet landscape. The Internet is being popularized by a new wave of services based on web technologies and with community culture, presented to end-users as web-based services. This is called the 'web 2.0' phenomenon, with an emphasis on the '2.0' establishing an analogy with software application numbering and versioning: a new version of the web has been released to the market. Traditionally, the Internet has been a more open environment, more favourable to innovation. The different business models that govern service delivery in the Internet have been posing fewer restrictions to entrepreneurship. They have especially facilitated the proliferation of a great number of applications to be trialled by users in a not-completely stable version called 'beta'. The beta proposition is valuable in the sense that is more coupled to marketing basics. A new product can be released early to the market in order to validate not only functionality and operativeness but also, and most importantly, to test waters, i.e. if the product concept is valuable for the end-users. Besides all this, service creation in the web environment is traditionally less expensive than compared to the telecom environment, mainly due to two reasons:

- The development and delivery technologies are simpler, less intricate and more homogeneous across the Internet landscape. Thus, the number of professionals and enterprises with ability to develop services in the Internet are vast, which in turn decreases the cost of service development. In comparison to the telecom world, where there are fewer professionals with development skills in the complex technologies for producing value-added applications in the telecom environment. An immediate example would be a comparison between the number of professionals with expertise in Java and HTML as compared to the professionals with knowledge of CAMEL or the vendor-proprietary solutions for intelligent networks, such as AXE or CS1+. Besides this, the telecom technology specifics required for the development of telecom services are mostly reserved to an operator, and normally are neither disclosed easily nor accessible to developers outside the strict circle of partnership of a telco. Figure 1.11 depicts this concept in terms of granularity of complexity in service development versus the number of developers. It clearly shows that in the native

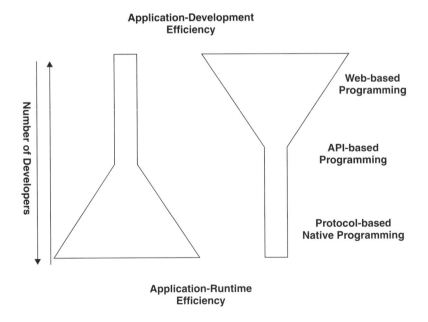

Figure 1.11 Programming complexity versus number of developers.

telecom environment, where the development is protocol-based, the complexity of development is high and the number of skilled developers is low. This greatly limits the innovation and increases the cost of service creation, while in the Internet environment, the web-based development is low on complexity and attracts a vast majority of developers. This proportionately increases the number of services developed and enhances the chances of innovation proportionately while reducing the cost for service creation.

- On the Internet, the client platform has played a key role in the virility of the expansion of web applications. Browsers are compatible, flexible and powerful, with great ergonomics as compared to the myriad of telecom devices with a varying range of features from model to model, from vendor to vendor. Even with the latest efforts to homogenize the mobile terminal platforms, the existing telecom terminals in the market are still less powerful with poor ergonomics. The pricing of devices is also a detrimental factor. Powerful and technology-rich terminals are only accessible to 10% of the mass market, and the technology upgrade implies a change of the terminal, while in the web environment, that is mostly solved with software downloads.

These factors seriously limit the possibilities for innovation at a reasonable cost in the telecom environment, while the Internet is witnessing progressive trends. Freedom in the business aspects and in the technology sphere is driving the Internet towards the web 2.0. Typically, all the features of the web 2.0 can be reduced to two paradigms:

- the user at the centre, with a deep socialization of services and a participative spirit, which encourage users to engage in service development;
- the Internet as a platform, providing the means to create new services rapidly by utilizing features of preexisting applications or functionalities.

The web 2.0 phenomenon and concepts as such will be explored in Chapter 5. From a mere user perspective, the web 2.0 represents a new wave of applications that are more appealing to the mass end-users. The services are more connected to user expectations than ever, by making the users as part of those services and by making the user an engine in the service delivery.

The web 2.0 principles, the spirit of the 2.0 change, could be used by adjacent domains of application. The era of the 2.0 has spread to Education 2.0, Government 2.0, Tourism 2.0, etc. Two principles that can be extracted as the common characteristics of the so-called 2.0 revolution are: the involvement of the user or receiving party, who is no more a passive actor, and the exposure of technology to the application domain in such a way that tasks are simplified to the user, rather than making them more complex to understand. When the technology is simpler to use, it reduces the entry barriers. In brief, the democratization of both the application domain and the technology is witnessed in web 2.0. Within the telecom industry, the 2.0 principles are starting to be applied shyly to the core business. Initiatives such as telco 2.0 are pursuing the task of reformulating the business models governing the telecom sector. Telco 2.0 principles intend to influence the telecom business to refocus and incorporate those changes that foster innovation. This is done mainly by recognizing that the operators and their ecosystem of vendors may not be skilled enough at designing attractive applications appealing to the end-user. Thus, telco 2.0 is tightly coupled to the web 2.0 world of applications.

However, while the web 2.0 is a consolidated reality in the market, the telecom 2.0 is only a distant reality. As an example of the telecom 2.0's way of doing things can be quoted the service offerings of the Apple iPhone, Google-boosted G1 device and similar 'Internet phones'. These are devices optimized to offer the users their favorite services on the Internet while also providing some dramatically basic telephony functionalities and a few telecom added-value applications. Precisely, IMS may be a suitable candidate to enhance the 'Internet phones' telephony features up to the 2.0 user perception. Another example of a telecom 2.0 trend in the market is the open telecom communities. This is an effort by some telecom operators to incorporate their service capabilities to Internet mash-ups by providing simple-to-use APIs. Usually in these open telco communities, developers can access the offered APIs without commercial commitment and with zero or small prices to foster development. Some of the remarkable open telcos programmes are OrangePartner, Vodafone Betavine, BT Ribbit, O2 Litmus, Telefonica's Open movilforum, Telenor Playground, Sprint Ahead and some others.

These examples indeed reveal that, for the moment, the initial telecom 2.0 realizations are rather linked to the web 2.0 sphere. The next step shall be the creation of genuine telecom 2.0 applications and services not necessarily supported or based on a web 2.0 service. In this very challenging context is where the IMS technology, its service features and business propositions can demonstrate the capability of the telcos, to provide impressive new ways of interpersonal communication and leisure services similar to the web 2.0 services.

1.5.2 Immediate Future: Web–Telecom Convergence

While looking for their specific path towards the 2.0 revolution, telcos must engage with the web 2.0 momentum and get positioned in an ICT landscape that is gaining popularity and, specially, closeness to the end-user. The mass-market typically classifies emerging services either as 'cool' or out-fashioned services. Behind the 'cool' qualification, there is an implicit perception of sophistication and usability that should be targeted when designing new added-value applications. More importantly, telecom can compensate for problems that exist with the Internet related to the Internet's architecture as well as with the business model.

Listed below are a few such problems that exist in the Internet. It also lists the consequent opportunities that arise to the telecom industry (Caida, 2003).

- *Beta releases* are used as an excuse for low quality. Internet products and services are rushed out to the users as guinea pigs. There exist minimum commitment towards maintenance and support for each service.
  Telco opportunity. Telecom has a reputation for provisioning high quality services and they can leverage this opportunity.
- *Proprietary solutions masquerading as open ones.* The major applications are 'open' only to the extent that they provide a trapdoor into their closed worlds, e.g. MSN IM, Skype, Flash, Google search. Interoperability is not possible.
  Telco opportunity. To be more open than the Internet players. If not for killer applications, telecoms could offer killer application creation environments.
- *Lack of clear ownership of user-generated content.* This hinders porting of applications across platforms.

Telco opportunity. To offer portability of data and identity as a virtue and not an imposition.

- *Lack of accountability.* Lack of ownership mentioned above also results in lack of accountability.
  Telco opportunity. To create services as well as the social institutions to support them and regulate their (ab)use.
- *Poor multilanguage and localization.* The Internet does not given enough priority for multilanguage representation and localization.
  Telco opportunity. Globally available services could be tailored to local needs.
- *Spam, viruses, phishing and scammers.* The big Internet companies making huge profits have not taken the responsibility of cleaning the Internet.
  Telco opportunity. Could sell products based on security, not necessarily on new features.
- *A new site, another identity.* No identity solution available on the Internet.
  Telco opportunity. The phone numbers and account credentials could be authenticated using SIM-card credentials and could be shared among all applications. It could be a market-maker for exchange of identity data. What Verisign does for assuring the ownership and identity of servers, telcos can do for users. Similar to what Paypal does for hiding the bank and credit card details of the users, telcos can do for user identity.
- *Lack of support.* The Internet as mentioned earlier lacks accountability and hence support. The culture of 'free' means you too often get what you pay for.
  Telco opportunity. Could make the quality of the support a differentiator.
- *Advertising everywhere.* The Internet is flooded with ads and pop-ups.
  Telco opportunity. Could position itself as the premium alternative. Offer the users to skip the ads.
- *The real cost is hidden in the Internet.* End-users supply their own equipment, draw on the technical support of friends, spend time researching what applications are compatible with which devices and peripherals and OS, etc.
  Telco opportunity. Could take responsibility for making the pieces work together and market the simplicity to the end-user.

As a summary, telecom could make hay by compensating for the problems that exist in the Internet by offering some or all of the following features:

- flat-rate voice bundle and flat-rate data bundle, which has become a reality with some providers;
- unlocked/open handsets;
- open SIP stacks and other application APIs such as location API;
- SMS/MMS API;
- broad adoption of Internet IM on mobiles;
- API access to own data (call history, contacts, etc.).

Telecom services are usually criticized as being distant from what the users need. On the other hand, telecom skills to manage the lifecycle of a service commercialization including CRM and quality of experience are renowned. Telecom services are perceived by end-users as trustworthy and reliable. Telecom branding provides this perception, but lacks the proximity and appeal to users. This might be achieved by converging features coming from the web with features coming from the telecom world. From a telecom operator's viewpoint, convergence, in order to be successful, needs to provide a means of getting closer to the user while still exploiting most of the telecom services and capabilities, or be relevant to the operator business. The operator can realize this convergence in the following manner:

- Telecom brands are associated with new web 2.0 services which follow the mainstream of applications in the Internet, while pure telecom service features in these web-based services are marginal or not the core platform functionality. Examples of these would be Orange's Pikeo and Bubbletop or Telefonica's Ketek social network. In brief, they enhance the telco positioning in the Internet market by providing pure internet-based services.
- Telecom service features are the basis of new applications in the web domain, usually by integrating telecom applications into the web or by constituting the key elements in the service delivery of a web 2.0 application to the user. Some examples of this might be the Zyb mobile social network or the Shozu mobile application for user-generated content. In brief, they enhance the telco positioning in the Internet market by providing hybrid telecom-based services.
- By exploiting contents and information in the service delivery of telecom services or mashing them up with the normal execution of telecom services such as those involving personal communications. An example of these would be the Movial Social communicator, which combines regular telephony features with mobile access to social networks and web 2.0 content-oriented sites. In brief, they enhance the telecom position in the telecom market by providing hybrid Internet-rich applications.
- By applying concepts and technology coming from the web 2.0 to telecom service delivery. For instance, the widget type of organization and rendering of services is being applied to the delivery of specific applications to IPTV and mobile handsets. In brief, they enhance the telecom position in the telecom market by providing applications technically inspired by Internet.

Chapter 5 describes the WIMS 2.0 initiative principles of convergence based on IMS services and technology, which follows the above-mentioned approaches.

1.5.3   Future Internet: The Next Landscape for Both Web and Telecom Worlds

The Internet is now critical for a whole variety of social and economic functions with over a billion users worldwide. However, today's Internet was designed in the 1970s and mismatches between the original purposes and its current usage scenarios are beginning to limit Internet development. The questioning and possible need for a redesign of the Internet architecture is nowadays more and more widely accepted. The Future Internet, also known as the future of the Internet, is the vision of the evolution of the Internet as known today, driven from the application and economy domains of the super-network. Future Internet thus is not only dealing with technology evolution of the network plane of the Internet, i.e. the Internet protocol and complementary mechanisms, but rather comprises a holistic view on the potential features of an Internet of the future. Two phenomena are causing the ICT industry to start formulating a vision beyond the current state of the art of the Internet:

- From the pure telecom side, the increasing capacity of the accesses to achieve a truly broadband connection to the Internet, both based on fixed technology as well as wireless: fibre in the last mile in the wireline environments and the long-term evolution for evolved 3G in the mobile environment are the representative technologies in the roadmap of most telecom operators. All this permits traffic-intensive applications to emerge and augment user expectations with regards to media-rich services.
- From the pure application side, the web 2.0 revolution has demonstrated that the Internet can serve as the platform to support a variety of innovative applications more and more feature-consuming.

All this has led the ICT industry stakeholders to start activities to pave the way towards the Future Internet. This has been specially noted in the research space of the countries with a technological lead worldwide. For instance, USA, Japan and the European Union are coordinating and modelling their research efforts in various directions towards the Future Internet.

Early research work on the Future Internet concentrated only on networking issues, while forgetting about the application domain, social and socioeconomics dynamics of the Internet. The first ideas of a Future Internet come with the past efforts in substituting the networking mechanisms and introducing IPv6 as a replacement of IPv4. Difficulties in such a deployment at the very heart of the Internet architecture has postponed the work around technically enhancing the Internet. The IMS architecture – designed as the keystone for merging the cellular and the Internet world by adopting Internet protocol technologies and deploying them in the telecom core business, i.e. telephony – adopted the future Internet protocol IPv6 as the networking support for guaranteeing true peer-to-peer experiences in a world populated by millions of telephony devices.

After numerous trials and some green roll-outs of IPv6 with a clear Asiatic leadership, most of the Internet fabric today is still utilizing IPv4, as endpoints and user equipment will be for a long time retaining the old IP protocol. Even IMS deployments and 3G terminals are neglecting IPv6 for the moment; increasing patches to the IPv4 networking and solutions for achieving peer-to-peer connectivity over the well-known IPv4 infrastructure seems to be preferable by operators and vendors rather than acquiring expertise, design and roll-out backbone and transmission IP networks based on IPv6.

In the next stage, the second range of trials for a Future Internet were leaded by the US in the last ten years to redesign completely the whole architecture of the Internet, beyond the pure layer three protocol. The so-called clean state approach focuses on unconventional, breakthrough and long-term research that tries to break the network's ossification. To this end, the research programmes led by the USA can be characterized by two research questions: 'With what we know today, if we were to start again with a clean slate, how would we design a global communications infrastructure?' and 'How should the Internet look in 15 years?' (University, n.d.). This next wave of Future Internet concepts was launched by the end of 2006 and results are still not disclosed.

Today's conceptions on the drivers of a Future Internet widely admit that a comprehensive and holistic outlook is to be considered beyond the pure protocol problematic, networking aspects or even beyond the technological issues.

According to the European approach (Loyola, 2008a), numerous new usage scenarios and a great number of requirements not foreseen when the Internet was designed justify a fresh look at the Internet architecture. Greater robustness and security, better handling of privacy matters and improved management capabilities are just some of the emerging requirements. A multiplicity of innovative applications – of both an economic and social nature – was enabled by sensors, 3D media, virtualization of resources and ultra-high end-to-end through-put and required dynamic service architectures and platforms. The key trend towards wireless communications on the move is also emerging as a key requirement to be supported locally. The European Technology Platforms (ICT, 2008) have devoted considerable effort to defin-ing options and concepts for the Future Internet to support a sustainable society. The four key pillars are:

(1) the Internet by and for People;
(2) the Internet of Contents;
(3) the Internet of Services;
(4) the Internet of Things.

All of the above, supported by a powerful future network infrastructure, will form the Future Internet.

A telecom operator contributes with the network infrastructure and application enablers to construct this vision of the Future Internet. When examining the role of the IMS-enabled service architecture of a telecom network, contributions become especially valuable to the Internet of services. This notion comprises the set of concepts that envisage the Internet as a gigantic distributed service platform that permits new services to be developed as mash-ups of existing and reusable services, which become available through well-defined program-matic service interfaces reachable by means of the Internet service fabric, i.e. the World Wide Web and its core protocol, HTTP. The Internet of services would be not only a cloud of online service functionality to be integrated into richer applications but also encompasses all elements involved in service creation and delivery to users, including service hostage and management, dynamic adaptation of capacity to load, performance, etc. These are the current trends denominated as platform-as-a-service feature, which will be consolidated in the Internet of services to ease the complete lifecycle management of a service, reduce its development costs and allow a true online process for all steps in the service lifecycle. Thus, the key factors of an Internet of services will include the online availability of service functionality for mash-ups and the online availability of platform features. Beyond this, different levels of expertise will be required to construct and deploy services in the Future Internet, ranging from corporations and professionals to end-users, facilitating a true democ-ratization of the service creation of delivery. The Internet of services will foster innovation and participation, following the user-generated services and user-controlled services.

How will an IMS-enabled telecom network contribute? The key for the success of an Internet of services will broadly depend on the rich functionality offered and available online. The current situation in the web 2.0 era is that only information and media-oriented services are available for mash-ups with an open programmatic interface, together with some specific providers of platform-as-a-service features. Interpersonal communication services, where the strength of a telco resides in terms of services, are not included today, due to technological and business considerations. As part of the latter, IMS provides feature-rich interpersonal communications with a clear multimedia emphasis, thus providing a richer portfolio of service features to be exported to the online cloud of services of the Internet of services. As a clearly defined standard oriented to service construction, the IMS services can be reduced to minimum service feature building blocks or enabling features that can be regarded as ingredients in a service mix, easily divisible, which allows a rich commercialization of these 'ingredients for making services' rather than exposing monolithic and complex services into the cloud in the Internet, which probably would not attract developers to integrate the latter into their mash-ups.

Focusing on the technology domain, IMS is closer to the Internet technologies and will ease considerably the process of exposing the telecom service assets into the cloud of services available on the Internet. The programmatic interfaces in the Internet of services will be based either on a service orientation (e.g. WebServices) or resource orientation (e.g. REST) paradigms, which are intrinsically synchronous technologies with a request–response pattern in the underlying protocols, while telecom services are based on asynchronicity and its underlying protocols are usually event-driven. IMS, as the merger of both paradigms – as an infrastructure built upon Internet protocols to provide telecom services – will help the smooth transition from the telecom world of services into the Internet of services.

2

IMS Technology

The success of packet-switched technology in the Internet resulted in the proliferation of multimedia services. Multimedia services triggered the demand for higher bandwidths. From observing the fixed network environment and the growth of the Internet in the past decade, it can be stated that users have an inexhaustible demand for bandwidth. The mobile networking environment is currently experiencing a similar phenomenon. End-user's demand for higher bandwidths is driven by their need to access multimedia services and Internet applications on the mobile devices connected to mobile access technologies. This can be summed up as the 'Mobile Internet' paradigm, which has driven the evolution of originally circuit-switched mobile networks to the current all-IP network architecture. Therefore, it can be inferred that the evolution witnessed in the mobile telecommunication world is driven by two important factors: the evolution of radio access technologies and that of the supporting core network architecture, as depicted in Figure 2.1.

The usage of mobile data services was originally limited due to a lack of available bandwidths. The low bit rates inherent in GSM technology made the data applications over GPRS not widely used, and the bandwidths available for the end user become a key factor. The introduction of 3G radio technology enabled the launch of data services from a mobile terminal. The HSDPA enhancements are expected to popularize multimedia and broadband services on mobile devices. This led to the evolution of radio technologies to make available higher bit rates for delivering multimedia-rich applications. However, radio technology-centric evolution needs to be well supported by the core network and hence a parallell network-centric evolution is essential to provision multimedia services with appropriate end-to-end quality of service and in a cost-effective manner.

The current status of IMS technology is best understood by tracing the transition phases of mobile networks towards the all-IP converged network. This chapter begins with an investigation of transition stages of the cellular networks from 2G to 2.5G to 3G and into the long-term evolution technologies. The discussion is focused on the evolution of the radio technologies enabling higher bit rates and other features for the Internet-like multimedia applications to be provisioned on the mobile access networks. Following the radio

IMS: A Development and Deployment Perspective   Khalid Al-Begain, Chitra Balakrishna, Luis Angel Galindo and David Moro
© 2009 John Wiley & Sons, Ltd

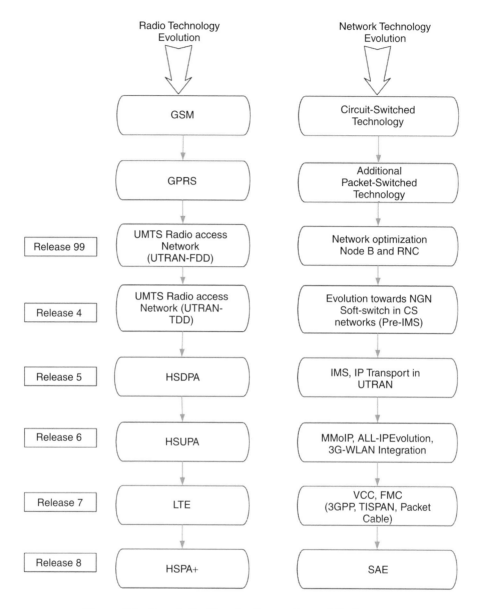

Figure 2.1 Snowball effect leading to the mobile Internet.

technology evolution, core network evolution to the horizontally layered IMS architecture, its functional operation and its distinctive features are presented.

IMS was originally conceived to make the mobile Internet paradigm come true by offering multimedia and Internet services anywhere and anytime on the cellular access technologies. Hence the first release of IMS was tailored to the GPRS/UMTS cellular network architecture. However, later releases of IMS aimed at true access independence by incorporating WLANs, fixed access networks such as xDSL technologies as well as cable access networks. This chapter also investigates the integration of the fixedline access technologies and cable access technologies with the all-IP IMS architecture. In this context the ETSI TISPAN architecture and the PacketCable architectures are discussed.

2.1 Evolution of Mobile Network Architecture

2.1.1 Transition Phases of Cellular Radio Access Technologies

The origin of the commercial cellular telephone system began with the second generation (2G) mobile cellular systems. The 2G systems comprise different technologies, among which the most important one is the global system for mobile communications (GSM) (3GPP, 1999a; ETSI, 1993), which rolled out commercially in 1991. The GSM allows sharing of a single 200 kHz radio channel among eight users by allocating a unique timeslot to each user. GSM is used in the 900 and 1800 MHz bands all over the world except North America (1900 MHz band). It offers voice on circuit-switched technology and a very limited data rate of 9.6 kbps. This opened a new market for mobile data communications through the short messaging service (SMS). The SMS is a connectionless packet service limited to messages containing less than 160 characters. GSM also has a facilitated data transfer service using circuit-switched data (CSD), offering throughput up to 14.4 kbps.

Code division multiple access (CDMA) (ETSI, 2002), which is also referred to as IS-95, is the other important technology in 2G mobile systems. It is mainly used in the America and Asia-Pacific regions. It is a parallel technology to GSM and was recognized by the Telecom Industry Association (TIA) in 1993 and commercialized in 1998. CDMA supports 64 users to share a common 1.25 MHz channel by attaching a pseudo-random code to each user, permitting the decoders to separate traffic at either end. It enables data rates up to 64 kbps by assigning seven supplementary codes to the fundamental code. The other technologies that are covered by the 2G systems are: *IS-136* (TDMA), which is used in North and South America and *personal digital cellular* (PDC), which is used only in Japan. All the 2G systems used circuit-switched technology to offer voice services and a limited data rate.

The low data rate limitations led to the standardization of the high speed circuit-switched data (HSCSD) (3GPP, 1999c; ETSI, 1996) and general packet radio service (GPRS) (3GPP, 1999b, 2001a; ETSI, 1998). HSCSD enables higher rates (up to 57.6 kbps), but it is again circuit based. Therefore, it is inherently inefficient for bursty traffic. Most operators use GPRS due to the drawback of the HSCSD technology. GPRS offers packet-switched transmission at bit rates of about 40 kbps by allocating several timeslots of a frame to the same data transmission. It has kept the GSM radio modulation, frequency bands and frame structure, and is designed to have enhanced features. It permits sending and receiving data at any time (Always-On) and offers higher bit rates by sharing the available radio resources between several users. GPRS is designed to allocate separate uplink and downlink channels, enabling simultaneous voice calls and data transfer, and also introduced billing based on volume. The enhanced data rate for global evolution (EDGE) improves GPRS by the introduction of the 8-PSK modulation, multiplying by 3 the online data rate compared to GPRS. EDGE is included in the 3G–IMT-2000 family of systems. Introducing a new radio modulation scheme triples the bandwidth offered by GPRS. The further evolution of the GSM standard is handled now by the GSM EDGE radio access network (GERAN) group of 3GPP (3GPP, 2002). This group covers in particular the connection of GSM/EDGE to 3G core networks and support of real-time services.

2.1.1.1   Third Generation (3G) Mobile Systems

These are also known as IMT-2000 and were promoted by the ITU to provide higher data rates to offer multimedia services. Since the work started in the standardization bodies ITU Task Group 8/1 (ITU TG8/1) for IMT-2000 and Special Mobile Group 5 (SMG5) subtechnical committee in ETSI for the universal mobile telecommunications system (UMTS), the third generation activities have formed an umbrella for advanced radio system developments. The 3G family is composed of five systems:

- wideband code division multiple access (W-CDMA) includes TDD and FDD modes;
- CDMA 2000 1X;
- time division synchronous code division multiple access (TD-SCDMA);
- EDGE (also called UWC-136);
- digital enhanced cordless telecommunications (DECT).

Among the above technologies, two technologies that are promising are:

- the W-CDMA standard, also called the universal mobile telecommunication system (UMTS), which operates in FDD and TDD modes and is standardized by the 3GPP;
- CDMA-2000 standardized by 3GPP2.

The list of official targets for third generation systems is long and diverse. Small low cost terminals, high spectrum efficiency, seamless roaming and provision of services for mobile and fixed users are a few of the targets.

UMTS mobile systems were introduced with the 3GPP Release 3, also known as Release 99 (3GPP, 2001b) and it required the installation of a completely new radio subsystem called the UMTS radio access network (UTRAN) (3GPP, 2001c). The ITU set guidelines for 3G systems in the IMT-2000 framework to support a data rate of 144 kbps for high mobility and 2 Mbps for stationary nodes. Although the UMTS system met the IMT-2000 guidelines from the start when Release 99 was standardized, the need for improved spectral efficiency, network optimizations and new services have motivated continued enhancements of UMTS. The terrestrial radio interface of UMTS, also called the universal terrestrial radio access network (UTRA), has been developed through a series of releases, from Release 3 to Release 7, which have progressively included a number of features as the FDD mode in Release 3 and the TDD modes high chip rate (HCR) and low chip rate (LCR also called TD-SCDMA) in Release 4.

High speed downlink packet access (HSDPA) with a downlink data rate of 14.4 Mbps was introduced in Release 5. The main objectives of HSDPA was to achieve a substantial increase in network capacity, an increase in peak throughputs (up to 14 Mbps) and a reduction in latency in the downlink. These objectives were achieved by implementing a number of new physical layer and MAC layer techniques, such as adaptive modulation and coding (AMC), fast scheduling and hybrid ARQ, all within a new suitable architecture.

High speed uplink packet access (HSUPA) for uplink was introduced in Release 6 to improve uplink spectral efficiency and further reduce the latency. HSUPA includes techniques such as fast node B scheduling, HARQ and shorter frame size. An uplink peak data rate of 5.76 Mbps is achieved by the introduction of larger code block sizes and/or a shorter frame length of 2 ms. HSDPA and HSUPA together form the high speed packet access (HSPA) technology and provide an efficient UMTS packet system.

In order to enhance the HSPA performance further, the evolution and optimization continued in Release 7 to include continuous packet connectivity (CPC) and multiple-input multiple-output (MIMO) for the downlink, achieving a 28.0 Mbps peak data rate. HSPA evolution (HSPA+) introduced higher order modulation for an uplink equivalent to 16-QAM (quadrature amplitutude modulation) and an equivalent to 64-QAM for the downlink. HSPA+ is characterized by enhanced spectrum efficiency, peak data rate, latency, improved layer 2 protocols, continuous packet connectivity and enhanced uplink to exploit the full potential of UMTS technology. These characteristics met the immediate and mid-term needs of the end-users. However, the operator and end-user expectations are growing rapidly and alternative competitive access technologies are emerging continuously. To ensure long-term competitiveness of 3G technology, it became essential to introduce a new radio access technology.

Long-term evolution (LTE) was launched by the 3GPP in order to ensure 3G competitiveness in a 10-year perspective and beyond. In this context, 3GPP included the 'Evolved UTRA and UTRAN' work item in 2004 (3GPP, 2006d, 2008e). The aim of the work item is to investigate the means of achieving enhanced service provisioning by improving data rates, capacity, spectrum efficiency and latency, thereby providing optimum support for packet-switched services (3GPP, 2007o, 2007p). The overall target of the long-term evolution (LTE) of 3G is to arrive at an evolved radio access technology that can provide service performance on a par with or even exceeding that of current fixedline access, at substantially reduced cost compared to current radio access technologies.

The requirements for the design of the 3GPP LTE system is prescribed in the 3GPP specification TR 25.913 (3GPP, 2006d) and is summarized as follows:

1. Providing significantly higher data rates compared to the existing technology such as the HSDPA and enhanced uplink, with target peak data rates up to 100 Mbps for the downlink and up to 50 Mbps for the uplink.
2. The capability to provide three to four times higher average throughput and two to three times higher cell-edge throughput when compared to systems based on HSDPA and enhanced uplink as standardized in 3GPP Release 6.
3. Increased spectral efficiency up to fourfold compared to 3G technology.
4. Improved architecture and signalling to reduce significantly control and user plane latency, with a target of less than 10 ms user plane RAN round-trip time (RTT) and less than 100 ms channel setup delay.
5. Support scalable bandwidths of 5, 10, 15 and 20 MHz and including bandwidths smaller than 5 MHz for more flexibility. In order to protect the investments already made by the operators, updates and modifications to the existing radio network architecture is being proposed. This involves a smooth migration into other frequency bands, including those currently used for second generation (2G) cellular technologies such as GSM and IS-95.
6. Support for operation in paired (frequency division duplex/FDD mode) and unpaired spectrum (time division duplex/TDD mode) is possible.
7. Support for end-to-end quality of service.

8. Support for interworking between the existing UTRAN/GERAN and other non-3GPP systems. The handover delay between them is to be less than 300 ms for real-time services and less than 500 ms for non-real-time services.
9. An enhanced multimedia broadcast multicast service (E-MBMS) shall be supported.
10. Reduced capital and operational expense shall be ensured.
11. Optimized support for low mobile speeds (0–10 mph) as well as support for high mobile speeds (10–30 mph).

The following technological building blocks enable to be met the LTE system requirements as prescribed by the 3GPP:

1. *Radio interface technology.* In order to meet the requirements of higher data rates, a new radio transmission technology called the orthogonal frequency division multiplexing (OFDM) has been selected for the downlink and single carrier frequency division multiple access (SC-FDMA) for the uplink. In an OFDM system, the available spectrum is divided into multiple carriers, called subcarriers, which are orthogonal to each other. Each of these subcarriers is independently modulated by a low rate data stream. Different bandwidths are realized by varying the number of subcarriers used for transmission, while the subcarrier spacing remains unchanged. In this way operation in spectrum allocations of 1.25, 2.5, 5, 10, 15 and 20 MHz is supported. OFDM enables transmission adaptation in the frequency domain in E-UTRA. OFDM has several benefits including its robustness against multipath fading and its efficient receiver architecture. It is used in WLAN, WiMAX and broadcast technologies.
2. In order to achieve higher throughputs and increased spectral efficiency so as to meet the coverage, capacity and data rate requirements, multiple-input multiple-output (MIMO) antenna solutions are used by the LTE systems. MIMO refers to the use of multiple antennas at the transmitter and the receiver sides. MIMO beam forming could be used to increase coverage and/or capacity, and spatial multiplexing, sometimes referred to as MIMO, can be used to increase data rates by transmitting multiple parallel streams to a single user (3GPP, 2007t).

The *CDMA-2000* family also comprises a series of standards, from CDMA 2000 1xRTT (using the same RF bandwidth as IS-95) to CDMA 2000 Evolution Data Optimized (EV-DO), which supports downlink data rates up to 3.1 Mbps and uplink data rates of up to 1.8 Mbps for Revision A, and much higher rates for Revision B (exceeding 10 Mbps DL obtained by bundling of RF channels), comparable to those of UMTS HSDPA.

2.1.2   Transition Phases of Mobile Core Network Architecture

In an effort to understand the evolution of the mobile networks to an all-IP infrastructure, the previous section focused on the evolution of mobile radio access technologies. This section investigates the transition stages of the mobile network architecture with focus on the evolution of the core network through the various releases starting from Release 99 to Release 8, as depicted in Figure 2.2

For the sake of completeness and better understanding, the discussion begins with the GSM/GPRS networks. A simplified GSM/GPRS network architecture is depicted in Figure 2.3. The base station subsystem (BSS) and network subsystem (NSS) in the GSM

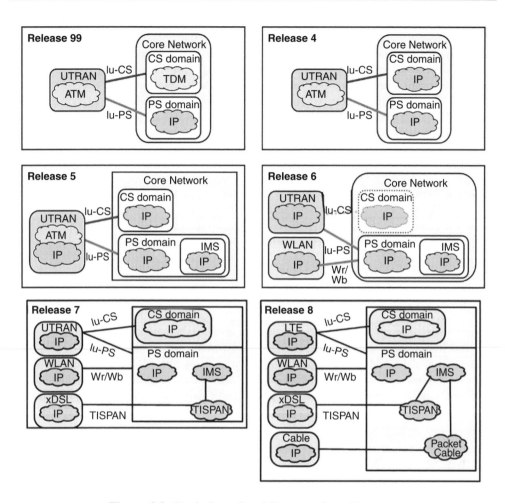

Figure 2.2 Evolution of mobile network architecture.

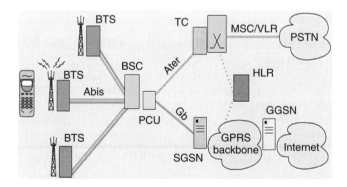

Figure 2.3 GSM/GPRS network architecture.

architecture support the voice and circuit-switched data services. The BSS consists of a base transceiver station (BTS) and a base station controller (BSC). BTS handles the radio physical layer and BSC deals with radio resource management and handover. The NSS for circuit-switched (CS) services consists of the mobile switching centre (MSC), the visitor location register (VLR) integrated in the MSC and the home location register (HLR). GPRS provides packet-switched services over the GSM radio. The packet control unit (PCU) is the additional network entity in the BSS that manages packet segmentation, radio channel access, automatic retransmission and power control. The major new element introduced by GPRS is an NSS that processes all the data traffic. It comprises two network elements:

- The *serving GPRS support node* (SGSN) keeps track of the location of individual mobile stations and performs security functions and access control.
- The *gateway GPRS support node* (GGSN) encapsulates packets received from external packet networks (IP) and routes them towards the SGSN.

The Gb interface between the BSS and the SGSN is based on the frame relay transport protocol. The SGSN and GGSN are interconnected via an IP network.

2.1.2.1   3GPP Release 3 Network Architecture

The 3GPP Release 3 network architecture shown in Figure 2.4 mostly reuses the components from the GSM/GPRS network architecture. However, UMTS is based on a new radio technology and has a big impact on UTRAN. It has additional equipments such as multimedia gateways for interconnecting packet and circuit-switched domains, allowing for the convergence of services based on these two different technologies. The first version of

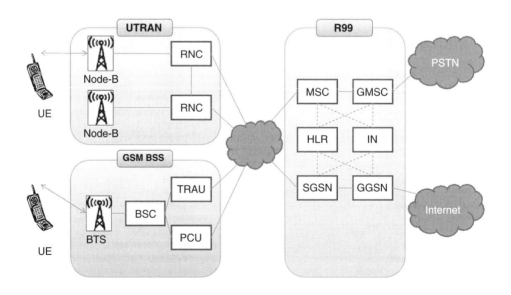

Figure 2.4 UMTS Release 3.

UMTS offered higher data bit rates so that multimedia services, such as video-telephony, could be proposed to the subscribers. Variable bit rates on the radio interface is another advantage offered by UMTS compared with 2G and 2.5G systems. Apart from the common features with the GSM/GPRS architecture, UMTS Release 3 mainly consist of two independent subsystems: the UMTS terrestrial radio access network (UTRAN) and the core network connected over a standard Iu interface:

- The *UMTS terrestrial radio access network* (UTRAN) is composed of node B and a radio network controller (RNC). The RNC is equivalent to the BSC in GSM. It controls the radio resource within its coverage through node B. Node B is the equivalent of the GSM base transceiver station (BTS) and converts the data flows through the Iub and Uu interfaces. Its role is mainly to perform physical layer functions such as modulation, coding, interleaving, rate adaptation, spreading, etc.
- The *core network* of UMTS is responsible for call establishment and handling, data transmission, mobility management, etc. It is based on the same technology as the GSM/GPRS subsystem (3GPP, 2001a) and is equivalent to the network subsystem (NSS) in the GSM/GPRS network. It reuses most components from the GSM/GPRS core network such as the MSC, VLR and HLR for the circuit-switched domain and the SGSN and GGSN for the packet-switched domain. However, there are a few enhancements in the core network:
  - *Packet-switched* (PS) is an evolution of the GPRS SGSN/GGSN with a more optimized functional split between the UTRAN and core network.
  - *Circuit-switched* (CS) is an evolution of the NSS with the transcoder function moved from the BSS to the core network.

As can be seen in Figure 2.4, the UTRAN consists of several possibly interconnected radio network subsystems (RNSs). An RNS contains one RNC and at least one node B. The RNC is in charge of the overall control of logical resources provided by the nodes B. RNCs can be interconnected in the UTRAN (i.e. an RNC can use resources controlled by another RNC) via the Iur interface. Node B provides logical resources, corresponding to the resources of one or more cells, to the RNC. It is responsible for radio transmission and reception in the cells maintained by this node B. A node B controls several cells.

2.1.2.2   3GPP Release 4 Network Architecture

3GPP started the work towards an all-IP architecture with Release 4. This evolution was typically driven by two main objectives, the first being to achieve independence of the transport and control layer to ease the implementation of new applications and the second being to achieve optimization for operation and maintenance for the access networks. Figure 2.5 depicts a Release 4 architecture.

3GPP Release 4 has an unchanged radio part. However, Release 4 triggered the evolution towards a next-generation network (NGN) type of architecture. In this context, it proposed a major enhancement in the circuit-switched domain. The most important enhancement of UMTS Release 4 is a new concept called the bearer independent core network (BICN). Up to and including Release 99, all circuit-switched connections have been routed through the core network via E-1 connections inside 64 kbps timeslots.Instead of using circuit-switched 64 kbps timeslots, the standards prescribed the traffic to be carried inside ATM or IP packets.

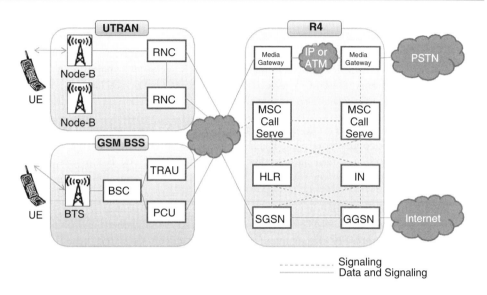

Figure 2.5 UMTS Release 4.

In order to achieve this, the MSC has been split into an MSC server and the media gateway (MGW). The MSC server is responsible for call control, communications, user mobility and mobility management. The media gateway is responsible for handling the actual bearer traffic and routing functions. MGW is also responsible for transcoding of the user data for different transmission methods. In this way it is possible, for example, to receive voice calls generated from the GSM A interface via E-1 64 kbps timeslots at the MSC media gateway, which will then convert the digital voice data stream into a packet-switched ATM or IP connection. The MSC server can manage many media gateways, which allows a better separation between control functions and routing functions. This evolution allows a better separation between control and processing functions in the network. Therefore it eases the introduction of new features and consequently new services. Hence the introduction of this new architecture was driven by network operators who realized the benefits of combining the circuit- and packet-switched core networks into a single converged network for all traffic.

The *3GPP Release 5* specification is a crucial step towards the development and implementation of multimedia services in the mobile environment. Firstly, high speed downlink packet access (HSDPA) (3GPP, 2005b) technology allows a significant increase in the bit rates up to 14 Mbps on the downlink channel. Services such as real-time video, web access and FTP, etc., would receive improved throughputs and, thus, the end-user perceives an enhanced quality of experience. However, at the network level, 3GPP extended the NGN concept by specifying IMS over the PS domain in Release 5. It introduces IP transport technology within the UTRAN, as an alternative to the ATM-based UTRAN. By relying on a multiservice IP backbone, IMS enables the introduction of innovative multimedia services over mobile networks. It introduces the capabilities to support IP-based multimedia services such as voice-over IP (VoIP) and multimedia-over IP (MMoIP), and makes use of the packet-switched network for the transport of control and user plane data. It eases the introduction of new multimedia services and applications by the operators. IMS is a subsystem that controls the provision of services between the servers and the core network. It allows integration of real-time applications and services. The PS domain deals with the mobility and handover aspects.

The most important evolution introduced in UMTS R5 is the IP multimedia subsystem (IMS). Figure 2.6 shows the proposed simplified version of the 3GPP all-IP UMTS core network architecture. New elements in this architecture are:

- *Call state control function* (CSCF). The CSCF is an SIP server that provides/controls multimedia services for packet-switched (IP) terminals, both mobile and fixed.
- *Media gateway* (MG). All calls coming from the PSTN are translated to VoIP calls for transport in the UMTS core network. This media gateway is controlled by the MGCF using the H.248 protocol.
- *Media gateway control function* (MGCF). The first task of the MGCF is to control the media gateways via the media gateway control protocol H.248. Also, the MGCF performs translation at the call control signalling level between ISUP signalling, used in the PSTN, and SIP signalling, used in the UMTS multimedia domain.
- *Home subscriber server* (HSS). The HSS is an extension of the HLR database with subscriber multimedia profile data.

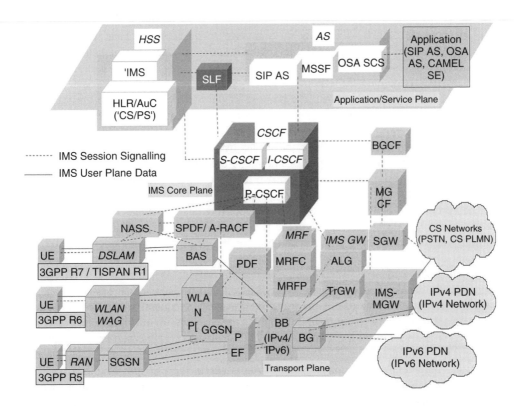

Figure 2.6 Network architecture evolution from Release 5 to Release 7.

2.1.2.3   3GPP Release 6

Evolution of the UMTS technology continues with 3GPP Release 6, where significant uplink performance enhancements (similar to HSDPA) are introduced through the enhanced dedicated channel (E-DCH), as explained in Section 2.1.1. From the core network perspective, new broadcast capabilities are introduced through the multimedia broadcast/multicast services (MBMS) feature. Release 6 also proposed enhancements to support WLAN integration and QoS improvements.

MBMS

The MBMS (3GPP, 2006c, 2007i, 2008c) is one of the key features introduced in Release 6. It defines capabilities to address the same information to many users in one cell using the same radio resources. The MBMS is a unidirectional point-to-multipoint service in which data are transmitted from a single source entity to multiple recipients. Transmitting the same data to multiple recipients allows network resources to be shared. By this, the MBMS architecture enables the efficient usage of radio network and core network resources, with an emphasis on radio interface efficiency. The MBMS is provided over a broadcast or multicast service area which can cover the whole network or be a small geographical area such as a shopping mall or sports stadium, allowing for region-specific content distribution. Examples of broadcast services are advertisements for upcoming or ongoing multicast services or localized advertisements such as ads for attractions or shops within the broadcast area. An example of a service using the multicast mode could be near real-time distribution of video clips from national and regional sports events for which a subscription is required.

The MBMS is defined in 3GPP TS 22.146 (Stage 1), 23.146/25.346/43.246 (Stage 2) and various Stage 3 specifications. The purpose of the MBMS is to enable the transfer of broadcast content to the user equipment within a service area with efficient use of radio resources from a central media server within the operators network, called the BM-SC. The MBMS architecture does not describe the means by which the BM-SC obtains the service data. The data source may be external or internal to the mobile network, e.g. content servers in the fixed IP network. 3GPP has defined two modes of operation: the broadcast mode and the multicast mode.

The introduction of MBMS in the radio interface requires techniques for optimized transmission of the MBMS bearer service in the radio interface such as point-to-multipoint transmission, selective combining and transmission mode selection between point-to-multipoint and point-to-point bearers. Radio access network signalling has been extended to enable user equipment to identify when MBMS transmissions for specific services are to take place and the bearers that are used for serving cell and neighbour cell transmission. The radio access network can decide to transmit the MBMS content by the most efficient means, point-to-point or point-to-multipoint bearers, or not to transmit within a cell at all based on an evaluation, made at the time of transmission, of the numbers of users that require a transmitted session. To reduce the cell capacity occupied by multicast transmission, user equipment can combine transmissions made in the multiple neighbouring cells. The MBMS is a best-effort service with no guarantee of continuous reception. Recovery mechanisms can be provided at the application layer but not within the radio access network.

Interworking

The IMS core network supports, at least from an architectural point of view, the interworking with IP external networks, which might not be fully compatible with IMS protocols through the Mb interface. That being the case, an appropriate gateway needs to be deployed. Besides, IMS also supports interworking with circuit-switching networks for voice call services based on SS7 protocols like ISUP or its bearer-independent counterpart, BICC. Protocol interworking between SIP and BICC or ISUP has also been addressed to support basic voice calls between the IM CN subsystem and the legacy CS networks.

2.1.2.4   3GPP Release 7

Release 7 introduces IMS/core network enhancements related to multimedia telephony, combining of circuit- and packet-switched services (CSI), policy and charging, and voice call continuity. It is proposed to keep the current radio network and core network separation to enable fast roll-outs and easy implementation to terminals. Listed below are the most prominent core network enhancements prescribed in Release 7.

Fixed Broadband Access to IMS (FBI)

It is recognized that several standards organizations are in the process of defining NGN (next-generation network) session control using IMS as a platform. Fixed broadband access provides a means of extending IMS services to wireline subscribers. The work on fixed broadband access (FBI) will embed IMS as the framework for advanced services for many types of operators. IMS is extended to act as a common core and services platform for fixed and mobile networks. Under the fixed broadband access (FBI) work, 3GPP is considering requirements from a number of different communities of operators including those based on DSL and cable technology. IMS is now accepted as the de facto basis for session oriented services in next-generation telecommunications networks. A number of features have been introduced primarily in order to extend the IMS towards providing services for fixed broadband access:

- NAT traversal;
- optional use of SIP preconditions for wireline access;
- PSTN bridging the use of IMS as a transit network with PSTN round the edge.

Combining CS and IMS (CSI)

CSI provides a means for a mobile station to combine a CS voice call with PS-based IMS services between the same two users. The motivation is to introduce multimedia telephony, without the immediate requirement to upgrade the radio to support VoIP. This is done by using legacy CS voice for the voice component of the multimedia session. In addition, CSI introduces an SIP-based mechanism for users to exchange terminal capabilities. Capability exchange provides a means by which party A can know which IMS services/media types can be included as part of a multimedia session with party B. Capability exchange can be done independent of CSI and is expected to be used for detecting if a voice call can be successfully upgraded to video. CSI provides a path to introduce subscribers to IMS services and in particular the concept of rich calls. The main usage scenario is to start each session with a CS voice call, then to detect what add-on services are possible and to display these available services to the user in order to encourage their use. By reusing CS calls for the voice, there is no risk of VoIP performance issues impacting IMS launch. CSI also allows operators to manage their investment in rolling out IMS while establishing the level of market demand for new services. CSI is the basis for video sharing services (adding video to an established voice call in progress) being launched on some networks.

Voice Call Continuity (VCC)

VCC is a home IMS application, which enables seamless continuity of voice services between the CS domain and IMS. A VCC subscriber's calls are anchored in the home IMS when roaming across the CS domain and IMS. VCC provides functions for call originations, terminations and call continuity across the CS domain and IMS. The VCC application is inserted in the control signalling path of the VCC subscriber's CS and IMS sessions for employment of a 3pcc (third party call control function) controlling these sessions. Calls established over the CS domain are redirected to IMS for insertion of a VCC application in the call control signalling path using standard CS domain techniques available for redirecting calls at call establishments; these calls are then processed according to standard IMS procedures. This allows for voice service continuity across the CS domain and IMS. VCC provides selection of the domain for delivery of incoming calls to VCC subscribers simultaneously registered in the CS domain and IMS. VCC provides interdomain mobility while maintaining active session(s), manifested by transfers between the CS domain and IMS, with the capability to transition multiple times in both directions. Figure 2.6 shows the network architecture evolution from Release 6 to Release 7.

2.1.2.5   3GPP Release 8

For LTE that offers a high-performance radio interface, a high-performance core network is required in order to experience commercial success. Impact on the overall network architecture including the core network is being investigated in the context of 3GPP system architecture evolution (SAE). It aims at optimizing the core network for packet-switched services and including the IP multimedia subsystem that supports all access technologies. The combined evolution of LTE and SAE forms the basis for 3GPP Release 8. As of today (3GPP, 1999d), 3GPP has approved to freeze the functional requirements of LTE as well as SAE as part of Release 8. There is proof of substantial industrial commitment towards LTE deployment in the form of contributions and intellectual inputs to the 3GPP LTE specification groups. Also, many recent press announcements from vendors and operators indicate the same.

The future migration or evolution of the GSM network architecture is SAE, or system architecture evolution, which has been renamed by 3GPP as the evolved packet system (EPS). The terms SAE or EPS may be used interchangeably. 3GPP defines EPS in Release 8 as a framework for an evolution or migration of the 3GPP system to a higher data rate, lower latency packet-optimized system that supports multiple radio access technologies. The focus of the work is on the packet-switched domain, with the assumption that the system will support all services including voice in this domain. In order to meet the improved

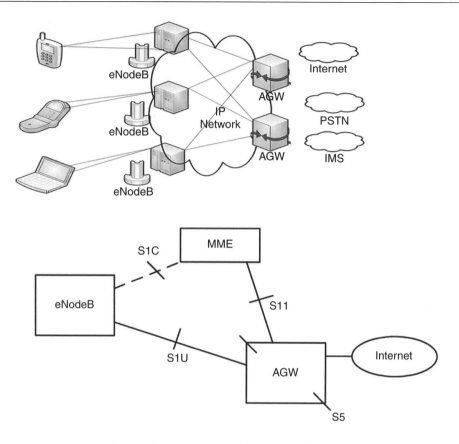

Figure 2.7 A simplified SAE architecture.

latency requirement, it was required to reduce the number of network nodes involved in data processing and transport. A flatter architecture as prescribed by the standards would lead to improved latency and transmission delay. Figure 2.7 depicts a simplified LTE system architecture and consists of two types of network nodes one at the user plane and the other at the control plane:

1. Evolved node B (eNodeB) is the enhanced BTS that provides the LTE air interface and performs radio resource management for the enhanced LTE radio interface.
2. Access gateway (AGW) provides the termination of the LTE bearer and acts as the mobility anchor point and packet date network gateway for the user plane.

Evolved Packet System (EPS) Architecture

In its most basic form, the EPS architecture consists of only two nodes in the user plane, a base station and a core network gateway (GW). The node that performs control plane functionality (MME) is separated from the node that performs bearer plane functionality (GW), with a well-defined open interface between them (S11), and by using the optional interface S5 the gateway (GW) can be split into two separate nodes (serving gateway and the PDN gateway). This allows for independent scaling and growth of throughput traffic and control signal processing and operators can also choose optimized topological locations of nodes within the network in order to optimize the network in different aspects. The basic EPS architecture is shown in Figure 2.7, where support nodes such as AAA and policy control nodes have been excluded for clarity.

One important performance aspect of EPS is a flatter architecture. For packet flow, EPS includes two network elements, called evolved node B (eNodeB) and the access gateway (AGW). The eNodeB (base station) integrates the functions traditionally performed by the radio network controller, which previously was a separate node controlling multiple node Bs. Meanwhile, the AGW integrates the functions traditionally performed by the SGSN. The AGW has both control functions, handled through the mobile management entity (MME), and user plane (data communications) functions. The user plane functions consist of two elements: a serving gateway that addresses 3GPP mobility and terminates eNodeB connections, and a packet data network (PDN) gateway that addresses service requirements and also terminates access by non-3GPP networks. The MME, serving gateway and PDN gateways can be collocated in the same physical node or distributed, based on vendor implementations and deployment scenarios.

The basic architecture of the EPS/SAE contains the following network elements:

- The mobility management entity (MME) manages mobility, UE identities and security parameters. MME functions include:
  - NAS signalling and related security;
  - inter-CN node signalling for mobility between 3GPP access networks (terminating S3);
  - idle mode UE tracking and reachability (including control and execution of paging retransmission);
  - roaming (terminating S6a towards home HSS);
  - GW selections (serving GW and PDN GW selection);
  - MME selection for handovers with MME change;
  - SGSN selection for handovers to 2G or 3G 3GPP access networks;
  - HRPD access node (terminating S101 reference point) selection for handovers to HRPD;
  - authentication;
  - bearer management functions including dedicated bearer establishment.
- Serving gateway is the node that terminates the interface towards EUTRAN. For each UE associated with the EPS, at a given point of time, there is one single serving gateway. Serving GW functions include:
  - the local mobility anchor point for the inter-eNodeB handover;
  - mobility anchoring for inter-3GPP mobility (terminating S4 and relaying the traffic between the 2G/3G system and PDN gateway), which is sometimes referred to as the 3GPP anchor function;
  - EUTRAN idle mode downlink packet buffering and initiation of network triggered service request procedure;
  - transport level packet marking in the uplink and the downlink, e.g. setting the DiffServ code point, based on the QCI of the associated EPS bearer;
  - accounting on the user and QCI granularity for interoperator charging;

- lawful interception;
- packet routing and forwarding.
- The PDN gateway is the node that terminates the SGi interface towards the PDN. If a UE accesses multiple PDNs, there may be more than one PDN GW for that UE. PDN GW functions include:
  - mobility anchor for mobility between 3GPP access systems and non-3GPP access systems, which is sometimes referred to as the SAE anchor function;
  - policy enforcement;
  - per-user based packet filtering (by, for example, deep packet inspection);
  - charging support;
  - lawful interception;
  - UE IP address allocation;
  - packet screening.
- The evolved UTRAN (eNodeB) supports the LTE air interface and includes functions for radio resource control, user plane ciphering and packet data convergence protocol (PDCP).

It can be summarized that the future network architecture is evolving to be flatter than the previously existing hierarchical layered architecture. From a four-node architecture in previous releases, an LTE/SAE network proposes a two-node only architecture. Comparing the functional breakdown with existing 3G architecture:

- Radio network element functions, such as the radio network controller (RNC), are distributed between the AGW and the enhanced BTS (eNodeB).
- Core network element functions, such as the SGSN and GGSN or PDSN (packet data serving node) and routers, are distributed mostly towards the AGW.

Latency is a key factor in the cellular data services. Sometimes, the most disturbing factor for the failure of a service is the high latency in radio applications, even if there are no real-time requirements. Consequenty, HSPA+ and SAE/EPC-derived enhancements, like the one-tunnel solution, target at reducing latency in packet data connections.

Other IMS-Related Enhancements of Release 8

- *Common IMS.* Since Release 7, 3GPP's definition of IMS has been open to access by noncellular technologies. This has generated cooperation with groups specifying IMS for wireline applications (e.g. ETSI TISPAN and Cablelabs). In Release 8, 3GPP's organizational partners (OPs) have decided that 3GPP should be the focus for all IMS specifications under their responsibility. The 'common IMS' work is an agreement between the 3GPP OPs to migrate work on the IMS and some associated aspects to 3GPP for all access technologies. This will simplify the deployment of fixed mobile convergence (FMC) solutions, minimize the risk of divergent standardization and make the standardization process more efficient.
- *The multimedia priority service* enhances IMS to provide special support for disaster recovery and national emergency situations. The multimedia priority service allows suitable authorized persons to obtain preferential treatment under a network overload situation. This means that essential services will be able to continue even following

major incidents. It is intended that users provided with the multimedia priority service
will be members of the government or emergency services. The multimedia priority
service provides IMS functions similar to those already available in the CS network.
When this feature is deployed, disaster recovery will be assisted by the multimedia
capabilities of IMS. This feature is also an enabler to the eventual replacement of CS
networks by IMS.

- *IMS enhancements for support of packet cable access* is a work item in Release 8 that
  introduces specific enhancements to IMS that are primarily of interest to the packet
  cable community. However, it is anticipated that some of the aspects will also be of
  interest to other IMS users. This work item consists of three main topics:
  - Security. The cable environment requires a specific security approach driven by
    its particular architecture for home networking. This work item will enhance IMS
    security to fit in the packet cable architecture.
  - Cable client deployment. Operational procedures in the cable industry typically
    involve the deployment of a blank client, which is one customized by commands
    sent from the network. This work item will provide the tools needed to support this
    deployment model for IMS.
  - Regulatory. Cable networks are often used for residential primary line support. This
    means that they must comply with regulatory features covering this aspect. This
    work item will provide the necessary regulatory features for cable deployment in
    North America and other regions. This will include support for equal access.
- *IMS service brokering* aims to enhance the existing service deployment technology in
  IMS to simplify further the deployment of services and to make the system more effi-
  cient. In particular, IMS service brokering considers the possible interactions between
  several developed services and how these will impact the network.
- *VCC-related enhancements.* There are two main areas of work related to VCC for
  Release 8: IMS centralized services (ICS) and VCC for SAE/LTE access and the CS
  domain:
  - The IMS centralized service (ICS) is an approach to the provision of communi-
    cation services wherein all services and service control are based on IMS mecha-
    nisms and enablers. IMS services are delivered over 3GPP CS, VoIP capable and
    non-VoIP capable 3GPP PS, and non-3GPP PS access networks, with provision of
    user transparent service continuity between these access networks. ICS users are
    IMS subscribers with a supplementary service subscription in IMS. ICS user ser-
    vices are controlled in IMS based on IMS mechanisms with the CS core network
    basic voice service used to establish voice bearers for IMS sessions when using
    non-VoIP capable PS or CS access. Centralization of service control in IMS provides
    consistent user service experience across disparate access networks by providing
    service consistency as well as service continuity when transitioning across access
    networks.
  - Release 7 VCC requires simultaneous activation of CS and PS radio channels for
    enablement of service continuity between CS and PS systems. This is not possible
    when transitioning between SAE/LTE access and CS access and with transitions
    involving some other combinations of 3GPP radio systems such as 2G CS and
    3G PS. Studies are being conducted to enable service continuity between such
    systems.

2.2   IMS – A Standardized All-IP Infrastructure

The telecommunication networks are evolving to an all-IP-based core network that has enabled the convergence of different access technologies as presented in the previous section. The session initiation protocol (SIP) (Rosenberg *et al*., 2002) and IP multimedia subsystem (IMS) (3GPP, 2007g) are the two most important driving forces towards this convergence. Prior to IMS, services were tightly coupled with a specific transport network and signalling protocol. With the introduction of IMS, the communication network architecture paradigm has transitioned from being stove-pipe vertically integrated to a horizontally integrated, layered architecture. The layered architectural design ensures that the transport and bearer services are separated from the signalling and session management services. It aims for minimum dependency between layers. It can be inferred that IMS is designed to separate service-related functions from the underlying transport-related technologies. This independence between the service, session and transport layers makes the underlying technology invisible to the user regardless of where in a multiservice, multiprotocol, multivendor environment the user resides. This concept of nomadicity enables seamless communication between fixed and mobile users. IMS architecture provides many common functions at the control layer and service layer that can be reused for fast service creation and delivery. A high level of abstraction at the service layer enables the service developers to develop applications quickly and focus more on the logic of the application rather than lower layer complexities. This also ensures a large developer base for IMS services, yielding to more scope for innovation. Finally, the end-users are catered for by a variety of blended multimedia services (blend of telecom and Internet services) which otherwise would not have been possible. A comprehensive service-centric view of the IMS architecture is presented below. However, the 3GPP standards have always focused on a network-centric view of the IMS architecture. In order to understand the service capabilities of IMS and its ability to act as a comprehensive service delivery framework, it is essential to study the IMS core network components, protocols and their functional operation. This section presents the IMS core that forms the session or service control layer in the horizontal-layered IMS representation depicted in Figure 2.8

IMS was introduced as a major architectural change in 3GPP Release 5 (3GPP, 2007j) and releases beyond (3GPP, 2006b, 2007f, 2007g). It has introduced new capabilities into the architecture through the addition of new network elements called the call state control functions (CSCFs), which are added to allow delivery of basic IP multimedia services, such as initiation of multimedia sessions between two subscribers and which has complete control of the sessions. The CSCFs are limited only to the session control layer of IMS and they depend heavily on an adequate application and service layer above. The session control capabilities of the CSCFs are abstracted and hooked on to the application plane in the IMS. The application plane stimulates new and innovative IP multimedia services, taking advantage of basic multimedia session control and services capabilities provided by the core network.

Figure 2.9 shows a simplified IMS core network architecture (3GPP, 2007j, 2007k). It shows the essential network elements used in providing real-time IP multimedia services. Since the IMS uses only the packet-switched domain for transport and local mobility management, it can be deployed without a circuit-switched domain. Hence, circuit-switched domain elements like mobile switching centres are not shown in the figure. The network

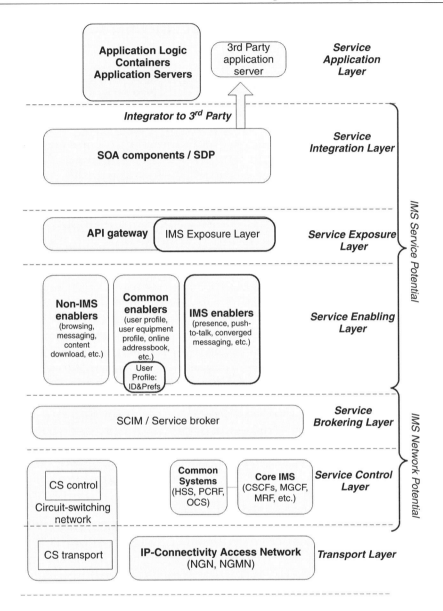

Figure 2.8 Service-centric horizontally-layered representation of IMS.

components that are included in the IMS architecture and their functionalities are listed below.

2.2.1 IMS Components

A central system component of the IMS network infrastructure is the call session control function (CSCF). One of the main purposes of the CSCF is the signalling routing function.

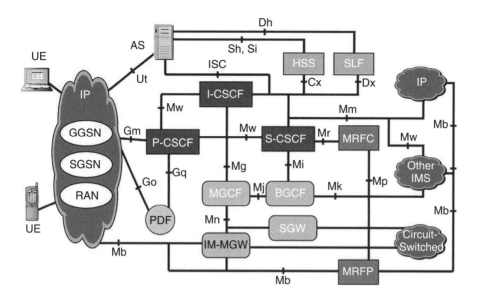

Figure 2.9 IMS core network components.

CSCF proxies all SIP signalling traffic and provides the following network services:

- session control services including subscription, registration, routing and roaming;
- combination of several different media bearers per session;
- central service based charging;
- secure authentication and confidentiality based on the ISIM/USIM; and
- quality of service control (policy decision function (PDF)).

Going deeper into the IMS architecture, the CSCF splits into three controlling functions, each differing in the functionality they provide:

- *P-CSCF (Proxy-CSCF)*. This node acts as the first point of contact for call sig-nalling coming from the user equipment (UE) to the IMS network. From the SIP point of view, the P-CSCF is acting as an outbound/inbound SIP proxy server. This means that all the requests and/or responses initiated by or destined to the IMS terminal will go through the P-CSCF. The P-CSCF is allocated to the IMS terminal during IMS registration and does not change for the duration of the reg-istration (i.e. the UE communicates with a single P-CSCF during registration). For a roaming subscriber, the P-SCSCF will be located in the visited network or, more specifically, the P-CSCF for a given user is located in the same network as the GGSN or the gateway of the generic IP transport system, from which service is received. Authentication of IMS users is another P-CSCF function. Once a user is authenticated, the P-CSCF asserts the user identity to the rest of the nodes in the network, so that they do not need to authenticate the user further. Based on the user's identity, the other nodes can provide personalized services and generate account records.

Behaviour and Interactions.

Incoming signalling to the IMS reaches the P-CSCF via the Gm interface. The P-CSCF forwards the call signalling on to the S-CSCF via the Mw interface, which shall be fully supported as well. Communication with the I-CSCF is also done via the Mw interface. The P-CSCF communicates now with the PDF via the Gq interface.

- *I-CSCF (Interrogating-CSCF).* The I-CSCF provides the functionality of an SIP proxy server. It also has an interface to the SLF (subscriber location function) and HSS (home subscriber server). This interface is based on the diameter protocol (Calhoun *et al.*, 2003). The I-CSCF retrieves the user path or the route through which the user could be communicated. Based on the path information, it routes the SIP request to the appropriate destination, typically an S-CSCF. It is located at the boundary of the IMS network and acts as an entry point for SIP signalling coming from another network. I-CSCF is usually located in the home network 1 and is responsible for assigning an S-CSCF to the subscriber. This node is an SIP proxy located at the boundary of the IMS network and acts as an entry point for SIP signalling coming from another network. The signalling could be: an SIP call setup request destined to a subscriber of the operator's network; an SIP call setup request destined to a roaming subscriber within the operator's network; a registration request.

  The I-CSCF is usually located in the home network. For incoming registration requests, the I-CSCF is responsible for assigning an S-CSCF to the subscriber. The choice of S-CSCF can be made dependent on the subscriber's identity (SIP address or international mobile subscriber identity (IMSI)), handled on a load-sharing basis or using a main server/backup server arrangement. The address of the I-CSCF is listed in the domain name system (DNS) records of the domain. When an SIP server looks for the next hop for a particular SIP message, it obtains the address of an I-CSCF of the destination domain. Additionally, the I-CSCF may optionally encrypt parts of the SIP messages that contain sensitive information about the domain, such as the number of servers in the domain, their DNS names or their capacity – this functionality is known as the topology hiding internetwork gateway (THIG). A network will typically include a number of I-CSCFs for the sake of scalability and redundancy.

Behaviour and Interactions.

The I-CSCF communicates with the home subscription server (HSS) via the Cx interface. In 3GPP specifications, this interface is based on the diameter protocol

- S-CSCF (serving-CSCF). The S-CSCF is an SIP server that performs session control. It maintains a binding between the user location (in terms of IP addressing and the path of network nodes that provide reachability to the user) and the users SIP address of record (also known as the public user identity). Like the I-CSCF, the S-CSCF also implements a diameter interface to the HSS. The S-CSCF provides SIP routing services. All the SIP signalling that the IMS terminal sends and all the SIP signalling the IMS terminal receives traverses the allocated S-CSCF. The S-CSCF inspects every SIP message and determines whether the SIP signalling should visit one or more application servers 'en route' towards the final destination – those application servers would potentially provide a service to the user. A network will typically include a number of S-CSCFs for the sake of scalability and redundancy. Each S-CSCF serves a number of IMS terminals, depending on the capacity of the node.

This is the central node of the signalling plane. The S-CSCF is essentially an SIP server, but it performs session control as well. In addition to SIP server functionality the S-CSCF also acts as an SIP registrar. This means that it maintains a binding between the user location (e.g. the IP address of the terminal the user is logged on) and the user's SIP address of record, also known as the IMS public user identity (IMPU). It also carries out the call/session and accounting control for a given subscriber. The S-CSCF is always located within the subscriber's home network. This means that all mobile-originated call signalling is routed via the user's home network. This might not be an optimal routing strategy, but it covers only signalling traffic; call traffic is forwarded using standard IP routing between the involved IP transport networks. The S-CSCF provides SIP routing services. All the SIP signalling that the IMS terminal sends and all the SIP signalling the IMS terminal receives traverses the allocated S-CSCF. The S-CSCF inspects every SIP message and determines whether the SIP signalling should visit one or more application servers 'en route' towards the final destination. Those application servers would potentially provide a service to the user. The S-CSCF also provides SIP translation services based on DNS E.164 Number Translation.

Policy enforcement is also a function of the S-CSCF. A user may not be authorized to establish certain types of sessions – the S-CSCF keeps users from performing operations unauthorized by the network operator. A network will typically include a number of S-CSCFs for the sake of scalability and redundancy. Each S-CSCF serves a number of IMS terminals, depending on the capacity of the node.

Behaviour and Interactions.

Like the I-CSCF, the S-CSCF also implements the Cx interface to the HSS, as already mentioned. This interface allows the S-CSCF to: download from the HSS the authentication vectors of the user who is trying to access the IMS and download the user profile from the HSS. The user profile includes the service profile, which is a set of triggers that might cause an SIP message to be routed through one or more application servers. Inform the HSS that this is the S-CSCF allocated to the user for the duration of the registration.

The S-CSCF also communicates with the application servers via the ISC interface.

Other components of the IMS core network are:

1. *Policy Decision Function.* The PDF (policy decision function) node makes policy decisions based on policy setup information obtained from the P-CSCF via the Gq interface. The PDF module could be defined as an extension to P-CSCF to handle the policy decision function. PDF exchange policy information with a policy enforcement point (PEP) on the gateway side enables media path bandwidth resources going through media gateway to be controlled. Some of the behaviours of this module are explained in the following.

Behaviour and Interactions

The tasks of the PDF are as follows:

(a) Policies. The PDF is responsible for maintaining the network operator's policy and user profile-based policy. Policies are in form of rules that tell how the call shall be handled. PDF thus maintains a policy repository which includes network and user profile-based policies.

(b) Mapping of policies into QoS parameters. It is the duty of the PDF to map these policies into QoS parameters so that these policies can be enforced at PEP.

(c) Services domain. The PDF is responsible for maintaining various services and the QoS required by each service. Thus, it maintains service-based local policy (SBLP) for a session.

(d) Define QoS parameters. The PDF needs to define values of QoS parameters to PEP. All these QoS parameters for various services need to be maintained in PDF and can be passed on to PEP during a bearer resource reservation.

(e) Information about PEP. The PDF needs to maintain a database about various PEPs it is handling, the PEP's capabilities and all the media resources that PEP would be managing.

(f) Manage media resources. The PDF needs to manage media resources to ensure that adequate resources are there for services. In addition, the PDF needs to ensure that resources are efficiently allocated and utilized. Hence, it is responsible for allocating, enabling and revoking resources for a session.

(g) Authorize the QoS decision. It is the responsibility of the PDF to make a decision to authorize QoS depending on policies, services a user is subscribed to and availability of resources to handle the services. If the session is not according to policies or if there are not enough resources available for the service, it rejects the authorize QoS request giving the reason for rejection.

(h) Authorization token generation and handling. The PDF generates an authorization token on request and includes session ID and PDF to contact in the authorization token. On receiving this authorization token from any other entity like PEP, the PDF can uniquely identify the session using the session ID.

(i) Gate information. The PDF would also maintain gate control information about various flows being monitored and filtered at PEP. The PDF sets up gate control at PEP.

(j) SBLP information. The PDF is responsible for passing SBLP information to PEP, like the destination IP address, destination port number, media direction information, media type, bandwidth and various other parameters.

The PDF's top-level behavioural requirement identifies six areas of responsibility, each of which has a specific role to fulfill, and is assigned to a logical subcomponent. The six responsibilities are listed below:

(a) session database – internal database;
(b) P-CSCF interfacing – communication manager for the PCSCF side;
(c) policy controlling – central processing of policy decision management;
(d) PEP interfacing – communication manager for Media Gateway side 3.0;

(e) COPS message parsing – COPS message translator and constructor;

(f) PEP manager – database for PEP (media controller information).

However, from the later releases the functions provided by the PDF are transferred to the policy control and charging rules function (PCRF). The PCRF encompasses the policy control decision and flow-based charging control functionalities. It provides network control regarding the QoS and flow-based charging (except credit management) towards the gateway. When the PCRF receives service information from the P-CSCF(AF), depending on the network operator's configuration, the PCRF may check whether the P-CSCF is allowed to pass the application/service information to the PCRF. The PCRF shall control how a certain service data flow that is under policy control shall be treated in the GW, e.g. discarded, etc., and ensure that the GW user plane traffic mapping and treatment is in accordance with the user subscription profile. The PCRF may check that the service information provided by the P-CSCF is consistent with the operator-defined policy rules before storing the service information. The service information is used to derive the QoS for the service. The PCRF may reject the request received from the P-CSCF. The PCRF indicates in the response the service information that can be accepted by the PCRF. The PCRF interacts with the GW by either responding to a request from the GW or by sending an unsolicited update of policy and charging rule information. An unsolicited update (i.e. the addition, modification or removal of policy and charging rules) may be triggered by a change of PCRF internal policy configuration or by receiving updated session information from a P-CSCF. The PCRF shall check whether the new policy configuration or session information changes the allocation of service data flows to bearers. If the binding of service data flows to bearers is not affected, the PCRF may send the unsolicited update, e.g. in the case of a change of QoS or charging key. Otherwise, the PCRF shall wait for a request from the GW with new information for the mapping (e.g. an updated TFT) and may only send unsolicited decisions that do not change the installed set of service data flow filters (e.g. a change of gate status or QoS). For new policy and charging rules or service data flow filters, the PCRF may also be configured or instructed by the AF to send the unsolicited update. Then, the PCRF shall bind these new rules or filters to the bearers based on the mapping information that was previously received from the GW (3GPP, 2008b, 2008d).

2. *Policy Enforcement Point.* In the IMS domain, the policy enforcement point (PEP) is responsible for enforcing the policies of the network. Broadly, PEP functions are as follows:

(a) communicating with PDF for registration, policies and information for each session;

(b) storing the operator's policy applicable to all the sessions in the local policy database;

(c) acting as a policy enforcement point for the operator's policies;

(d) establishing a session between users and allowing media to flow between users and the media controller (PEP);

(e) opening and closing the gate so that media can traverse between users on receiving the appropriate decision message from the PDF;

(f) monitoring sessions between the users and verifying that they are in accordance with the policies;

(g) notifying the PDF on detection of violation of policies by the user.

Behaviour and Interactions

PEP is generally responsible for enforcing the policies of the operator. It polices the incoming traffic and screens QoS usage and alerts PDF about violation of the rules that take place during the call. Resource allocation and bandwidth reservation for outgoing and incoming traffic on both sides of the media controller are also taken care of by the PEP.

3. *IM-SSF (IP Multimedia Services Switching Function).* The IM-SSF acts as an application server on one side and on the other side it acts as an SCF (service-switching function) interfacing the gsmSCF (GSM service control function) with a protocol based on CAP (CAMEL application part, defined in 3GPP TS 29.278 (3GPP, 2005a)).

4. *MRF (Media Resource Function).* The majority of next-generation services require media processing that can be delivered by general purpose media servers. For example, these media servers are able to play audio prompts, convert text to speech, mix audio for conference calls, etc. They are often delivered with VoiceXML (VXML) technology. The IMS standard includes the system component, called the media resource function for this purpose. It is further divided into a signalling plane node called the MRFC (media resource function controller) and a media plane node called the MRFP (media resource function processor), thereby separating the functions of media control from media processing. MRFC accepts instructions from an application service and directs the media server to handle the media stream. The MRFC controls the media server using the standard H.248/Megaco protocol. In practice the media server and the media gateway could merge into one component, as long as it is permitted by a system design for intended services. It acts as an SIP user agent and contains an SIP interface towards the S-CSCF. The IMS specifications also give MRFC other responsibilities, like charging of the media server, conference call control and control of subscriber roaming. These can be equally accomplished by application services themselves or by the CSCF. Moreover, since VXML-enabled media servers and application platforms often facilitate the development and integration of new services, the future IMS architecture could evolve towards a VXML-based interface standard between media servers and application services. This approach is without MRFC involvement. The MRFP implements all the media-related functions and provides a wide range of functions for multimedia resources, including provision of resources to be controlled by the MRFC, mixing of incoming media streams, sourcing media streams (for multimedia announcements) and processing of media streams. In general, the MRFP provides the following functions:

- provides resources to be controlled by the MRFC;
- mixes incoming media streams (e.g. for multiple parties);
- sources media streams (for multimedia announcements): text, files, within the bearer path, etc., with various options (language, customs, etc.);
- processes media streams (e.g. audio transcoding, media analysis);

- provides tones and supports DTMF within the bearer path;
- notifies the MRFC when an event has occurred, e.g. AS/CSCF may have directed it to collect DTMF digits.

A few MRF-related standards are listed below:

- VoiceXML makes the implementation of voice response (VR) applications (audio dialogues) productive and independent of call control and other application logic. VoiceXML scripts are stored on an HTTP server and supplied to the media server VoiceXML browser.
- CCXML (call control XML) is designed to provide telephony call control support for VoiceXML. CCXML scripts are stored on an HTTP server and supplied to the media server voice browser. A CCXML script might support an outdial notification function in a messaging system.
- NETANN (basic network media services with SIP), now on its eleventh version, is an SIP-based AS-MS protocol that supports the implementation of network announcements (Im sorry, the number) and other basic functions. NETANN can also invoke VoiceXML scripts.
- MSCML (media server command markup language) is an SIP-based AS-MS protocol used to implement advanced conferencing and fax applications.
- MSML/MOML (media sessions markup language/media objects markup language) are SIP-based AS-MS protocols used to implement advanced conferencing and fax applications.
- Media server control is a new initiative of an informal IETF working group called 'Network Working Group', which specifies an 'SIP control framework'. This approach separates SIP-based AS-MS control into various XML packages, which cover basic IVR, VoiceXML, fax and conferencing
- MRCP (media resource control protocol, RFC4463) is an SIP-based protocol to utilize text-to-speech (TTS) and automatic speech-recognition (ASR) functions from VoiceXML. Using a network protocol to allow these resources to be separated from the primary media server allows them to be independently developed and for the service provider to be independent of any particular TTS or ASR implementation.

5. *BGCF (Breakout Gateway Control Function).* The BGCF controls the transfer of calls to and from the PSTN. In general, it determines the network in which the interworking with the PSTN is to occur. The request for doing so will be received by the BGCF from the S-CSCF. The breakout gateway control function (BGCF) therefore performs the following functions:

- selects the network in which PSTN breakout is to occur;
- selects local MGCF or peer BGCF;
- provides security through authorization of peer networks.

IMS standards do not specify the criteria for the BGCF to use when selecting the PSTN/PLMN access point. Some possible factors include the following:

- current location of the calling UE;
- location of the PSTN/PLMN address;
- local policies and business agreements between the visited and home network.

6. *SGW (Signalling Gateway).* SGW performs lower layer protocol conversions.

7. *MGCF (Media Gateway Control Function).* The MGCF implements a state machine that performs protocol conversion and maps SIP to either ISUP (ISDN user part) over IP or BICC (bearer independent call control) over IP. The protocol used between the MGCF and the MGW is ITU-T Recommendation H.248 (Groveo and Lin, 2007). The MGCF communicates with CSCF and controls the connections for media channels in an IMS-MGW. The MGCF performs protocol conversion between ISUP and the IMS call-control protocols. MGCF performs the following functions:

   - controls the parts of the call state that pertain to connection control for media channels in a MGW;
   - communicates with the S-CSCF;
   - communicates with the I-CSCF and BGCF;
   - performs protocol conversion between ISUP and SIP;
   - out-of-band information received in MGCF and may be forwarded to the CSCF/MGW.

8. *MGW (Media Gateway).* The MGW interfaces the media plane of the PSTN (public-switched telephone network) or CS (circuit-switched) network. On one side the MGW is able to send and receive IMS media over the real-time protocol (RTP). On the other side the MGW uses one or more PCM (pulse code modulation) timeslots to connect to the CS network. Additionally, the MGW performs trans-coding when the IMS terminal does not support the codec used by the CS side.

   The IMS-MGW may terminate bearer channels from a switched circuit network and media streams from a packet network. The IMS-MGW may support media conversion, bearer control and payload processing (e.g. codec, echo canceller or conference bridge). The MGW performs the following functions:

   - interacts with the MGCF for resource control;
   - terminates bearer channels from the circuit-switched network and media streams from the packet network (e.g. RTP streams in an IP network);
   - supports media conversion, bearer control and payload;
   - processing (e.g. codec, echo canceller, Conference Bridge);
   - performs transcoding and interworking.

9. *HSS (Home Subscriber Server).* IMS specifications include a definition of the subscriber profile database, a central repository for subscriber information called the home subscriber server (HSS). The HSS maintains all subscriber information necessary for establishing sessions between users and for providing them with services. The HSS is an equivalent of HLR in GSM, but it goes even further in its standardization. The HSS supports many interfaces, out of which the most important ones are Cx and Sh interfaces. The interface between the application server and the HSS is denoted as Sh, while the Cx interface is used by the HSS to communicate with the I-CSCF and the S-CSCF. The HSS includes information about:

   - subscriber registration (name, address, subscribed services, etc.);
   - subscriber preferences (baring information, forwarding settings, etc.) which are included in the service profile that resides in the HSS;
   - subscriber location;

- HSS offers the AS a transparent data storage, per subscriber, for information of the specific service hosted by the AS. The Sh interface is built upon the existing DIAMETER standard used for accessing subscriber information. Note that DIAMETER is an extension of the RADIUS standard.

The IMS standards take into account previous operators' investments in their network infrastructures and moderate future technology deployments in that extent. The HSS shall be viewed as a server that provides a consolidated interface to multiple sources of information, better than another big repository for subscriber data. By this, IMS can rely on subscriber information from multiple sources like the HSS, network HLRs, location servers, business repositories, etc. The HSS can pull the data together under one interface, rather than replacing already existing repositories. For more information regarding the IMS HSS subscriber data requirements refer to the 3GPP specification TS 23.008 (3GPP, 2007q). For details of the Cx and Sh interface specifications refer to 3GPP TS 29.228 (3GPP, 2007d) and TS 29.328 (3GPP, 2007r) respectively.

10. *SLF (Subscription Location Function)*. The SLF is a simple database that maps users addresses to HSSs. Both the HSS and the SLF implement the diameter protocol.

11. *AS (Application Server)*. The AS is an architectural piece that accommodates service-related entities, which may be dedicated platforms, like enabler servers, conferencing/MRF or gateways to other service environments, like CAMEL/IN, or brokering functionalities, like a service brokering/SCIM. They also act as exposure entities following the OSA model. They may be generic sevice platforms such as a service container with a middleware that usually adheres to the JAIN SLEE of SIPServlet specs, together with other Java components. It provides value-added IP multimedia services and resides in the user's home network or in a third party location. The AS can also provide service capability interaction manager (SCIM) functions to manage interactions. In general, the application server provides the following functions:

- provides service control for IMS;
- may be directly connected to S-CSCF or via the OSA gateway for third party security;
- interacts with the HSS to obtain subscriber profile information;
- may support applications such as presence, conference control, online charging, etc.

Different options of application servers exist, as shown in Figure 2.10:

- CAMEL services via the CAMEL support environment (CSE) are intended for the support of existing IN services (provides service continuation).
- OSA services via the open service access service capability server are intended for the support of third party application providers. OSA SCS provides access and resource control.
- IMS services on the SIP application server is intended for new services. A multitude of widely known APIs such as CGI, CPL and SIP servelets are available. An SIP AS hosts and executes IP multimedia services based on SIP. It is the heart of the application plane and provides end-user service logic. It could include a variety of telephony and nontelephony application servers to cater for a broad

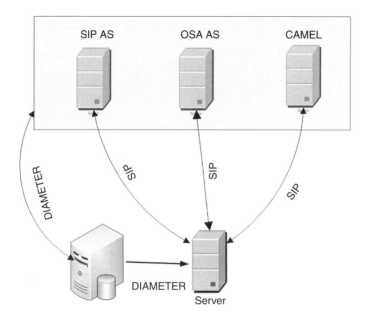

Figure 2.10 Application server options for IMS.

range of voice and multimedia services. The SIP AS provides (i.e. hosts and executes) value-added services to the subscribers. These could be anything from a streaming video service (Video-on-Demand, VoD) to providing voice and video mail services. Depending on the actual service, the AS can operate in the SIP proxy mode, SIP user agent mode (i.e. endpoint) or sip back-to-back user agent (B2BUA) mode (i.e. a concatenation of two SIP user agents).

• IMS services directly on the CSCF (similar to SIP AS), such as SIP AS co-located on the CSCF, are considered to be useful for simple services. These may also be beneficial for the service availability and the service performance. This is not prescribed in the standards, but it is a vendor-specific proprietary solution.

CAMEL supports:

• legacy IN services in 2G and 3G networks;
• services based on proven and reliable IN technology (reuse!);
• but CAMEL is expensive and limited in evolution.

SIP AS supports:

• multimedia conferencing services, integrated with HTTP;
• exploitation of cheaper Internet technology;
• easier service creation, but not yet proven for carrier grade services.

OSA is an API that could be mapped to both CAMEL and SIP:

• OSA is enabling EAI in telecoms;
• proven technology (reuse of existing services in NGN);
• support implementation of different business model (walled versus open garden);
• best support of third parties.

Behaviour and Interactions

The AS interfaces the S-CSCF using the SIP protocol via the ISC interface.

12. *Charging Collection.* There are two interfaces specified by IMS for charging for next-generation services. The interface Rf allows for off-line charging, whereas the Ro interface provides online charging capabilities to support real-time charging models. Both interfaces use a common format of DIAMETER message, the ACCOUNTING REQUEST. Either both or one of the interfaces should always be supported by an IMS compliant application service. The main function of the IMS charging entity is to adopt a flexible structure of the call data records and to integrate with existing back-end business systems and online charging technologies.

3GPP has also defined interfaces that are standardized means by which the above-mentioned components interface and talk to each other. Listed below are the interfaces as specified by the 3GPP:

- The *Gm* interface supports the communication between the UE and the IMS network (i.e. the P-CSCF) related registration and session control. The protocol used for the Gm interface is SIP.
- The *Mw* interface allows the communication and forwarding of signalling messaging between CSCFs, e.g. during registration and session control.
- The *Cx* interface supports information transfer between the CSCFs (I-CSCF and S-CSCF) and the HSS. Procedures related to S-CSCF assignment, accounting and authorization and routing information retrieval require information transfer between the S-CSCF/I-CSCF and HSS. The procedures that require information transfer between CSCF and HSS are:
  - procedures related to the S-CSCF assignment;
  - procedures related to routing information retrieval from the HSS to CSCF;
  - procedures related to authorization (e.g. checking of the roaming agreement);
  - procedures related to authentication: transfer of security parameters of the subscriber between the HSS and CSCF;
  - procedures related to filter control: transfer of filter parameters of the subscriber from the HSS to CSCF.

  For further information on the Cx interface refer to TS 23.228 (3GPP, 2007g).
- The *ISC* interface is between the serving CSCF and the service platform(s). An application server offering value-added IM services resides either in the user's home network or in a third party location. The third party could be a network or simply a standalone AS. The serving CSCF to the AS interface is used to provide services residing in an AS. Two cases were identified: S-CSCF to an AS in the home network and S-CSCF to an AS in the external network (e.g. third party or visited).

  The SIP Application Server may host and execute services. The SIP application server can influence and impact the SIP session on behalf of the services and uses the ISC interface to communicate with the S-CSCF. The ISC interface shall be able to support subscription to event notifications between the application server and S-CSCF to allow the application server to be notified of the implicit registered public user identities, registration state and UE capabilities and characteristics in terms of SIP user agent capabilities and characteristics. The S-CSCF shall decide whether an application server is required to receive information related to an incoming initial SIP request to ensure appropriate service handling. The decision at the S-CSCF is based on (filter) information received from the HSS. This filter information is stored and conveyed on a per application server basis for each user. The name(s)/address(es) information of the application server(s) are received from the HSS. For an incoming SIP request, the S-CSCF shall perform any filtering for ISC interaction before performing other routing procedures towards the terminating user, e.g. forking, caller preferences, etc. The S-CSCF does not handle service interaction issues. Once the SIP application server has been informed of a SIP session request by the S-CSCF, the IM-SIP AS shall ensure that the S-CSCF is made aware of any resulting activity by sending messages to the S-CSCF. When the name/address of more than one application server is transferred from the HSS, the S-CSCF shall contact the application servers in the order supplied by the HSS. The response from the first application server shall be used as the input to the second application server. Note that these multiple application servers may be any combination of the SIP application server, OSA service capability server or IM-SSF types. The S-CSCF does not provide authentication and security functionality for secure direct third party access to the IM subsystem. The OSA framework provides a standardized way for third party secure access to the IM subsystem. If an S-CSCF receives an SIP request on the ISC interface that was originated by an application server destined to a user served by that S-CSCF, then the S-CSCF shall treat the request as a terminating request to that user and provide the terminating request functionality as described above. Both registered and unregistered terminating requests shall be supported. It shall be possible for an application server to generate SIP requests and dialogues on behalf of users. Such requests are forwarded to the S-CSCF serving the user and the S-CSCF shall perform regular originating procedures for these requests.

More specifically the following requirements apply to the IMS service control interface:

– The ISC interface shall be able to convey charging information as per 3GPP TS 32.200 (3GPP, 2005c) and TS 32.225 (3GPP, 2006e).
– The protocol on the ISC interface shall allow the S-CSCF to differentiate between SIP requests on Mw, Mm and Mg interfaces and SIP requests on the ISC interface.

• The *Sh* interface allows the application server to communicate to the HSS. For this interface, the following criteria shall apply:

– The Sh interface is an intraoperator interface.
– The Sh interface is between the HSS and 'SIP application server' and between the HSS and the 'OSA service capability server'.

The HSS is responsible for policing what information will be provided to each individual application server:

– The Sh interface transports transparent data, for example, for service-related data, user-related information, etc. In this case, the term 'transparent' implies that the exact representation of the information is not understood by the HSS or the protocol.
– The Sh interface also supports mechanisms for transfer of user-related data stored in the HSS (e.g. user service related data, MSISDN, visited network capabilities, user location (cell global ID/SAI or the address of the serving network element, etc.)).

- Note that before providing information relating to the location of the user to an SIP application server, detailed privacy checks frequently need to be performed in order to meet the requirements in TS 22.071 (3GPP, 2007s). The SIP application server can ensure that these privacy requirements are met by using the Le interface to the GMLC instead of using the Sh interface.
- The Sh interface also supports mechanisms for transfer of standardized data, e.g. for group lists, which can be accessed by different application servers. Those application servers sharing the data shall understand the data format. This enables sharing of common information between application servers, e.g. data managed via the Ut interface.

- The *Gq* interface allows for dynamic QoS-related service information to be exchanged between the policy decision function (PDF) and the P-CSCF. This information is used by the PDF for service-based local policy decisions.

2.2.2 IMS Protocols

The protocols that are employed in the IMS architecture are the protocols that are necessary for multimedia communication, such as the session control protocol (SIP), session description protocol (SDP) and media transport protocols such as RTP and RTCP. Further, IMS uses the DIAMETER protocol for AAA functionality and is also utilized by the policy and charging architecture within the 3GPP system. Although introduced by IMS, PCRF is not exclusive for IMS services. The MEGACO protocol is used for media control. The IMS protocols are classified into the following categories : session signalling protocol, media control protocol, authentication protocol and media delivery protocol.

2.2.2.1 Session Signalling Protocol

The session protocols include the session control protocol and session description protocol.

Session Initiation Protocol

The session initiation protocol (SIP) (Rosenberg *et al.*, 2002) was originally proposed by the Internet Engineering Task Force (IETF) to establish, modify and terminate multimedia sessions in all-IP networks. In November 2000, the SIP was accepted as a 3GPP signalling protocol; later it was chosen as the session control protocol for the IMS. The SIP is not a standalone integrated solution for multimedia sessions in a communications system; it is rather a component that can be used in conjunction with other protocols to build a complete multimedia architecture able to provide a broad range of services for users. However, the SIP works independently of any of these protocols and without dependency on the type of session that is being established. Refer to Section 3.1.1 to learn about the potential of the SIP and the capabilities it lends to IMS-based multimedia sessions.

Session Description Protocol

The session description protocol (SDP), published by the IETF as RFC 4566 (Handley, Jacobson and Perkins, 2006), is a format for describing multimedia sessions and its streaming initialization parameters for the purposes of session announcement, session invitation and other forms of multimedia session initiation. The media and transport information included in an SDP session description is the following:

- the type of media (video, audio, etc.);
- the transport protocol (real-time transport protocol (RTP) / user data-gram Protocol (UDP) / internet protocol (IP), H.320 (Greene, Ramalho and Rosen, 2000), etc.);
- the format of the media (H.261 video (Turletli and Huitema, 1996), Moving Picture Experts Group (MPEG) video, etc.).

In addition, the SDP conveys address and port details to enable the streams to be sent, received, or both. Furthermore, information about the bandwidth to be used and contact information for the person responsible for the session are additional desirable information, due to the fact that resources necessary to participate in a session may be limited.

By means of a short structured textual description, the SDP is able to meet an important video streaming requirement, which is to convey the previously mentioned parameters (media details, transport addresses and other session description metadata) to the participants, since they allow them to gather sufficient information to discover and participate in a multimedia session.

The SDP provides a standard representation for such information, regardless of the transport protocol used (including, among others, session announcement protocol (SAP), SIP, real-time streaming protocol (RTSP), electronic mail using the multipurpose Internet mail extensions (MIME) extensions and the HTTP). It is important to keep in mind that the SDP is not intended to support negotiations of session content or media encodings; this is viewed as outside the scope of a session description. These tasks are typically performed by the service logic residing in the application servers.

2.2.2.2 Media Control Protocol

MEGACO/H.248 is the official industry standard protocol for interfacing between the media gateway controllers (MGCs) and media gateways (MGs) (Groves and Lin, 2007). The standard is the result of a collaborative effort between the IETF and ITU standards organizations and has been accepted industy-wide. MEGACO/H.248 has also been adopted by the 3GPP and TISPAN organizations for all media control operations in the IMS architecture. MEGACO/H.248 is logically derived from the MGCP protocol and enhances MGCP by providing the following functionalities:

- support for multimedia and multipoint conferencing;
- improved syntax for efficient semantic message processing;
- TCP and UDP transport options;
- formalized extension process for enhanced functionality;
- expanded definition of packages;
- improved security.

2.2.2.3 Authentication Protocol

The DIAMETER protocol was defined by IETF and performs authentication, authorization and accounting (AAA) in the IMS and NGN. It is defined in RFC 3588 (Calhoun *et al.*,

2003). Diameter security is provided by IPSEC or TLS. The DIAMETER protocol provides the following facilities:

- connection and session management;
- user authentication and capabilities negotiation;
- reliable delivery of attribute value pairs (AVPs);
- extensibility, through addition of new commands and AVPs;
- basic accounting services.

2.2.2.4 Media Delivery Protocol

A media session requires a protocol responsible for transporting media; in this case, the protocols involved are the real-time transport protocol (RTP) and the real-time transport control protocol (RTCP). RTP (RFC 3550) (Schulzrinne *et al.*, 2003) allows real-time media transport, playing out the media at the proper time over unreliable transports (UDP and data-gram congestion control protocol (DCCP)) and jitter presence, i.e. when the data are being transported by the network without keeping a timing relationship. The arrival times of the packets cannot be used to establish a relationship between the media streams; such constraint is resolved by means of time stamps with which the receivers can recover the data timing order. Receivers use buffers and interpolate techniques (to fill possible gaps, if a packet has not been received in time). Since there is a direct relationship between delay and quality, it is a challenge is decide appropriately when to start playing media to the user. If the receiver starts as soon as the first packet is received it is possible that some packets do not arrive when they have to be played; on the other hand, waiting before playing the media would make the maintenance of a normal conversation difficult. RTP packets also carry sequence numbers, binary sender identifiers and the payload type (to identify the current speaker and the encoding and transport format of the data carried in the RTP packet, respectively).

RTCP (Huitema, 2003) is always used together with RTP to monitor QoS and convey information about the participants in an ongoing session (media synchronization and mapping between RTP binary sender identifiers and human-readable names). To enable QoS statistics it implements a feedback on the quality of reception of data by means of RTCP reports, where senders report the number of packets they sent and receivers report the number of packets received. RTCP also provides a mapping between RTP timestamps and a wall clock, allowing synchronization of different media streams, useful, for example, in lip synch (audio-video synchronization). When SDP is used, RTP packets are normally sent to a port with an even number and RTCP messages are sent to the consecutive uneven port.

2.2.3 Distinct Features of IMS

This subsection presents some of the intrinsic features of IMS other than the radio technology features and core network components already discussed. Among them, 3GPP user identities and their relationships are presented first followed by device-related IMS features.

2.2.3.1 3GPP User Identities and Relationships

Normal 3GPP networks use the following identities:

- international mobile subscriber identity (IMSI);
- temporary mobile subscriber identity (TMSI);

- international mobile equipment identity (IMEI);
- mobile subscriber ISDN number (MSISDN).

IMSI is a unique phone identity that is stored in the SIM. To improve privacy, a TMSI is generated per geographical location. While IMSI/TMSI are used for user identification, the IMEI is a unique device identity and is phone-specific.

IMS also requires an IP multimedia private identity (IMPI) and an IP multimedia public identity (IMPU). Both are not phone numbers or other series of digits, but uniform resource identifiers (URIs), which can be digits (a tel-uri, like tel:+1-555-123-4567) or alphanumeric identifiers (a sip-uri, like sip:john.doe@example.com). There can be multiple IMPUs per IMPI (often a tel-uri and a sip-uri). The IMPU can also be shared with another phone, so both can be reached with the same identity (e.g. a single phone number for an entire family). The HSS user database contains the IMPU, IMPI, IMSI and MSISDN as well as other information. The format for IMPUs and IMPIs is standardized in 3GPP TS 23.003 (3GPP, 2007I).

1. IMPI:
   - This is a network access identifier (NAI) as specified in IETF RFC 4282 (Aboba *et al.*, 2005), e.g. username@realm.
   - RFC specifies that NAI should be at least 72 octets; it recommends 253 octets.
2. IMPU:
   - This is an SIP URI as specified in IETF RFC 3261 (Rosenberg *et al.*, 2002) and should take the form sip:user@domain.
   - Tel URL as specified in IETF RFC 3966 (Schulzrinne, 2004) e.g. tel:+4112345.

2.2.3.2 IMPI (IMS Private User Identity)

Each IMS subscriber is assigned at least one private user identity (IMPI). Unlike IMPU, the IMPI takes the format of a network access identifier (NAI). The IMPI is used exclusively for subscription identification and authentication purposes. In short, we can say that an IMPI performs a similar function in the IMS environment as an IMSI does in GSM.

The IMPI has the following purposes:

- It is used as a key for user authentication credentials stored in the HSS.
- Depending on the authentication methods, it may be stored in the device and used in SIP signalling or not.
- It may be used only in SIP messages related to IMS registration, as IMS registration implies user authentication.
- It may be included in CDRs (not mandated by standards)
- In a mobile cellular context (3GPP):
  - For SIM-based authentication (SIM application) it is not used in IMS registration messages and must be derived from the IMSI stored on the SIM accessed by the terminal. In addition, the UE must register with IMS using an IMPU, which is also derived from the IMSI.
  - For USIM-based IMS authentication (USIM Application) it is derived from IMSI stored on the USIM accessed by the terminal and used in IMS registration messages.

- For ISIM-based IMS authentication (ISIM application card) it is effectively trans-
  ported in SIP messages and can be any arbitrary ID defined by the operator (may
  be based on the IMSI).
- It is unique to an application (SIM, USIM, ISIM), i.e. unique to a device registered
  with IMS.
- In a fixed and mobile context (IETF approach):
  - It corresponds to the user name in (IETF) digest authentication (initially defined
    for HTTP).
  - It is therefore used in IMS registration messages
  - It may be shared between different devices, more especially when IMS is accessed
    through soft clients. These devices can concurrently register with IMS using this
    shared IMPI.

If there is no ISIM application, the private user identity is not known. If the private
user identity is not known, the private user identity shall be derived from the IMSI. The
following steps show how to build the private user identity out of the IMSI:

- Use the whole string of digits as the username part of the private user identity.
- Convert the leading digits of the IMSI, i.e. MNC and MCC, into a domain name. The
  result will be a private user identity of the form

```
"'<IMSI>@ims.mnc<MNC>.mcc<MCC>.3gppnetwork.org"'
For example: If the IMSI is 234150999999999
(MCC = 234, MNC = 15),
the private user identity then takes the form:
234150999999999@ims.mnc015.mcc234.3gppnetwork.org
```

2.2.3.3   IMPU (IMS Public User Identity)

An IMS user is identified in a deterministic way. An IMS user is allocated with one or
more public user identities. The IMPUs are allocated by Swisscom. An IMPU can be either
a SIP URI15 or a Tel URI14. When the IMPU contains an SIP URI, the following format
will be used: Sip: first.last@operator.com. Furthermore, it is possible to include a telephone
number in an SIP URI using the following format:

$$\text{Sip:+1-555-123-4567@operator.com;user=phone}$$

The TEL URI is another format that the IMPU can take. The format of the TEL URI is
the following:

$$\text{Tel:+1-555-123-4567}$$

In short, IMPUs have the following purposes:

- Identify the IMS user for communication and service access purposes.
- They are addresses used for SIP routing in the IMS network.
- They can be dynamically and simultaneously associated with multiple devices through
  IMS registration, thus defining a clear separation between the user and the devices it
  may use.

- They can be defined as a phone number (e.g. MSISDN, fixed number) or be similar to an email address (user@domain).
- They may be multiple for the same user, permitting it to have aliases or different persona associated to different services.
- They may be associated with an individual or to a group of individuals.

2.2.3.4 Relationship between Private and Public User Identities

The home network operator is responsible for the assignment of the private user identities and public user identities. Other identities that are not defined by the operator may also exist.

Public user identities may be shared across multiple private user identities within the same IMS subscription. Hence, a particular public user identity may be simultaneously registered from multiple UEs that use different private user identities and different contact addresses. If a public user identity is shared among the private user identities of a subscription, then it is assumed that all private user identities in the IMS subscription share the public user identity. The relationship for a shared public user identity with private user identities and the resulting relationship with service profiles and IMS subscription are depicted in Figure 2.11. An IMS subscription may support multiple IMS users.

The public user identity 1 and public user identity 2 cannot belong to the same alias public user identity group since they do not share the same service profile. The public user identity 2 and public user identity 3 may belong to the same alias public user identity group depending on whether they have the same service configuration for each and every service.

Public Service Identity

The public service identity shall take the form of either an SIP URI or a Tel URI. A public service identity identifies a service or a specific resource created for a service on an

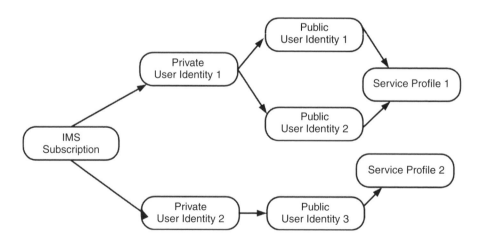

Figure 2.11 Public and private user identities.

application server. The domain part is predefined by the IMS operators and the IMS system provides the flexibility to create the user part of the PSIs dynamically. The PSIs are stored in the HSS either as a distinct PSI or as a wildcard PSI. A distinct PSI contains the PSI that is used in routing, while a wildcard PSI represents a collection of PSIs. Wildcard PSIs enable optimization of the operation and maintenance of the nodes. A wildcard PSI consists of a delimited regular expression located either in the user info portion of the SIP URI or in the telephone subscriber portion of the Tel URI. The regular expression in the wildcard PSI shall take the form of extended regular expressions (ERE). The delimiter shall be the exclamation mark character ('!'). If more than two exclamation mark characters are present in the user info portion or telephone subscriber portion of a wildcard PSI then the outside pair of exclamation mark characters is regarded as the pair of delimiters (i.e. no exclamation mark characters are allowed to be present in the fixed parts of the user info portion or telephone subscriber portion). When stored in the HSS, the wildcard PSI shall include the delimiter character to indicate the extent of the part of the PSI that is wildcarded. It is used to separate the regular expression from the fixed part of the wildcard PSI. For example, the following PSI could be stored in the HSS: 'sip:chatlist!.*!@example.com'. When used on an interface, the exclamation mark characters within a PSI shall not be interpreted as a delimiter. For example, the following PSIs communicated in interface messages to the HSS will match the wildcard PSI of 'sip:chatlist!.*!@example.com' stored in the HSS:

sip:chatlist1@example.com

sip:chatlist2@example.com

sip:chatlist42@example.com

sip:chatlistAbC@example.com

sip:chatlist!1@example.com20

Private Service Identity

The private service identity is applicable to a PSI user and is similar to a private user identity in the form of a network access identifier (NAI), which is defined in IETF RFC 4282 (Aboba *et al.*, 2005). The private service identity is operator defined and although not operationally used for registration, authorization and authentication in the same way as the private user identity, it enables public service identities to be associated with a private service identity, which is required for compatibility with the Cx procedures. Refer to Table 2.1 for the role of different identifiers as specified by the standards.

2.2.3.5 Device-Related IMS Features

User Profile and Service Profile

The user profile is downloaded over the Cx interface from the HSS to the S-CSCF. Each user profile typically consists of IMS user subscription information such as the private user identitiy of the user. Each instance of the IMS user subscription contains one or more instances of the service profile. Figure 2.12 represents the relationship between the user profile and service profile.

Table 2.1   Role of different identifiers.

Identifier	3GPP role	Provisioning in 3GPP	Provisioning in fixed access
IP address	Support media and signalling stream	Downloaded into terminal from DHCP in access network and uploaded into S-CSCF as part of registration process	Associated with line card as part of the service provisioning process
Private identifier	To identify terminal to system as part of registration/ identification process; also used for billing	Held in ISIM explicitly or derived from USIM, then loaded into S-CSCF via registration process	'Pseudo'-IMSI provided and held in the UPSF as part of the provisioning process
Public identifier	Used to identify required terminals on incoming calls; also used as CLI on outgoing calls	Programmed into ISIM and loaded into HSS as part of the service provisioning process	Programmed into USPF as part of the service provisioning process

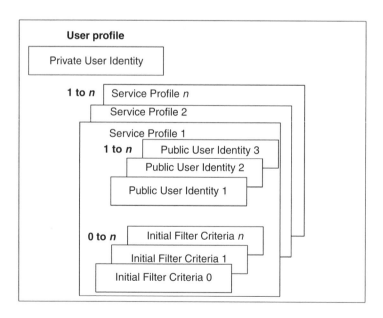

Figure 2.12   Relationship between the user profile and service profile.

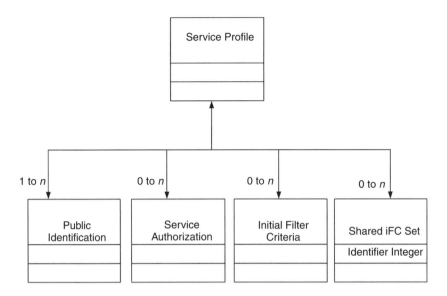

Figure 2.13 Detailed representation of the service profile.

Figure 2.13 shows that a service profile is associated to possibly several IMPUs or PSIs. In the case of IMPUs, they are part of the same IMS subscription, i.e. they correspond to the same subscriber. Each instance of the service profile consists of one or more instances of public identification. Public identification contains the public identities associated with that specific service profile. The information in the core network service authorization, initial filter criteria and shared iFC set apply to all public identification instances that are included in one service profile. Each instance of the service profile contains zero or one instance of the core network service authorization. It identifies a policy through an integer that is to be used by the S-CSCF to decide the types of media that are authorized inside an SIP session associated with the IMPU or PSI. In the latest version of the specification (3GPP, 2007d), core network service authorization also includes a list of communication service identifiers that are authorized for the IMPU/PSI. If no instance of the core network service authorization is present, no filtering related to subscribed media or restriction on IMS communication service identifiers applies in S-CSCF.

Each instance of the service profile contains zero or more instances of the initial filter criteria (iFC) and shared iFC set (3GPP, 2007d). In order to optimize the provisioning and storage of service profiles, the concept of shared iFC sets were introduced. These are sets of initial filter criteria that are shared among multiple subscribers. Instead of being integrally defined in the service profile, they are referenced through an integer. According to the specification, this integer points at iFC sets that are locally stored in the S-CSCF. Shared iFC sets may be shared by several service profiles.

The iFCs form the most important part of the service profile and are composed of the following elements as represented in Figure 2.14:

- A *priority* is defined as an integer. Each iFC in a service profile must have a different priority, which permits the S-CSCF to process the iFCs in a deterministic order.

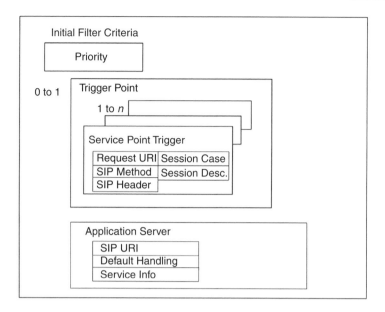

Figure 2.14 Detailed representation of initial filter criteria.

- A *trigger point* consists of a set of criteria that has to be met by the SIP request to be routed to an AS. These criteria are defined as a set of service point triggers (SPT) that are linked through logical Boolean operators (AND, OR, NOT). The service point trigger is basically a logical condition based on the characteristic of an SIP message such as the SIP URI of an application server, to which the request should be routed if the trigger point is true.
- A *default handling element* indicates to the S-CSCF the default handling of the SIP request in case the AS does not respond – either to continue with processing or reject the SIP request.
- An optional *service information element* is a string provisioned in the service profile that the S-CSCF should place in the body of the SIP request before it is sent to the AS. Originally, it was assumed that a service information element could be added in every SIP request that the S-CSCF would forward (proxy) to an AS. This could have been interesting, for instance, to provide a kind of identification of the iFC that led to the forwarding (based on this ID, the AS could determine what services need to be executed). However, it soon appeared that tempering with the body of an SIP request is incompatible with the behaviour of an SIP proxy. The possibility to use service information is therefore limited to the only SIP request issued to an AS that does not result from a proxy behaviour: REGISTER. Consequently, the use of service information should remain very limited in the future.

Globally Routeable User Agent URIs (GRUU)

It is common for individual users to have multiple devices that they use for different purposes. One user may carry a mobile phone, a wireless PDA and a PC with wireless

Table 2.2   Distinct features of IMS.

Capability	Description
Subscriber nomadicity	Decoupling the subscriber from specifc access and specific terminal equipment
Application ubiquity	Application availability from any access network; content *tuning* to match access and terminal capabilities
Resource control	Authorization and availability; accounting in forms of measuring resource usage and resource assurance, policy resource usage and fraud prevention
Subscriber identity and authentication	Common model for all devices, access and applications
Service blending	Service brokering enables applications to provide adaptive behaviours based upon subscriber events and states
Billing	For scenarios crossing multiple provider boundaries

capabilities. In this environment it has become important that the system provides tools to allow the different devices to be addressed efficiently and to harmonize the service presented to the user. The GRUU feature allows the network to specify that a particular IMS transaction is related to a particular device belonging to the user. This feature enhances the experience of IMS users who wish to share a single public identity among multiple devices.

Other Features

Other distinct features of IMS are listed in Table 2.2.

2.3   Evolution of Fixed Mobile Convergence

The term 'fixed mobile convergence' (hereafter referred to as FMC) is currently used in different contexts. Its use is spans from interworking between mobile networks and the PSTN domain to convergence in dual mode devices, which enable WLAN/2G/3G voice roaming, and to FMC as a service delivery platform common to fixed and mobile networks. However, in the context of this book fixed mobile convergence (FMC) is a technology that brings in mutual advantages of the fixed broadband access, the cable operators and mobile operators. This convergence, in the initial phase, allows the users of each domain to benefit from services that are developed in the other domain. For instance, mobile users may receive video streaming offered in cable technology on their cell phones or users can send and receive SMS/MMS on their fixed phones and PCs.

The goal is to create combined services by horizontal convergence of services provided in mobile, fixed PSTN, broadband and cable access domains. Horizontal convergence is a concept that contrasts with traditional vertical interconnection of service-specific network technologies. In vertical convergence, the integration is only at the transport level and not at the service level. The horizontal convergence of fixed mobile domains creates innovative multimedia services such as unique numbering, common contact lists and remote private video recorder (PVR) programming by cell phones. Moreover, new capabilities such as receiving the bandwidth consuming services on cell phones via broadband access or content sharing over fixed and mobile devices can be provided for the users.

To reach this goal, the need for a standardized service control overlay that provides a clean split between data transport and service level is indispensable. The role of this control overlay will be to make the services provided in a domain reachable to the users of other network technologies. Based on the session initiation protocol (SIP), IMS is the most complete IP-based service control plane to set up an overlay on the underlying transport infrastructure and provides the possibility of end-to-end IP-based services. All the services that are developed by service providers will be connected to IMS via a standard and SIP-based interface called IMS service control (ISC).With such an architecture, the new services can be developed independently from underlying infrastructure technologies by different service operators and be connected to IMS via ISC. This reduces the time to market for emerging services and is more profitable for operators. IMS was standardized by the 3GPP for the 3G mobile technology in 3GPP Release 5, but the later releases have integrated other wireless and wireline network technologies such as WiMAX, WiFi, xDSL and broadband cable into the IMS. This integration is being done through next-generation network initiatives by TISPAN and the Cable Television Laboratory (CableLab) Project initiative called PacketCable 2.0 (2007a).

The TISPAN Project on telecommunication and Internet converged services and protocols for advanced networking, by the European Telecommunications Standards Institute (ETSI), in the field of the next-generation networks (NGN) project, considers part of the work on IMS done by 3GPP as their service layer model.

PacketCable (2007a) is introduced and standardized by CableLab to provide IP multimedia services over HFC (hybrid fibre coax) access networks. VoIP is the only major IP-based service considered in (PacketCable 1.5 (2005a)). In contrast, CableLab in PacketCable 2.0 is seeking to develop such an architecture to provide advanced SIP-based services in cable networks by adopting IMS as the service control overlay.

IMS Facilitates Horizontal Fixed Mobile Convergence

In the model proposed by IMS, services become independent of network infrastructure technology. IMS is another significant step (after the intelligent network) to eliminate the cost and complexity of building a separate smokestack network for each new service. In IMS, each service will be connected to the network via standardized interfaces and will become available to the users in any access technology. This is an architecture that makes it simpler and more cost effective for operators to roll out new and personalized services. IMS enables the development and deployment of converged IP infrastructure in which services are shared across multiple access technologies. IMS localizes the users in different domains and access technologies and triggers the required services for the users. Deploying IMS by operators of fixed and mobile networks means implementation of a uniform service control overlay that uses a unique IP-based signalling (SIP). In this model, as depicted in Figure 2.15, CSCFs deployed in different domains are interconnected. This allows a domain access to services of other domains. A uniform application service interface (ISC) and accessibility of services of other domains enables service convergence between different domains. According to all of these features, IMS is considered as the dominant solution for providing fixed mobile convergence.

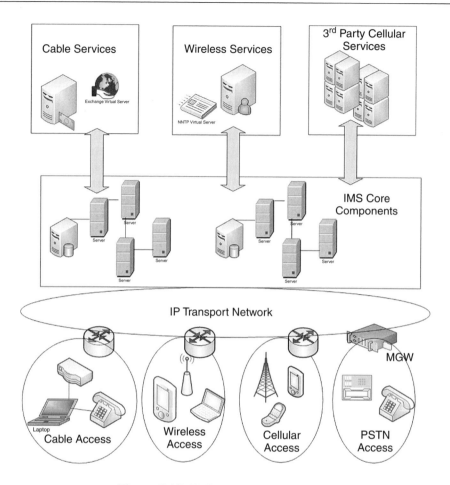

Figure 2.15 IMS – an FMC facilitator.

2.3.1 TISPAN NGN

The 3GPP IMS prescribes a converged IP core network architecture perceived entirely by the mobile operators. The European Telecommunication Standards Institute (ETSI) extended the IMS architecture in the scope of its work on next-generation networks (NGN). TISPAN, which is the standardization body within the ETSI, produces its own releases of standards to incorporate the wireline and fixed operator perspective. Although TISPAN works in the fixed networks environment, it works in close partnership with the 3GPP to bring in the fixed mobile convergence (FMC) to NGNs. NGN Release 1 specifications were finalized in December 2005. It essentially adopts the 3GPP IMS (IP multimedia subsystem) standard for SIP-based applications, but also adds further functional blocks and subsystems to handle non-SIP applications and other requirements of fixed networks not addressed by IMS (NGN 2007). TISPAN makes specifications for several non-IMS subsystems, like the network attachment subsystem (NASS) and the resource admission control subsystem (RACS). NGN Release 2 was finalized in early 2008, which has added a key element to the NGN such as

IMS and non-IMS-based IPTV, home networks and devices, as well as the NGN interconnect with corporate networks.

The TISPAN efforts are focused on two main aspects: one is to improve the subscriber experience and the other is to deliver new services that do not work well over the existing best-effort Internet, or that currently need separate networks. However, from the service provider's perspective, the TISPAN NGN is all about consolidation and cost savings as much as it is about introducing new services and better end-user experience.

The TISPAN Release 1 architecture is based on the 3GPP IMS Release 6 architecture but is intended to address the fixed network operator and provider issues. The architecture uses cooperating subsystems sharing common components, which allows the addition of new subsystems over time, to cover new demands and service classes. It also ensures that the network resources, applications and user equipment (mostly inherited from IMS) are common to all subsystems, thereby ensuring user, terminal and service mobility to the fullest extent possible.

A high level view of the TISPAN NGN architecture is depicted in Figure 2.16, which hides the complexity but highlights the layered approach of the architecture. At the top is the applications domain, which is really a placeholder for the specific application services that a service provider wants to deliver to its customers, and which are not specified by

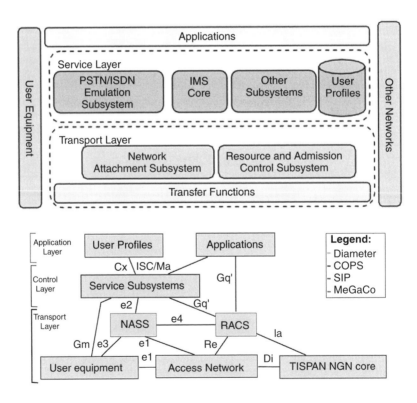

Figure 2.16 TISPAN NGN architecture.

TISPAN. It essentially consists of the application servers (ASs), which host the IMS services and the home subscriber server (HSS). Underneath is the control layer, which consists of a number of reusable service subsystems, among them the 3GPP's IMS core. These service subsystems underpin the IMS applications and enable them to be delivered to the end-users. They include the RTSP-based streaming services, the IMS SIP-based services and the PSTN emulation services, among others. A key feature is that all these services share a variety of other common components. In particular, they are linked to a common database, which means that they all share subscriber identity information and subscriber profile information, rather than having their own separate silo versions of that data. At the bottom is the transport layer, which consists of the NGN core, end-user equipments, access networks and the two most important non-IMS blocks of TISPAN: the NASS and the RACS.

2.3.1.1 NGN Transport Plane

The network attachment subsystem (NASS) provides registration at access level and initialization of user equipment for access to TISPAN NGN services. NASS provides network-level identification and authentication, manages the IP address space of the access network and authenticates access sessions. NASS also announces the contact point of the TISPAN NGN service/applications subsystems to the user equipment. Network attachment via NASS is based on implicit or explicit user identity and authentication credentials stored in the NASS. The NASS can be divided into several functional entities (3GPP, 2008a), as depicted in Figure 2.17:

- The network access configuration function (NACF) is responsible for the IP address allocation to the user equipment and could be implemented by a dynamic host configuration protocol (DHCP) server.
- The access management function (AMF) is the interface equipment between the access network and the NACF.

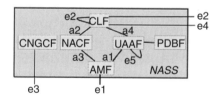

NASS internal structure and interfaces

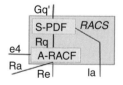

RACS internal structure and interfaces

Figure 2.17 NASS and RACS internal architectures.

- The connectivity session location and repository function (CLF) is used to associate the user IP address to the location information. In addition, it may store further information about the user (profile, preferences, etc.).
- The profile data base function (PDBF) stores the user profiles and authentication data.
- The user access authorization function (UAAF) performs authentication for network access, based on the user profile stored in the PDBF.
- The CNG configuration function (CNGCF) is used to configure the customer network gateway (CNG) when necessary.

The resource and admission control subsystem (RACS) provides applications with a mechanism to request and reserve resources from the access and aggregation networks. To achieve this, real-time multimedia services (VoIP, video-conferencing, VoD and online gaming) must be able of trigger the QoS resource reservation and admission control and policy control capabilities of the network. RACS provides the means for an operator to enforce admission control and set the respective bearer service policies. It provides the means for value-added services to obtain network resources that are needed to offer end-user services. RACS is session aware but service agnostic. It can be divided into two functional blocks: the serving policy decision function (S-PDF) and the access resource and admission control function (A-RACF), as described in 3GPP (2007h). The S-PDF performs policy decisions and is able to send resource requests to an A-RACF and/or border gateway function (BGF). It communicates the policy decisions back to the application function (e.g. the PCSCF). The A-RACF performs admission control, which means that it checks if the requested resources may be allocated for the involved access. It returns the results of admission control to S-PDF.

2.3.1.2 NGN Service Subsystems

Service Subsystems

TISPAN consists of multiple service subsystems to cater for different services that will be delivered to the end-users, and IMS is just one of those subsystems. TISPAN, as already stated, takes the core of 3GPP-specified IMS, the SIP control plane, and reuses it within the TISPAN NGN architecture, but replaces the original IMS access network and policy and admission control capabilities that were defined specifically for the wireless and mobile networks. A major driver for TISPAN is PSTN emulation, given the huge PSTN user base. PSTN/ISDN emulation mimics a PSTN/ISDN network from the point of view of legacy terminals (analogue or ISDN) by an IP network accessed through a gateway. All PSTN/ISDN services remain available and identical (with the same ergonomics for the user), so that end users are unaware that they are not connected to a conventional TDM-based PSTN/ISDN. There are two different forms of PSTN emulation service documented in TISPAN.

Non-IMS PSTN Emulation Service

This is a very simple architecture with a monolithic softswitch controlling an access gateway (for the line side) and trunking gateways (for the trunk side). The protocol used between the softswitches is SIP-I, which encapsulates the national variant of ISUP over SIP. Many vendors are developing this architecture. PSTN emulation service over IMS reuses the full IMS to emulate PSTN/ISDN services, and introduces a new functional entity for the line side – the access gateway control function (AGCF).The protocol used between the IMS functional entities is SIP, which means that it maps all PSTN/ISDN supplementary services into SIP. The PSTN emulation subsystem thus contains all the call routing and switching infrastructure, as well as gateway functions that bridge between that emulated IP-based environment and the legacy SS7 and circuit-switched environments that end stations are used to. The PSTN environment shares all the same resources as IMS-based services, and hence the PSTN services and IMS services can coexist on top of a shared IP network.

TISPAN is also considering several other service subsystems. In particular, Release 2 focuses on video streaming services in the RTSP subsystem. This will enable both IPTV and on-demand TV services to be delivered over the converged access network.

(i) Application Components

The application servers are SIP entities that execute the services. TISPAN defines two types of application: one that uses the service subsystems and that directly interacts with the RACS and the second that uses the service subsystem capabilities. The latter interface with the subsystem and use the services from the various subsystems to get their requirements pushed down through the network.

(ii) Additional Components

The home subscriber server (HSS) or user profile server function (UPSF) is a secure database storing user profile information. It can be accessed by the S-CSCF using the diameter protocol. It is to be noted that the HSS can be seen as an evolution of the former home location register (HLR). In case several HSS are used in a domain, a subscriber location function (SLF) is required. The SLF is a simple database indicating in which HSS a user profile is located (3GPP, 2007e).

TISPAN Interworking

Several components are used as gateways for interworking with legacy circuit-switched networks (e.g. SGW, MGCF, MGW, BGCF) (3GPP, 2007b). The signalling gateway (SGW) interfaces the signalling plane of the circuit-switched (CS) network. In addition to controlling the media gateway, the media gateway control function (MGCF) is used for protocol conversion between SIP and ISUP7 or BICC8. The media gateway (MGW) is able to send and receive media over packet-switched (PS) and CS protocols.

The functional architecture for TISPAN interworking with other networks is described in Figure 2.18 and the various components are briefly described below:

- *Interconnect border control function (IBCF).* IMS entities use the IBCF to perform interconnect procedures at the boundary between and to operator domains. There may be multiple IBCFs within an operator's network. The use of the IBCF is optional in the architecture and depends on the operator's interconnect policy. IMS entities insert the IBCF along the SIP signalling path if required by local policy. It provides application-specific functions at the SIP/SDP protocol layer in order to perform

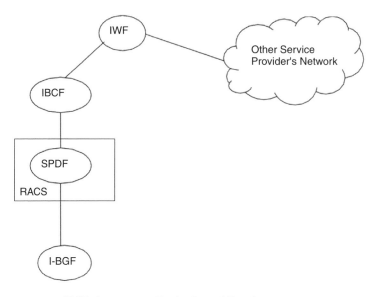

IBCF   Interconnect Border Control Function
IBGF   Interconnect Border Gateway Function
IWF   Inter - Working Function
RACS   Resource and Admission Control Subsystem
SPDF   Server Based Policy Decision Function

Figure 2.18 TISPAN interworking components.

interconnection between two operator domains. It provides network topology hiding, transport plane functions as well as facilitating interoperation between IPv4 and IPv6 SIP applications.

- The *interworking function* provides signalling protocol interworking between the SIP-based IMS network and other service provider networks using H.323 or other SIP profiles.
- *Interconnect border gateway function (IBGF).* This provides a pinhole firewall and NAT to protect the service provider's IMS core. It controls access by packet filtering on IP address/port and opening/closing gates into the network. It uses NAPT to hide the IP addresses/ports of the service elements in the IMS core.

TISPAN interfaces and protocols are listed in Table 2.3.

2.3.2   PacketCable and IMS Integration

PacketCable is a project started by Cable Television Laboratories Inc. (CableLabs) and its cable operator members with the goal of enabling a wide variety of IP-based multimedia services over two-way hybrid fibre coax (HFC) cable access systems (PacketCable, 2007a). PacketCable specifications have been released in four project phases: PacketCable 1.0 (PacketCable, 2007b), PacketCable 1.5 (PacketCable, 2005a), PacketCable multimedia (PCMM) (PacketCable, 2005b) and PacketCable 2.0 (PacketCable, 2007a). Before going through the architectural details of PacketCable, it is worth mentioning that the PacketCable

Table 2.3 TISPAN interfaces and connections.

Interface	Between	Protocol
Gm	UE and a P-CSCF	SIP
Mw	P-CSCF and another P-CSCF	SIP or SIP-I
Mr	CSCF and an MRCF	SIP
Mg	MGCF and a CSCF	SIP or SIP-I
Mi	CSCF and a BGCF	SIP or SIP-I
Mj	BGCF and an MGCF	SIP or SIP-I
Mk	BGCF and another BGCF	SIP or SIP-I
ISC	CSCF and application server	SIP or SIP-I
Ic	IBCF and other IP networks	SIP or SIP-I
Iw	IWF and other IP networks	H.323 or SIP
Le	MGCF and SGF	SSURN/ SSURN over SIGTRAN
Lb	IWF and IBCF	SIP-I or SIP
T1	At MGCF	SSURN/ SSURN over SIGTRAN
T2	At IBCF	SIP or SIP-I
T3	At IWF	H.323 or SIP
Ut	UE and application server	HTTP/XCAP
Dh	AS and an SLF	DIAMETER
Cx	CSCF and a USPF	DIAMETER
Dx	I-CSCF and an SLF	DIAMETER
Ro	Charging function and AS/CSCF	DIAMETER
Rf	Charging function and AS/CSCF	DIAMETER
Gq	IMS and a P-CSCF	DIAMETER

architecture utilizes the services of three underlying networks: the HFC access network, the managed IP network and the PSTN. PacketCable 1.0 defines the subscriber environment and its interfaces to other network components (Figure 2.19) including the cable modem (CM), multimedia terminal adapter (MTA), cable modem termination system (CMTS) and call management server (CMS). CM connects the user device to the cable access. MTA provides for the user codecs and all signalling and encapsulation functions required for media transport and call signalling on top of IP. It may be implemented as a standalone device or embedded in the cable modem. The CMTS provides data connectivity and complementary functionality to cable modems. The CMTS is located at the cable system head-end or distribution hub on the operator side of the HFC access networks. The CMS provides call control and signalling-related services for the MTA and CMTS. It is a trusted network element that resides on the managed IP part of the PacketCable network. The CMS consists of a call agent and a gate controller. The call agent (CA) refers to the control component of the CMS that is responsible for providing signalling services. To control a call, the NCS (network call signalling) protocol is used between the CMS as the server and the MTA as the client (PacketCable, 2004). The NCS is a revision of the media gateway control protocol (MGCP).The gate controller (GC) is a logical QoS management component within the CMS that coordinates all quality of service authorization and control. Finally, the RKS (record keeper server) collects all the charging information from the CMS and CMTS to provide

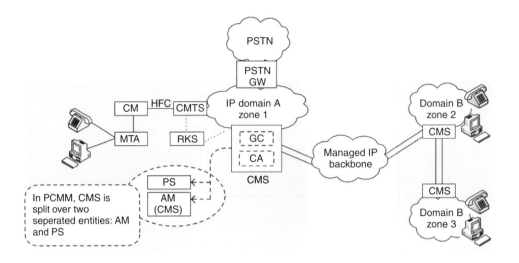

Figure 2.19 PacketCable 1.0 architecture.

billing information. PacketCable 1.5 extends the definition of two concepts, the zone and the domain introduced in the PacketCable 1.0. A PacketCable zone is defined as a single CMS and the endpoints it manages and a PacketCable domain is the set of security realms managed by a single administrative and/or legal entity. The PacketCable 1.5 has introduced interconnection of different operator zones. The SIP is considered as the signalling protocol between different CMSs. In the next release, CableLab specified PCMM to provide advanced IP-based services in cable accesses. According to the new requirements, the functional elements are enhanced. In this release, CableLab has defined the functional components and interfaces necessary to provide quality-of-service (QoS) and resource accounting to any multimedia-based application. The policy-based admission control architecture defined in IETF RFC 2753 (Yavatkar, Pendarakis and Guerin, 2000) is adopted. CMS is divided into two separate components: an application manager (AM) and a policy server (PS). It can be said that the application manager is the advanced version of the CA and the policy server is the advanced version of the gate controller. The policy server acts as the policy decision point (PDP) defined in RFC 2753 (Yavatkar, Pendarakis and guerin, 2000). According to the available application resources, user profiles and network policy rules, a PDP decides to accept or deny a requested multimedia session. The policy enforcement point (PEP) function is also supposed to be inserted in CMTS. According to the authorization token issued by the PDP (policy server in this case), the PEP allocates the resources for the media transfer of the accepted session. The policy and QoS signalling protocols, event message generation for resource accounting and security interfaces for multimedia architecture are defined in the PCMM specifications.

2.3.2.1   PacketCable–IMS Integration   (PacketCable, 2004)

To reach a scalable and reliable architecture for horizontal fixed mobile convergence, Cable-Labs has decided to adopt IMS as the service control overlay in the PacketCable 2.0 project.

The target architecture in PacketCable 2.0 will have essential differences from the current architecture. With the existing architecture in PacketCable 1.5, horizontal fixed mobile convergence is impossible. As shown in Figure 2.20, the IP services of one domain will not be accessible for other domains, since the connection of different IP domains is always via PSTN. Hence, the cost of the call between different IP domains will be considerable because of signalling translation and media packet transformation in the borders. To migrate to PacketCable 2.0 with IMS overlay, a step-by-step and evolutionary strategy to reuse the existing infrastructure would be adopted. Figure 2.20 shows these evolution phases in the PacketCable architecture. It is a reasonable strategy that (i) utilizes proven technology; (ii) generates new revenues and reduces the costs; (iii) incrementally adds applications and features. In

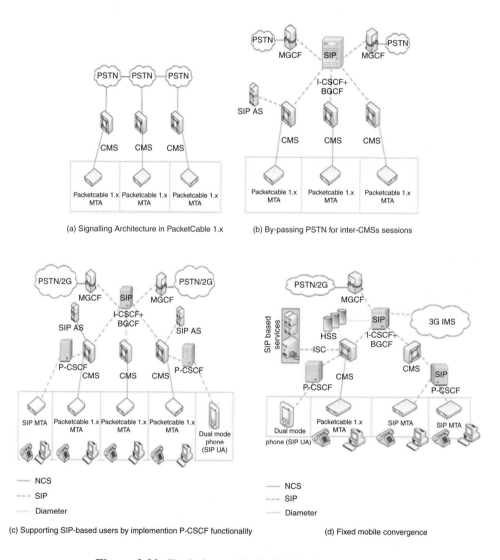

Figure 2.20 Evolution to PacketCable 2.0 architecture.

the first phase, as depicted in Figure 2.20(b), PSTN will be bypassed by adding the I-CSCF function in BGCF. Although in this phase SIP devices cannot be supported, some services that are more easily implemented on SIP, such as voice calling and click-to-dial, may be implemented for the conventional cable user endpoints. These are the services that can make a differentiation for MSOs from telephony provider competitors. In the second phase of migrating towards IMS, MSOs may support SIP clients by adding the P-CSCF functions and developing SIP MTAs. In addition, S-CSCF functions should be included in the CMS. Therefore, an IMS-like service control overlay will be set up with completion of this phase. Moreover, by expanding the customer space, MSOs are able to focus on their business goals and provide more advanced service features such as conference calling, call forwarding and integrated voicemail and messaging.

Finally, in the last phase of architecture evolution, the fixed mobile convergence is expected to happen. The IMS architecture is supposed to be implemented completely in this phase (see Figure 2.20(d)). Then different application servers of cable and 3G domains will be converged by using IMS standard interfaces. In this architecture, non-SIP-based end devices are supported as well as SIP-based devices (SIP endpoints or SIP-MTAs). There-fore, CMS is enhanced and is not replaced by S-CSCF. The three main enhancements in CMS are, firstly, interconnecting to HSS by deploying diameter interface; secondly, defin-ing the ISC interface access to SIP-based ASs; and, thirdly, setting up the SIP session on behalf of non-SIP-based end devices. In Mani and Crespi (2008), the details of the required architecture and signalling flow for a non-SIP-based device access to PacketCable IMS are introduced. If the user device does not support SIP (or the user does not own SIP-MTA), the description of the requested media in SDP will be negotiated by using NCS. The CMS receives this request and verifies whether the user has the right to the requested media according to its profile residing in HSS. If the user is eligible for the requested service in IMS, CMS will act as the user agent (UA) on behalf of the user and provide the required SIP messages for the next SIP proxy. When the user uses an SIP-based device, it sends the SIP messages to P-CSCF directly. Then P-CSCF forwards the SIP request to the proper CMS. PacketCable 2.0, by IMS integration, specifies a revolutionary architecture at the service layer and replaces NCS with SIP. NCS is a centralized signalling protocol to con-trol nonintelligent legacy telephony end devices. With NCS, CMS monitors the end device status and issues the corresponding signalling to establish or terminate a session. The end device has no intelligence to provide the required call control signalling. In contrast, SIP proposes a distributed signalling architecture with intelligent end devices. The end devices are able to provide call control signalling and react to the incoming requests. Consequently, the involvement of IMS components (including CMS) in session signalling is limited to AAA functions, session routing and service triggering. Hence, with elimination of respon-sibilities for end device control, supplementary services such as session forking/merging, caller ID presentation/restriction and call forwarding may be implemented on board in CMS. Moreover, PacketCable 2.0 is able to support a variety of advanced end devices to satisfy end-users demands for high quality experience of multimedia services. On the other hand, PacketCable 2.0 introduces essential enhancement from the application and service point of view. Based on SIP, IMS facilitates the implementation of advanced telephony services like voice/video conferencing, messaging, push-to-talk and so on. Furthermore, the flexi-bility of IMS in the application level and its unified service triggering interface allow the service components to be combined and new advanced services to be defined. Therefore, with PacketCable 2.0, this service combination can happen between cable and 3G domains and provide interesting services such as single fixed mobile numbering, unified messaging, video-on-demand (VoD) streaming to cell phone, content sharing between cell phones, personal video recorders (PVR) and other devices, remote PVR programming by cell phones and buddy list spanning fixed mobile devices. Single fixed mobile numbering means that a mobile fixed user, using a multimode terminal, will be able to roam in a cable/3G technology network and have access to the local services. The SIP signalling route and the session flow for registration are, respectively, in this scenario, a wireless cable modem from one side creates an 802.11 WLAN and from the other side is connected via HFC access to the cable IP domain. The 3G user is equipped with a multimode 3G/802.11 device. This device connects to the WLAN of the CM, obtains an IP address and discovers the P-CSCF in the cable domain. After this stage, this user is ready to register in the IMS level. To this end, it submits a register request. Upon receipt of the register request, the P-CSCF examines the 'home domain name' to discover the entry point to the home network (i.e. the I-CSCF) where the request should be transferred. Then, I-CSCF interrogates HSS by submitting Cx-Query/Cx-Select-Pull. The HSS checks whether the user is allowed to register via that P-CSCF according to the user subscription and operator limitations/restrictions. Then if the user was authorized for this registration, HSS returns the SCSCF name or capabilities corresponding to the user service profile by sending Cx-Query Resp/Cx-Select-Pull Resp to the I-CSCF. The register request arrives at the S-CSCF. The S-CSCF demands HSS for information about the indicated user including the service filter criteria (iFC). Based on the filter criteria, the S-CSCF sends register information to the service control platform and performs whatever service control procedures are appropriate. Finally, S-CSCF accepts the registration request of the user and replies with a 200-OK. From this step on, the user is able to establish a call session. Figure 2.20 demonstrates a call session establishment route as well as a session flow originated by a 3G user roaming in the cable domain. Adoption of IMS in PacketCable is a big step towards the horizontal convergence of cable access technologies and wireless mobile systems. However, to achieve this, there are important concerns in signalling compatibility, resource reservation and security. In the following chapters, these issues and corresponding solutions will be reviewed.

Part II

Convergence – Services and Deployment Perspective

3

IMS – A Service Perspective

We have seen in Chapter 2 that from the network architecture perspective, IMS offers a horizontally integrated, access-agnostic, open-standards framework. It is being adopted as the common core network by the TISPAN architecture of the fixed network operators as well as in the cable operator's PacketCable architecture, thereby enabling true fixed mobile convergence. The converged network architecture does offer business benefits such as reduced operational and capital expenses, and rapid service deployment, among other technological advantages. However, the 3GPP representation of the IMS service architecture is network centric and defines the IMS application layer only as an interface to the IMS core network. This representation is potentially misleading as it drives the telecommunication engineers to perceive the IMS service architecture as an IP replication of the intelligent networks (INs). In such an architecture, the core network supports the basic voice services, while the application servers complement the core network for the delivery of 'value-added' services. In such a scenario, the use of the SIP is limited to session control. This limits the potential of IMS and SIP as a whole. The true potential of IMS lies in the service layer and in the use of SIP as a generic service creation and service control protocol. A service-centric view of the IMS architecture not only demonstrates IMS as a service delivery framework but enhances the status of IMS architecture to a service creation and service integration framework, as shall be discussed in this chapter. This chapter presents the intrinsic features of the SIP protocol and the IMS service architecture that can be exploited by the operators, service providers, and the application developers. IMS service possibilities are discussed under the topics *blended services* and *end-user services*.

3.1 The IMS Potential

In an attempt to present the service-centric view of IMS, this section presents the horizontal-layered representation of the IMS service delivery architecture depicted in Figure 3.1 to demonstrate its service potential.

IMS: A Development and Deployment Perspective   Khalid Al-Begain, Chitra Balakrishna, Luis Angel Galindo and David Moro
© 2009 John Wiley & Sons, Ltd

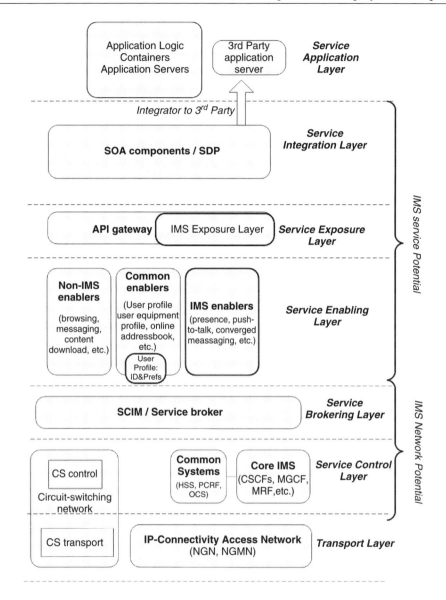

Figure 3.1 Service-centric horizontally layered representation of IMS.

Service potential of IMS is realized by studying the following aspects of IMS service delivery architecture:

- *IMS essentials.* These provide the reasons why IMS is considered a suitable candidate for service delivery.
- *SIP potential.* The power of SIP as a service control protocol and exploitation of the 3GPP-SIP headers for service control and delivery is the first proof to the IMS service potential.

- *IMS-enhanced service delivery framework.* IMS architecture by itself is limited to session control and heavily depends on an adequate service and application plane above. The IMS architecture typically provides a network overlay, offers security, authentication and authorization. It comes with an inbuilt charging mechanism, media handling and legacy networks interworking capabilities. However, a service delivery framework should provide service creation and execution capabilities, OSS/BSS interfaces and service exposure interfaces. The session control capabilities of the IMS core network could be abstracted and hooked on to a service delivery plane above the IMS architecture. The service delivery plane is depicted in the horizontal representation of IMS in Figure 3.1. The service plane stimulates new and innovative IP multimedia services, taking advantage of basic multimedia session control and services capabilities provided by the core network. Refer to Section 1.3 for definitions of the horizontal layers that form the service delivery plane of the IMS.

3.1.1 IMS Essentials

As described in previous sections, IMS provides a set of enablers as ingredients for a service mix, and powerful mechanisms to enable service creation by composing these ingredients. The IMS distinct features are reviewed in Section 2.3.1, which presents the technical characteristics of IMS technology. Even if those are breakthrough powerful features, by themselves they do not justify a technology-driven roll-out. The deployment of a new architecture over the existing network and systems should be driven by clear business motivations. That being the case, a question arises as to the necessity of IMS for a telecommunication network, when apparently most of the service features provided by this new system are already possible with a legacy infrastructure. At a first glance, there could be an immediate reasoning for voice and video calls, and messenger-like services. Voice calls are enabled by circuit switching, and in many cases they constitute a mature and cost-effective part of the network, which are well known, with a long expertise within the organization of a telco and mature O+M systems. Above all, voice-oriented services, especially mobile and corporate services around speech, are the most prominent source of incomes for any regular operator. When thinking of improving these communication services with some of the distinct features of IMS, like conversational video or instant messaging, the same discussion appears: for instance, video telephony is currently well deployed in mobile third generation networks, based on circuit switching and the multimedia protocol suite ITU H.324M adapted for this purpose by the 3GPP. As for the conversational messaging enhanced with a presence, the market is requesting, the most popular solutions are already in place. Even if proprietary, Windows Live Messenger and Yahoo! Messenger are being mobilized to reach the wide population of operators' customers. In some other cases, it seems that the second preferred option is the Jabber/XMPP-based presence and messaging. With such a perspective, it seems that IMS has been standardized to provide the most popular services already in place, provided either by the circuit-switched infrastructure or by popular proprietary solutions, with none or little off-the-shelf enhancement in terms of service features. Thus, apparently, questions like the following might easily arise: 'What is the value of IMS?', 'Why to deploy IMS?', 'Is it just a matter of migrating existing services to a standardized IP environment?' In that case, the vast majority of strategists would not recommend initiating such a migration; there is no point in a pure technological migration without a business rationale behind it. The following facts provide, in brief, the essence of the IMS value and why IMS is an essential infrastructure *in terms* of standards, future capacity, green ICT, integration ability and new services.

- Standards not linking to a vendor proprietary technology. When deploying external solutions, integration costs are high, and most importantly, from a business perspective, the future of that piece of business is to be partially controlled by the solution provider, which can increase costs and reduce benefits. Operator brand risks being diluted.

- Capacity increase and cost reduction for the new access technologies. Circuit switching has no evolution path within the broadband trend. The future technologies for accessing the network, like FTTH in the fixed case or the LTE/Super3G in the mobile case, are pure packet-switching access networks. In the near future, there will be a strong need to have in place a mature solution for providing speech and conversational services over this IP-based system, with at least the same portfolio offered today to the mass market and to the corporate segments. Some of these new technologies, like the LTE and the associated evolved 3G technologies, are designed from scratch with IMS as the default service control. Thus, IMS procedures and services are tremendously optimized in these future environments. As examples, the evolved 3G network will have a true always-on connectivity with default IMS registration as a must. When comparing service performance, an LTE cell can support 3 to 5 times of simultaneous IMS voice calls as compared to a circuit-switching 3G cell, when both cells are assigned the same amount of radio spectrum. Since LTE can support higher amounts of spectrum, this figure can easily exceed 14 times. This means that if a standard 3G cell can support up to 66 simultaneous calls with circuit switching, an LTE call might support up to 960 simultaneous calls with IMS.

- Green ICT. IMS contributes to the green principles by reducing the energy consumption due to various factors. Aside from capacity increase and the unified architecture and network convergence that enables power savings by offering several services of the portfolio with the same infrastructure, there are pure logistics facts. Because infrastructure is shared, the requirements for heating, lighting, air-conditioning, etc., are reduced. Furthermore, soft-switching like the one offered in IMS consumes around 35% ITU-TWR7 less power per subscriber than its predecessor hard switches. Less space and smaller equipment is required; less cabinets therefore implies minimizing the cooling and energy needs.

- Ability to integrate smoothly. In general, IP technology is prone to integration, or at least poses fewer requirements for the smooth integration. In the case of IMS, this chapter presents examples of the ability of IMS to integrate with other IP-based infrastructures and services, like the IPTV system, with web-based services either in the domain of the operator or in the wide Internet and with external conversational services also based in IP like those based on H.323 protocol or proprietary ones like Skype, Ribbit and Windows Live. The characteristics of the IMS architecture also permits integration ability with the circuit-switching domain of a telco network to hook on the maturity of these domains to produce feature-rich services like rich media circuit calls or the voice call continuity (VCC) feature.

- IMS new service features, to build appealing and useful services to the end-user. This is discussed in the rest of this section.

Even if the efficiency and higher capacity of IMS as compared to the legacy infrastructure justify the roll-out of IMS, from the service perspective, the IMS service enablers and capabilities also enable applications and end-user services that cannot be provided with the legacy infrastructure, and which might not be evident. IMS is essential for the technical realization of an increasing number of services, due to the technical constraints of the circuit-switching network in terms of media transport, flexibility and intelligent network limitations. The following service cases represent some of these IMS-essential services:

- *Ultra high video quality to enable high resolution video telephony from TV to TV, from TV to mobile.* Today's circuit-based H.324m video telephony offers low quality even for the small screens of the mobile terminals. Aside from sociological aspects, this has contributed to the low acceptance and market penetration of the 3G video telephony. The current trend considers the TV end system as a potential terminal for services beyond media consumption, for instance for interpersonal communications. Due to its ergonomics and positioning within the domestic piece of the digital style-of-life, TV terminals will be crucial for the acceptance of video telephony, regarded as a standalone service as well as integrated with IPTV services. Thus, IMS is utterly essential to be able to provide breakthrough high definition for video communication involving TVs and mobile devices, exploiting the full potential of larger screens.
- *Media split within a call.* In the specific case of a multimedia call, for instance a video call, a typical scenario that IMS enables is the media split to deliver the audio and video media to different devices, to adjust better the medium to the characteristics for the terminal equipment. For instance, this could be the automatic or user-requested media split of a video call in a home environment, where an IPTV terminal equipment acts as the video endpoint, while the user's mobile phone can be kept as the audio device.
- *Tele-presence calls* for 3D video and surround audio. As a trend in the IPTV environment, new TV and set-top box equipment will support 3D rendering of moving pictures and other content, making the most out of the high throughput of the fibre access link. As part of the future tele-presence services, a 3D video call can leverage the screen technology of the 3D TV and IMS flexibility to work with any number of media flows within the call session, any media type and any codec. This can be applied also for enhancing the audio of the standard call to include a rich definition, stereophonic, diverse audio channel or any other requirement for 3D or surround effects that enhance the tele-presence services.
- *Automatic update of presence based on home-zone, calendar and other contexts.* Presence is a powerful enabler that can be nurtured with numerous sources of information about the user context. However, the challenge for a successful service is limited by profitability, i.e. the cost of extracting that context information to be integrated into the IMS presence, rather than its deployment or usage. Location is a good example of a valuable asset when tailoring the delivery of a service to the user-specific geo-position. However, precise location calculation in cellular networks is an expensive operation due to the high consumption of network resources. If this datum needs to be continuously monitored or updated in the IMS presence, it becomes a nonrealizable presence

source. A different approach lies in realizing that a location datum can be profitable and useful for services even if not accurate. This would be the home zone case, which provides only binary information, whether the user is at the 'home zone' or not. Usually utilized for applying special tariffs when the user is located in a pre-determined area, the user's 'home zone' is also very valuable when integrated into the presence information held by IMS, and utilized for a huge variety of services. For instance, a simple one would be an automatic change of the configuration of call diverts when the user enters the 'home zone' or 'office zone' in the case of corporate users. In the latter case of corporate environments, another cost-effective presence source is the corporate calendar, which can feed the presence enabler of IMS with information about the user's limited availability for communicating when the user is at meetings.

- *Discontinuous high quality video share.* The current circuit switching enables bidi-rectional video-telephony designed for conversational communication between two peers. However, mainly due to psychosociological issues and poor video quality, this scenario is not demanded by users in the great majority of European and American countries. When engaged in a call, a user does not usually have the need to watch the correspondent peer, or does not feel comfortable being watched, aside from other considerations like ergonomics and the price of a long call with video enabled. In most cases, users are indifferent to the video-telephony service. Furthermore, today's deployment does not permit a video call to switch dynamically to a voice call or vice versa without hanging up the call. Therefore, if video communication is to continue in the market, more flexibility is requested and different usage scenarios fitting the users' need should be pursued. This is the case for a discontinuous, high quality, video share service. This IMS service, not realizable with circuit switching, presents a different usage scenario: a video stream is enabled in any moment during the voice call, and for a limited time, in a unidirectional basis. After all, the goal is to permit the user to share the device camera with the remote peer in the voice call, usually intended to provide a visual impression on the surroundings of the user, rather than for watching the correspondent peer involved in the call. Once more, since IMS does not restrict the specific codec to be used, video quality can be gradually enhanced with time, without risking interoperability, in contrast with the current video telephony service, which determines the specific video format and its quality to a limited nonevolution able set. Besides, since the video stream can be activated or deactivated, the price can be modulated per video burst or per the time the video stream is active, and thus the result could be more controlled and attractive to the user.
- *Media sharing for tele-commerce, tele-marketing and customer care.* Beyond speech, additional media during a phone call can assist helpdesks, customer care and order desks of companies to fulfil their task better. When talking to a customer, there are certain scenarios where the appropriate pictures of products or video clips shared from the customer care towards the customer user can be very efficient to complete a sale, place the adequate order or assist a customer appropriately. Tele-marketing and tele-commerce can benefit from the media-rich nature of IMS telephony to provide a multimedia experience to customers.
- *Customized alerting* with multimedia ring-forward and ring-back tones. Ring-back tones have proven to be a value-added service for mass market with increasing

profitability and CAPEX efficiency. Notwithstanding, evolution of today's services to more sophisticated service concepts is hampered by the circuit-switching limitations of voice telephony. For instance, tones are limited to audio, while other richer media would potentially interest consumers for a true multimedia experience, which is the natural ambience of IMS. Together with media restrictions, the direction of the customized alerting is ring-back in today's implementation. The forward direction, where the caller can chose the multimedia tone to be played in the callee's terminal as an announcement of an incoming call, is not technically feasible with today's SS7-based telephony, while IMS poses different technical possibilities for its realization. With IMS customized alerting, the multimedia nature allows the popular service to be applied beyond the leisure market to a corporate environment. A true multimedia ring-back tone potentially composed of video, moving pictures, animations, combined with advertisement or company information, business cards, etc., shall be of interest for corporate segments.

- *Corporate telecom services and its integration with the back-office.* One of the key fields to launch IMS services is Large Companies. IMS is an enabler to integrate Fixed and Mobile Capabilities and Communications with the IT back-office of the company. Integrations of business applications as CRM, ERP, Directories, etc., from the company with IMS capabilities as presence, instant messaging, voice calls, address book, etc., provide a great value to increase efficiency in those integrated companies. This situation joint with the clear possibility for a large company to buy the appropriate mobile phones with the IMS clients preconfigured ensures that this integration can be offered for the employees with a return of investment for the company in the way of increasing productivity.

- *Applications for logistics and integrated warehouse.* One of the sectors demanding applications where different flows and capabilities can be integrated is the logistics. This sector demands applications that increase efficiency for the integrated store. Applications where location, presence, instant messaging, telephony, catalogue directory, formularies, etc., would be integrated to control the store or to manage the workflow for logistics represents profitable applications for cross-docking, replacement or other activities key in this sector.

3.1.2 SIP Strength

The session initiation protocol (SIP): (Rosenberg *et al.*, 2002) is used as a service control and signalling protocol in the IMS. SIP sessions are not limited to voice sessions or just person-to-person communication. An SIP session could include one or more media components relating to communication such as voice, video or messaging text. SIP sessions could also include applications such as gaming, whiteboard or content such as text files, streaming video and web pages. An SIP session could be renegotiated at any time to add, remove or replace any media component using the session description protocol (SDP) (Handley, Jacobson and Perkins, 2006). An IMS service is delivered within an SIP session and a non-IMS service could become an IMS service when delivered within an SIP session. The SIP session may be set up between the user accessing the service and the service itself, or between the user and an IMS application server hosting service controller logic and interfacing with the actual service through an appropriate interface (e.g. web services, SIP).

SIP offers numerous advantages to the IMS service and a few important ones could be summarized as follows:

- The inherent multimedia nature of SIP permits multiple media components (e.g. video, text, application interface) to be combined/composed in the same service session, thus enabling service composition.
- The SIP session determines a context for the delivery of the service with a beginning implied by SIP-INVITE and an end implied by SIP-BYE. The duration of the session is measurable within the IMS control layer.
- The content of a service session can be personalized for each user and negotiated between the user and the service controller. Through normal SIP support in the device, the user has real-time control on the service session (e.g. termination, content).
- The service session can be stopped at any time, either by the user or the service controller.
- Through SIP session mobility procedures (under specification in the IETF), the user can transfer an ongoing service session from one device to another, without stopping it.
- The SIP signalling generated by the user's terminal reaches the SIP application server and transports meaningful service-related, media-related, session-related information, which can be exploited by the IMS core network as well as the SIP AS to optimize the service delivery. This is achieved by exploiting the 3GPP-SIP headers, which is discussed in detail in the following paragraphs.
- The establishment and renegotiation of the session permits the user's terminal and the SIP AS to reuse core network support to set the relevant QoS and security associations just like for a person-to-person voice or multimedia session (Handley, Jacobson and Perkins, 2006).
- The SIP session combines a traditional telecom service (person-to-person voice communication) with multimedia content delivery. The ability to combine core telecommunication (i.e. person-to-person communication) with service and content delivery is also a significant advantage. For instance, communication components can be added to what has started as a service session. Alternatively, person-to-person communication sessions between two or more parties can be extended seamlessly to content and/or application sharing.

3.1.2.1   Exploitation of 3GPP-SIP Headers

Within the IMS domain, the native SIP headers are enhanced by the 3GPP-specific SIP headers. Some of these headers may be of interest to the IMS core components or the SIP application server within the IMS operator's network or the media server when they process the incoming SIP messages.

A list of 3GPP-IETF headers are discussed below that are exploited by the IMS core network components and the SIP application server:

- *P-Access-Network-Info* (Garcia-Martin, Henrikson and Mills, 2003) is a session-related header and the most important of the SIP private headers that could be used by the SIP application server for content adaptation. It informs the session control logic about the access technology used by the client device, such as X-DSL, UMTS, WLAN, etc. This information is used by the media process and control units such as the media servers to encode the content at a bit rate that is best suited to the maximum bit rate provided by the access technology that the client is connected to. The same header also provides the information about the location of the client in the form of cell ID for cellular line or line ID for fixed lines. The service logic residing in the SIP application servers could make use of this information for optimizing the content delivery by choosing the optimal content server in terms of location, and the charging units could make use of this information for charging the user appropriately. The user agent that supports the P-Access-Network-Info header has the ability to insert it in any SIP request or response generated by it towards a trusted SIP proxy or SIP application server. Intermediate proxies, i.e. the CSCFs, ignore and bypass the header. The SIP application servers inspects the header without modifying the value and passes on the relevant information to the media control unit, which actually makes use of the information to adapt the media accordingly. A typical P-Access-Network-Info is as shown below:

```
The syntax of the header is shown below:
P-Access-Network-Info="P-Access-Network-Info"HCOLON
                        access-net-spec
access-net-spec        = access-type *(SEMI access-info)

access-type            = "IEEE-802.11a"/"IEEE-802.11b"/
                         "3GPP-GERAN"/"3GPP-UTRAN-FDD"/
                         "3GPP-UTRAN-TDD" /
                         "3GPP-CDMA2000" / token
access-info            = cgi-3gpp/utran-cell-id-3gpp/
                         extension-access-info
extension-access-info = gen-value
cgi-3gpp               = "cgi-3gpp" EQUAL
                         (token / quoted-string)
utran-cell-id-3gpp     = "utran-cell-id-3gpp" EQUAL
                         (token / quoted-string)
```

- *P-Asserted-Identity* and *P-Preferred-Identity* (Jennings, Peterson and Watson, 2002) allow the application server to check if the user has been authenticated by IMS, possibly in another operator's domain. A proxy server that handles a message can, after authenticating the originating user in some way (e.g. digest authentication), insert such a P-Asserted-Identity header field into the message and forward it to other trusted proxies. The P-Preferred-Identity header field is used from a user agent to a trusted proxy to carry the identity of the user sending the SIP message wishes to be used for the P-Asserted-Header field value that the trusted element will insert. It informs the SIP AS or the proxy about the public user identity with which the user is accessing the IMS service. This provides an IMS single-sign-on mechanism, which is also applicable to HTTP-based signalling through an HTTP authentication proxy. A typical P-Asserted-Identity and P-Preferred-Identity is as follows:

```
PPreferredID ="P-Preferred-Identity" HCOLON
PPreferredID-value
*(COMMA PPreferredID-value)
PPreferredID-value = name-addr/addr-spec

PPreferredID ="P-Preferred-Identity"HCOLON
```

```
                        PPreferredID-value
                      *(COMMA PPreferredID-value)
         PPreferredID-value = name-addr/addr-spec
```

- *P-Charging-Function-Addresses* and *P-Charging-Vector* (Garcia-Martin, Henrikson and Mills, 2003) provide the addresses of the online and/or offline charging nodes, respectively, to be used. It provides the contextual charging information supporting charging correlation between the media plane information such as the media type and volume of media and the duration of the media measured at the IMS control plane entities with application layer events such as the location of the terminal (e.g. cell ID) and information about the access technology used by the terminal (P-Access-Network-Info). A typical P-Charging-Function-Addresses and P-Charging-Vector headers are as follows:

```
   P-Charging-Addr    ="P-Charging-Function-Addresses"
      HCOLON charge-addr-params
                       *(SEMI charge-addr-params)
 charge-addr-params   =ccf/ecf/generic-param
 ccf                  ="ccf" EQUAL gen-value
 ecf                  ="ecf" EQUAL gen-value
 P-Charging-Vector    ="P-Charging-Vector" HCOLON
    icid-value
                       *(SEMI charge-params)
 charge-params        = icid-gen-addr / orig-ioi /
                         term-ioi / generic-param
 icid-value           = "icid-value" EQUAL gen-value
 icid-gen-addr        = "icid-generated-at" EQUAL host
 orig-ioi             = "orig-ioi" EQUAL gen-value
 term-ioi             = "term-ioi" EQUAL gen-value
```

- *P-Visited-Network-ID* header field is used to convey to the registrar or home proxy in the home network the identifier of a visited network. It identifies the network in which the user is currently roaming, allowing service level agreements with this network to be checked and applied accordingly. This information is used for charging.

3.1.2.2 SIP for Service Control

The service control ability to SIP is offered by the following intrinsic features of SIP that are being used by the IMS.

User Identities

Section 2.2.3 presents *IMS user identities* (Sparks, 2007) as the distinct feature of IMS. By definition, user identities can represent both users and services. In the context of IMS two identities are of relevance, the user identities termed as the *IMS public user identity* (IMPU) and the service identity termed as the *public service identity* (PSI). These identities are not statically tied to any particular location or a specific end-device. SIP registration procedures permit the dynamic association of an IMPU to one or more devices, which can be as diverse as mobile phones, fixed phones, personal computers, televisions or MP3 players. A request addressed to a specific IMPU can be routed to several devices based on a mechanism termed as *forking* (refer to Section 1.1.2). SIP also enables to be performed that intelligent routeing of requests favour a specific end-device or a specific location based on various criteria, such as the type of SIP signalling, the content of the session, the mobility capability of the device and preferences of the callee or the caller. This feature of SIP offers the possibility to accommodate varyious end-user devices with varying access technologies (i.e. both fixed and mobile). This leads to true user and service oriented convergence which is brought about by:

- the fact that IMPUs can be shared across end-users and end-devices;
- the fact that the routeing mechanism in IMS is based on two distinct features of IMS, the *service profile* and *initial filter criteria* (iFC).

User Profile and Service Profile

Refer to Section 2.2.3 for a detailed explanation on user profile and service profile and the relationship between them. However, the aspects that are relevant to demonstrate the service control capabilities are presented here:

- *User-orientend service convergence.* An important feature of SIP as a service control protocol is that it uses identities that can represent both users and services. In the context of IMS, user identities are called public user identities (IMPUs) and service identities are called public service identities (PSIs). These identities are symbolic as they are not statically tied to a specific location. SIP registration procedures permit the dynamic association of an IMPU to one or more devices, which can be as diverse as mobile phones, fixed phones, personal computers, televisions or MP3 players. A request addressed to a specific IMPU can therefore be routed (forked) to several devices, and intelligent routing mechanisms (Rosenberg, Schulzrinne and Kyzivat, 2004a, 2004b) exist to select or favour specific destinations over the others, based on certain criteria such as the type of SIP signalling, the content of the session, the caller and callee capabilities, etc. Apart from accommodating various fixed and mobile access technologies, the ability of IMS to unify end-users across devices through the sharing of IMPUs enable user-oriented convergence.
- *Service routing.* IMS service profiles (refer to Figure 3.2) consist of SIP routeing information stored in a network database called the home subscriber server (HSS). This SIP routeing information is associated with both types of public identities that exist in the IMS: the IMPUs for users and PSIs for services and service-related resources. An IMS service profile therefore influences the routing of SIP requests that are either originated or addressed to a particular IMPU or PSI. The IMS core network entity that utilizes service profiles for SIP routing is the S-CSCF, which retrieves them from the HSS over the Cx (diameter-based) reference point. The service profile is transferred over Cx in a standardized XML format and is composed of a list of initial filter criteria (iFC) (refer to Figure 3.3), which are processed one after the other by the S-CSCF. An iFC essentially consists of a condition to be met by the SIP request and the address of an application server to which the SIP request should be routed in that case.

The processing of a service profile by the S-CSCF may lead to the routeing of an SIP request to several application servers over the IMS service control (ISC) reference point. If no application server decides to serve as an endpoint to the SIP request, the S-CSCF

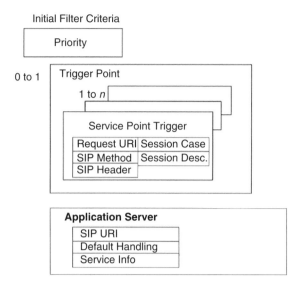

Figure 3.2 Detailed representation of initial filter criteria.

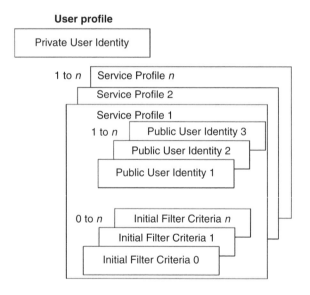

Figure 3.3 Relationship between a user profile and service profile.

then proceeds with normal SIP routeing procedures. Detailed service routeing procedures between the S-CSCF and ISC interface are presented as part of the IMS service architecture in Section 3.2.1.

The mapping between public service identities and a specific target address are stored in the HSS. This mapping could be one-to-one or one-to-many based on whether a service profile is associated with the PSI or is part of a service profile associated with an IMPU.

Service profiles can impact the routeing of requests addressed to a PSI in two ways:

- A service profile associated with the PSI may permit alternative target addresses to be defined based on the details of SIP messages addressed to the PSI. For instance, an SIP INVITE may be routed to one location while an SIP MESSAGE may be routed to another one.
- A service profile associated with an IMPU may include rules that relate to the PSI being used as the intended destination in SIP messages initiated by the IMPU. For instance, a SIP INVITE initiated by the IMPU and addressed to the PSI may be routed to one location while a SIP MESSAGE initiated by the IMPU and addressed to the PSI may be routed to another one.

The sequence of operation in a service profile-based routeing of a SIP request is as follows:

1. iFCs are evaluated at the S-CSCF when receiving an initial request in a dialogue such as INVITE, SUBSCRIBE or a standalone request such as MESSAGE, OPTIONS, etc. Note that S-CSCF does not evaluate iFCs on PRACK, NOTIFY, UPDATE or BYE.
2. Upon registration, the S-CSCF receives a user profile from the HSS.
3. Filter criteria determine the services that are applicable to the collection of public user identities of the profile.
4. S-CSCF assesses the criteria in the order of their priority: the lower the number the higher the priority.
5. Filter criteria contain trigger points, which are Boolean conditions.
6. When the trigger point is fired, the request goes to the corresponding AS.
7. After receiving the request back, the next criteria is checked.

The service point triggers (SPTs) that form part of the filter criteria check the following characteristics of a SIP request:

- The session case permits a definition of whether the request was originated or terminated by the public identity while it was registered or while unregistered.
- The SIP method element permits a check to be made of whether the SIP request is an INVITE, a SUBSCRIBE, a MESSAGE, a PUBLISH or any method that may be created in the future (the type is a string, not an enumeration).
- The request URI element permits a check to be made of where the SIP request is addressed to. This is typically used for iFCs that relate to private or restricted access to a PSI (the request URI is the PSI).
- The SIP header element permits a check to be made of whether a particular header exists in the SIP request, as well as the content of this header. It is possible to use a wildcard in the value of the content. Note that, once again, there is no preconception about the header, which can be any standard, nonstandard or future standard header.
- Finally, the session Description element permits a check to be made of whether the session description prototocol (SDP) body that may be attached to the SIP request is an INVITE. The SDP permits the details of the content of a session (e.g. media, application, codec) to be described.

Finally, in a complex mobile service landscape wherein the operator has deployed a large number of services, it is absolutely crucial that the operator is able to control the invocation of services and the interaction between the various service components. In the CS and PS domains, service execution is application controlled, which makes service interaction increasingly complex and reduces overall service transparency and control. IMS meets this challenge by providing efficient service provisioning functionality.

When a user registers on the mobile network operator's IMS network, his subscriber service profile (SSP) is downloaded by the CSCF from the HSS. The SSP contains a great deal of service-related information per end-user and enables the CSCF to:

- identify which service(s) needs to be executed, based on filter criteria held in the SSP;
- determine the order in which multiple services are executed (if applicable);
- determine the address(es) of the application server(s) that should execute the requested end-user service(s).

The filter criteria dictate the application server(s) of the order in which services should be executed in the case where multiple services need to be executed on the same application server(s). Service execution, control and interaction enable MNOs to use IMS as a reusable service infrastructure platform by allowing them to control and manage effectively the complexities involved in service filtering, triggering and interaction.

SIP Event Notification   (Gouraud, 2006; Roach, 2002)

SIP event packages can be used to distribute information and event notifications in an SIP network.The concept of event package permits SIP to be used to create, modify and distribute event states between SIP entities. SIP PUBLISH (Niemi, 2004) is used to create or modify an event state remotely, SUBSCRIBE to fetch or monitor it and NOTIFY to supply the event state and/or inform about changes. A large number of event packages exist, including packages related to presence, session states or user activity on a keyboard. However, the framework is extendible and event packages can be created for virtually anything. The service potential of event packages lies in the fact that the event states relate to data, commands or any type of events. For example (Gourraud, 2007), SIP event packages could be used to automate totally the process of service publishing, discovery and configuration. The event packages access or monitor user, device or application-related event states, wherever these event states are maintained (e.g. in the endpoint on the network).

Automatic Service Discovery and Configuration   (Gouraud, 2006; Gourraud, 2007)

A user can access IMS services from a variety of fixed and mobile devices, including a mobile phone, a TV set at home, a PC, a fixed phone or even an MP3 player. When a device is connecting to IMS either automatically or manually, depending on the device, the IMS client automatically retrieves from a service registry in the network the list of services a user is authorized to access. These services may either be IMS services (i.e. services accessed through an SIP) or non-IMS services, like streaming content.

When a new service is made accessible to the user through subscription, the service registry is automatically updated and the update is automatically notified to all of the user's currently registered devices. If a new service requires specific client update or configuration, the task is automated as well.

(i) Discovery of Services

User A's client issues an SIP SUBSCRIBE, which both originates from and is addressed to User A's public identity (e.g. sip:User A@operator.com) for a relevant event package (an IETF-defined 'UA Profile' event package might do it). As the SIP request originates from User A, it is routed to the S-CSCF that serves him. The service profile associated to sip:User A@operator.com determines that a SUBSCRIBE originated from User A, addressed to User A, and whose event package is 'UA Profile' shall be routed to User A's service registry. Upon the receipt of the SUBSCRIBE, User A's service registry issues a NOTIFY as an answer. The SIP NOTIFY may either transport in its body an XML document including the service list, include an HTTP URI to User A's service portal (accessed through content indirection) or both (refer to Figure 3.4).

Each and every one of User A's devices can issue the same request and discover the service. It is possible that differences exist between the list, based on the device, its capabilities or the configuration of each service.

User A gets authorization to access a new service. The discovery of this service is integrated into the automatic service discovery process, which is based on IMS. Upon provisioning of relevant subscription information, the application server hosting the service for User A issues a SIP PUBLISH for the 'UA Profile' event profile. The PUBLISH may either originate from the service's public service identity (e.g. sip:serviceX@operator.com)

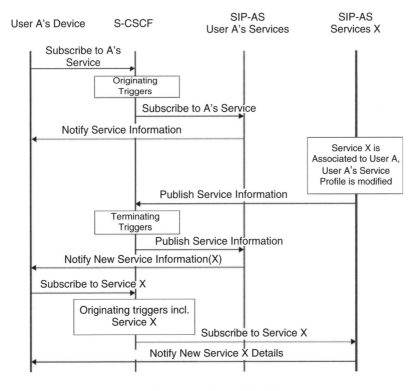

Figure 3.4 Automatic service discovery.

or from User A's identity (sip:User A@operator.com), and it is addressed to User A. Since it is addressed to User A, the request reaches an S-CSCF serving this user. User A's service profile determines that a terminating PUBLISH for the 'UA Profile' event package addressed to User A shall be routed to User A's service registry. User A's service registry updates the list of available services, following which the service registry issues a SIP NOTIFY towards all of User A's devices, which currently have an active subscription to the 'UA Profile' event package. These devices automatically and instantly discover the new service (either directly in an XML document or through new content indirection). Other devices will discover it the next time they subscribe to the 'UA Profile' event package.

(ii) Automatic Service Configuration   (Gouraud, 2006)

In order to be read-to-use, the new service must be configured in the device(s) and some software components might also need to be downloaded to the client. Details may vary depending on the device. The new service entry in User A's service registry includes information such as a description of service X, means to access it (e.g. HTTP URI, IMS PSI, RTSP URI, application-specific information) and means to configure/update the client if needed. This latter part includes a PSI to be used to access service X configuration information in the network. This PSI might be the same as the one to access the service, in case the service is accessed through a PSI, but it may also be specific to the service configuration, more especially if the service itself is not accessed through SIP. In order to retrieve service configuration information, User A's client issues a new SUBSCRIBE for the 'UA Profile' event package. This SUBSCRIBE still originates from User A, but this time is addressed to the service configuration PSI as included in the service registry (sip:serviceX@operator.com).

The SIP SUBSCRIBE is routed to the S-CSCF serving User A. User A's service profile determines that a SUBSCRIBE for the 'UA Profile' event package, originated by User A and addressed to sip:serviceX@operator.com, shall be routed to the application server hosting service X for User A. This entry in User A's service profile was provisioned when User A received authorization to access service X. The application server issues a NOTIFY as an answer, with either relevant service configuration information transported in the body or an HTTP URI to access it through content indirection. User A's client then receives relevant configuration information and gets ready to access the service. It may use any relevant protocol(s) for it.

SIP Content Indirection   (Burger, 2006)

SIP content indirection allows a SIP message to indicate to the recipient that it should retrieve a content identified by a URI included in the SIP message. The URI scheme determines the protocol to use (e.g. HTTP, XCAP, FTP, RTSP) and specific SIP headers provide additional information about the content, such as what to do with it (e.g. store, display, execute), its size or its expiry time.

A typical usage of content indirection permits a service to redirect a client to a web page offering an advanced user interface for the user to interact with the service. Another use is associated with SIP NOTIFY. NOTIFY is used to report an event state or a change to an event state to the client. This event state can actually consist of an XML document which is too voluminous to be included in the body of the NOTIFY (e.g. rich presence information, user profile data). The NOTIFY will therefore use content indirection and ask the client to retrieve the document via a more appropriate protcol, e.g. HTTP or XCAP.

A third use would enable content indirection for a user to provide access to a file (e.g. picture, video clip, love letter) stored in a network repository to other users by sending a text IM (SIP MESSAGE) with content indirection. Compared to a more intrusive push approach, content indirection permits the recipients to decide whether or not they want to retrieve the content, as well as when they feel appropriate to do it. SIP is the language of IMS and it can be used as an end-to-end service protocol between two devices, between a device and an application server, and between two application servers. In theory, every SIP message that can originate from or be terminated at a device can also originate from or be terminated at an application server. Therefore, just as a device can issue an instant message, send a SIP SUBSCRIBE or start a session towards another device, it can do the same towards an application server, and an application server can itself do the same towards a device or another application server. An IMS service can therefore act as if it was a user. In some cases it can impersonate a user, acting as an agent of this user in the network by processing or initiating SIP requests on the user's behalf. In other cases, the service can simply use its own PSI in order to interact with other services and users under its real identity. It can even have its own presence or be part of a user group. The usage relationship between a service and an enabler can cross network boundaries. For instance, a service owned by an operator can use the presence server from another operator as an enabler, simply by issuing a presence request (SUBSCRIBE for 'presence' event package) addressed to a subscriber of this operator.

SIP REFER Method   (Sparks, 2004)

SIP REFER permits Peer A to request Peer B to initiate a service interaction towards Peer C or a server. Peer B may decide to initiate the service interaction or not, and this service interaction may succeed or fail. REFER is implicitly associated with an event package, which means that Peer B will notify Peer A about the processing of the request (SIP NOTIFY).

While REFER was initially introduced to support SIP session transfer and ad hoc conferencing, which implies that the request is to use SIP towards another peer, the specification does not place any constraint on the URI Peer B is referred to. It can therefore be of any scheme, such as SMTP (email), HTTP or RTSP (streaming).

REFER is therefore another way for an SIP peer to ask another SIP peer to use another protocol to access a service.

SIP as a Service Transport Protocol   (Gourraud, 2007)

An SIP message has a body, and this body can include any type of content, such as documents defining service control semantic. The first example of this usage came with an SIP session initiation, with an SIP INVITE transporting an SDP (session description protocol) document to negotiate the content and conditions of the session.

However, SIP is not limited to transporting SDP documents, and SIP methods like PUBLISH, SUBSCRIBE, NOTIFY and MESSAGE can typically transport XML documents. The only limitation is that SIP is not supposed to be a transport protocol, and the size of the attached document should remain small. However, if it reaches a size threshold, content indirection can be used to provide alternative access to the service semantic.

SIP as Mediator for Other Service Protocols

A potential future in which SIP would be used in the process of delivering multiple services, but not merely as a service protocol, rather as a mediator between the user and the service, as the glue between multiple service protocols or as the integrator of service delivery into a coherent context, the SIP session.

User-Centric Service Orientation   (Gouraud, 2006; Gourraud, 2007)

To understand how SIP and IMS can support a user oriented architecture (UOA), it is first important to realize that SIP is not only a very generic session control protocol that can support application-level sessions (see previous entry). SIP also features an important set of non-session-related extensions, which make it a very generic service control protocol. The most important of these extensions is certainly the concept of 'event package', which is based on the methods SUBSCRIBE, NOTIFY and PUBLISH. An SIP-based presence relies on event packages, which permit a user (or application) to access, monitor or modify presence information. However, presence is only the tip of the event package iceberg. There already exist numerous event packages, which permit, for example, the activity of a user to be monitored on a keyboard or the changes to be made to a user profile, or for a server to receive and interpret stimuli from a graphical user interface (e.g. the user touched this area and then entered this and that key). Event packages permit any type of event, data and commands to be created, updated and distributed into an SIP network.

The user orientation of IMS is based on characteristics that are specific to the SIP protocol and on others that belong specifically to the IMS service architecture. In short, SIP addresses can relate to services, but are more usually associated with users. The IMS service architecture permits SIP requests to be routed to devices and to application servers based on the identity of the user. The routeing of SIP requests to devices associated with the user is based on normal SIP routeing procedures, while the routeing of SIP requests to specific application servers is based on 'service profiles' associated with each user and stored in the IMS user database, the HSS.

In practice, while SOA would permit an application to access the presence of user X by using a presence service with the identity of a user as parameter, SIP and IMS permit the presence of a user to be accessed via the generation of a SUBSCRIBE, which is addressed to the user identity. Instead of being routed to a unique presence service applicable to all users, the SIP request can be routed to an instance of the presence service that is specific to this particular user.

Such a reversal of the addressing and routeing approach (service for SOA and user for SIP) is very important, as a request addressed to an SIP user address in IMS can reach not only a user but also the devices associated with this user, and applications or application instances associated with this user, whether these applications reside in the network (e.g. presence) or in the device(s) of the user (e.g. a game).

Imagine a health monitoring device installed on the body of the user. A very convenient way for a centralized entity to access health indicators of the user and to monitor their changes over time would be to issue a SUBSCRIBE for a specific event package and to address this SUBSCRIBE to the user itself. The same SUBSCRIBE addressed to different users will reach the health monitoring devices associated with each of them.

Now, think of all the potential applications of SIP event packages and add to it all the potential applications of the generic session control mechanisms of SIP. Think that there exist other SIP extensions of interest to add to the pack. Then, imagine the potential of a service architecture that combines both SOA and SIP/IMS principles.

3.2   IMS-Enhanced Service Delivery Framework

To justify the investment on the IMS infrastructure by the operators, IMS should not be limited to a specific set of services and specific business model; instead it should act as a service delivery framework to enable the many services related to multimedia: communication, content and data. It should act not just as a host for some of these services but as a facilitator and add value to services provisioned by the Internet or third-party providers. IMS service architecture and the development architecture that includes the service creation, service composition and orchestration enhances IMS to a service delivery framework. Having understood the concepts of service profile and IMS service profile-based routeing in the previous sections, the IMS service architecture refers to the use of the ISC reference point by the IMS core network entity S-CSCF to route SIP requests to application servers. It should be noted that the ISC interface can also be viewed from the application server perspective, which is presented in the next section as the IMS development architecture. The IMS development architecture also includes application servers and service implementation techniques.

3.2.1   IMS Service Architecture – Network Aspect   (Gouraud, 2006)

The ISC interface is not a control protocol that permits an application server to instruct the S-CSCF and vice versa. ISC uses SIP and its role is to extend SIP reachability from the IMS core network to IMS application servers. When the S-CSCF receives a SIP message and applies iFCs to it, the potential consequence is that this SIP message may be forwarded to one or more application servers. Similarly, when an application server generates an SIP message, the role of the S-CSCF will be to route it to another SIP entity (application server or device). In this interaction, the S-CSCF only acts as a proxy between two SIP entities. The real service control is what transpires between these two SIP entities. Consequently, in the IMS service architecture, when normal SIP routing and iFC-based routeing are combined, the S-CSCF simply acts as an SIP service router, which can support SIP-based interactions between IMS clients and IMS application servers. These interactions may relate to session control. There are no chapters or a specification dedicated to ISC in the 3GPP, reiterating the fact that the 3GPP architecture is a network-centric standard. However, TS 24.229 (3GPP, 2007b) provides the detailed description of the S-CSCF features and embedded in there are the characteristics of the ISC. Figure 3.5 is used as an illustration to demonstrate the functioning of the IMS service architecture.

In the IMS architecture, a specific party in an end-to-end SIP interaction is either identified by an IMPU or a PSI. While the figure shows an example of an IMPU-to-IMPU interaction, the following alternatives could also exist:

- IMPU to PSI: an IMS client or an AS-based application acting on behalf of a user interacts with an AS-based application identified by a PSI.

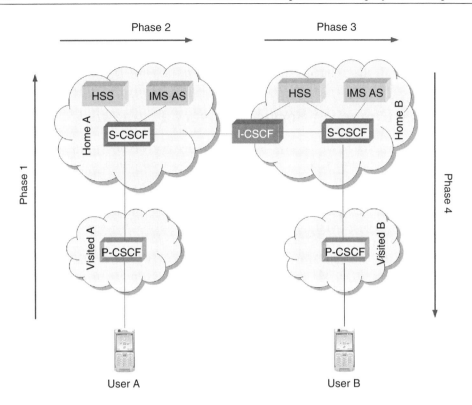

Figure 3.5 IMS service architecture in operation.

- PSI to IMPU: an AS-based application interacts with an IMS client.
- PSI to PSI: an AS-based application interacts with another AS-based application.

It is presumed in the illustration that User A is registered with IMS. This implies that User A's end-device has discovered the P-CSCF and is associated to it based on 3GPP or TISPAN procedures. The IMS registration procedure is performed leading to User A's authentication, the association of an S-CSCF to the user's end-device and the setting up of a security association between the client and the P-CSCF.

The figure shows direct interfaces between different operator networks. However, the GSM Association (GSMA) defined the possibility for third party transit networks to be used in between, in order to reduce the need for bilateral operator agreements. The end-to-end SIP interaction facilitated by the IMS service architecture is studied in the following phases:

- Phase 1: call initiation.   As per the standards, the P-CSCF removes any potential route defined by the IMS client and replaces it with a route to the S-CSCF serving the user in its home network. This route is defined during the IMS registration of the user. The request may traverse an I-CSCF and in case the home network prefers to hide its network topology in the potential visited network, a border gateway is used in common practice. When the S-CSCF receives an initial INVITE request destined for the served user, in the illustrated example for User A, the S-CSCF shall:

- Examine the SDP offer (the 'c=' parameter) to detect if it contains an IP address type that is not supported by the IM CN subsystem and processes the initial request when the IP address type is supported. It also processes the initial INVITE request without examining the SDP when it has prior knowledge of the terminating UE supporting IPv4 and IPv6 addressing. The processing of the initial request when the IP addressing type does not match is out of the scope of this section, which focuses on highlighting the service architectural features of IMS. Refer to the 3GPP specification (3GPP, 2007b) for details.

- Phase 2: service profile-based service routeing for the originating user. The S-CSCF serving the originating party applies the service profile for the originating IMPU. This service profile consists of a list of initial filter criteria. It corresponds to the set of services the originating party is associated with. In practice, this means that the service route for the SIP request originated by User A is defined by the operator and implemented by the S-CSCF. In this phase, the S-CSCF may route the SIP request to one or more application servers (one after the other in a signalling chain or pipe). It is possible that one of the application servers terminates the SIP request (i.e. acts as an SIP user agent). This may happen in three cases:
  - The request was addressed to a PSI hosted by the application server (e.g. the user tries to access content).
  - The originating party issued a request addressed to itself, which was actually targeted at one of its services in the network. For instance, SIP requests that update the presence information of a user (PUBLISH for presence event package) are addressed to the user's own IMPU.
  - The application server terminates the request on behalf of User B. This may happen, for instance, with services that block traffic to unauthorized destinations. Another example could be the case where a service, potentially after having accessed the presence of User B, decides to transform an IMS instant message (SIP MESSAGE) into an email, an SMS or an Internet IM. In this case, IMS SIP routeing is terminated, to be superseded by an alternative protocol outside IMS.

  Signalling path termination. The SIP interaction may terminate at phase 2 if one of the application servers decides to act as an endpoint for the SIP request.

  IMS exit point. The SIP interaction may exit IMS at phase 2 if an application server serves as a gateway towards a different protocol/domain (e.g. email, SMS, MMS, Internet IM protocol). Note that phase 2 takes place before the IMS core network tries to route the SIP request based on the request-uri. This means that the request-uri might not be an IMS routeable URI: it could be a URI specific to the protocol and/or domain to which the SIP request is actually targeted. In case no application server terminates the SIP request, the S-CSCF proceeds with the next phase.

- Phase 3: routeing to the destination network. The S-CSCF routes the SIP request according to its request-uri. In case the request-uri is a TEL URI, the S-CSCF may route the request to the circuit-switched network when the TEL URI does not resolve into an SIP URI through ENUM.Through DNS, the S-CSCF may also determine that the domain of the request-uri is resolved into a non-IMS network like the Internet. For an IMS exit point, the SIP request is routed outside IMS by the IMS core network if it is a TEL URI that does not resolve into an SIP URI or if it is an SIP URI whose domain is outside the IMS (e.g. in the Internet). Either User B is registered with IMS or not. If it is, the I-CSCF retrieves the location from the HSS and routes the request to the S-CSCF serving it. If it is not, there is a procedure to allocate an S-CSCF dynamically to User B and to route the request to this S-CSCF. The rationale is that User B's IMS services in the network might need to be reachable even when User B is not available. For instance, for services such as presence, a voice call continuity server that will route the call to the circuit-switched network, a call forwarding service needs to be reachable even when the user is not as an alternative, instead of originating from the IMS and having followed phases 1 to 3, the request received by the I-CSCF may have originated from the circuit-switched network. The request in such case is received from MGCF, i.e. the signalling gateway with CS, or from a non-IMS network like the Internet.

• Phase 4: service profile-based routeing for the terminating user.   The S-CSCF serving the terminating party applies the service profile associated with the IMPU (or PSI) corresponding to the request-uri. This service profile consists of a list of initial filter criteria. It corresponds to a set of services associated with the terminating party. In this phase, the S-CSCF may route the SIP request to one or more application servers (one after the other).

It is possible that one of the application servers terminates the SIP request (i.e. acts as an SIP user agent). This may happen in two cases:

– The request was addressed to a PSI hosted by the application server (e.g. the user tries to access content or the user posts messages in a chat room). This is the case where an operator offers services to the world. Note that there exist alternative mechanisms to route a request to a PSI.

– The application server terminates the request on behalf of User B. This might be the case for presence where the request is addressed to User B but was actually aimed at its presence server and any other service acting on behalf of the terminating party, such as voice call continuity, call or message blocking, or forwarding services.

Signalling path termination. The SIP interaction may terminate at phase 4 if one of the application servers decides to act as an endpoint for the SIP request.

IMS exit point. The SIP interaction may exit IMS at phase 4 if an application server serves as a gateway towards a different protocol/domain such as email, SMS, MMS, Internet IM protocol.

• Phase 5: delivery to terminating party.   In case there is still an SIP request to be delivered to User B, the S-CSCF routes the request to the P-CSCF(s) serving User B. In case several endpoints are registered with the same IMPU, there may be several requests, which in turn get routed to the device User B.

3.2.1.1   Other Service Interactions

The illustration addressed the situation where the SIP request is initiated by an IMS client corresponding to an IMPU and is destined to another IMS client also represented by an IMPU; this is an IMPU-to-IMPU service interaction. However, there exist two types of identities in IMS: public user identities (IMPUs) are SIP URIs (sip:user@domain) or TEL URIs (tel:+3312345) that identify users and public service identities (PSIs) are SIP URIs (sip:service@domain) or TEL URIs (tel:+3354321) that identity services, service features or user groups.

Based on the IMS service architecture that uses the service profile-based routeing procedures, there are few other possibilities of service interaction, as mentioned below:

- IMPU-to-PSI service interaction where PSI is a network-based service. The service may be explicitly identified by PSI or may implicitly be identified by the content of the SIP message such as in PUBLISH; it has a header called 'event' with the value 'presence'. In this case, the SIP request is likely to be addressed to User A himself (self-addressing). User A is a subscriber of the operator offering the service. He may be roaming or in a home network. Routeing to the application server hosting the service is based on User A's service profile in the HSS and the fact that User A originated the SIP service request. This service routeing mechanism ensures that User A is authorized to access the service. If not, the service profile would not route the request to the application server. Other SIP routing mechanisms would then be used, which would either route the request to an application server handling nonauthorized requests or would reject the request if the target address cannot be resolved to an IP address. It also ensures that the routeing to the application server can be specific to User A. Service examples:
  - User A publishes new presence information using the SIP method PUBLISH.
  - User A discovers the services he is subscribed to using the SIP method SUBSCRIBE.
  - User A records a greetings message for his multimedia mailbox using the SIP method INVITE.
  - User A starts a video-on-demand session using the SIP method INVITE.

  In all cases, User A must be authorized to access the service. In the first three cases, the service is personal to User A (User A's presence, User A's service registry, User A's mailbox). In the last case, the service can be shared with external entities such as VoD servers.

- IMPU-to-PSI service interaction where the PSI is a public network service. The service must be explicitly identified via a PSI. User A might be a subscriber of the operator offering the service or alternatively a subscriber of another operator, or simply a user accessing the service from the Internet. The routeing of the SIP request is based on the resolution of the PSI to an application server hosting the service. Either the PSI has a DNS entry or the PSI is associated with a service profile in the HSS, which determines that some SIP messages addressed to it have to be routed to this AS. Service examples:
  - public news service using the SIP method SUBSCRIBE or INVITE;
  - pay per view video on demand using the SIP method INVITE;
  - discussion forum using the SIP method INVITE for messaging sessions.

  If the public network happens to be the Internet, then the service might be explicitly identified by the SIP URI or might be implicitly identified in the SIP message, which is addressed to an Internet user. The routing of the SIP message to the Internet is based on standard SIP routeing procedures applied by the IMS core network. Service examples:
  - access to presence located in the Internet (SIP method is SUBSCRIBE);
  - VoIP session with Internet client;
  - chat/IM with Internet client.

- IMPU-to-PSI service interaction where PSI represents a service located in the user's device or a server connected to IMS. The service is located in a device or a server that belongs to User A and is connected to the IMS network like an IMS user. It could also be accessible through the Internet. It might be offered by User A's operator or by another service provider (possibly User A himself). The service is addressed through an identity perceived as a user identity from the IMS network. In a typical case the device or server is registered through User A's identity. User A therefore issues a SIP request addressed to himself, which will be routed to the device/server through normal SIP routing procedures. In order to route the request to a specific client (and avoid forking procedures), the request is likely to be addressed to the globally routeable user agent URI (GRUU) associated with the device or server. The GRUU permits the device associated with the user to be explicitly identified. Service examples:
  - User A retrieves a secured HTTP link to access private information on his PC (the SIP method is SUBSCRIBE and uses the fact that IMS is a secured network and that User A's identities are asserted both on the client and the server).
  - User A programs Windows Media server to record a specific program (the SIP method is PUBLISH).

3.2.1.2 Requests from Application Servers

The IMS application servers are capable of generating SIP requests. They use either an IMPU or PSI as the originating party and an IMPU or PSI as the request-uri. When an AS issues a request with an IMPU as the originating address, it is supposed to route it to an S-CSCF serving the IMPU. Starting with 3GPP R7, it is possible to allocate an S-CSCF to the IMPU even if it is not registered with IMS (i.e. the IMS AS issues a request on behalf of an IMPU that is not registered with IMS). IMS routeing then proceeds with phase 2, as discussed earlier.

When an AS issues a request using a PSI as the originating party, it may either reach a S-CSCF serving the PSI or route the request directly to an I-CSCF for the terminating party. Depending on the case, IMS routeing then proceeds with phase 2 where the service profile is associated with the PSI or directly with phase 3. ISC is not a control protocol that permits an application server to instruct the S-CSCF to do this or that. ISC is just SIP, and its role is to extend SIP reachability to IMS application servers.

When the S-CSCF receives a SIP message and applies iFCs to it, the potential consequence is that this SIP message may be forwarded to one or more application servers. Similarly, when an application server generates an SIP message, the role of the S-CSCF will be to route it to another SIP entity (application server or device). In this interaction, the S-CSCF only acts as a proxy between two SIP entities. The real service control is what may happen between these two SIP entities. Consequently, in the IMS service architecture, when you combine normal SIP routeing and iFC-based routeing, the S-CSCF simply acts as a SIP service router, which can support SIP-based interactions between IMS clients and IMS application servers. These interactions may relate to session control, but not necessarily.

3.2.1.3   Value-Added Service Interactions

In the previous sections, a combination of service interactions between IMS clients and IMS application servers where the origination of service is either a IMS client or an application server is presented. A more interesting capability of IMS service architecture is the addition of application servers that may add value to the end-to-end IMS service interactions. Listed below are the various possibilites of adding value to the end-to-end IMS service:

- The service may either act as an SIP proxy or as a back-to-back user agent (B2BUA). Without going into details, the proxy behaviour permits little control on the end-to-end SIP interaction and generally applies to services that simply monitor SIP traffic or perform actions that do not directly impact the end-to-end service interaction (e.g. the service initiates a new service interaction as a consequence of the one it is proxying). On the other hand, acting as a B2BUA is more adequate for services that intend to control the end-to-end service interactions (e.g. changing a leg, modifying a session description).
- The service either inserts itself in a service interaction initiated by a peer, or the service is itself the initiator of the service interaction between the peers. Third party call control (Rosenberg *et al*., 2004) is an ideal example for this use case.
- The service may be tranparent to the peers involved in the service interaction. This is the case when the service only adds value to an otherwise peer-to-peer service interaction. Alternatively, the service may be visible; i.e. it appears in SIP signalling as the recipient or initiator of SIP messages via its public service identity (PSI). For instance, a multimedia content service identified via a PSI actually aggregates individual media components accessible via SIP as part of its delivery session. Note that SIP is only one of the protocols that the front-end service can use towards back-end services.
- The service may either apply to one peer of the end-to-end service interaction or be shared by all participants in the peer-to-peer relationship. In the former case, the service is personal to a user and may be customized specifically for him/her, while in the latter case the service is shared between two or more users. A typical example of a shared service is one that supports the multiparty nature of an SIP service interaction, e.g. a multimedia conference server or a chat room for IMS.
- The service may either be the single intermediary between the peers or may be chained with other services. An example of chained services on a service interaction path of type messaging (SIP MESSAGE) could be an anti-spam service, an anti-virus service and an archive service. Another example applied to VoIP would be the chaining of a voice call continuity (VCC) server responsible for switching between WiFi/IMS and cellular/circuit-switched during the voice call, with a voice application server supporting typical telephony features for the user.
- Finally, a service may systematically act as an intermediary in the peer-to-peer service interaction or may sometimes decide to take over the role of the peer for the service interaction. In the VoIP example given above, the VCC server will always act as an intermediary, while the voice application server may sometimes act as an intermediary

when it decides to let call setup happen, or as a peer if it decides to reject the call attempt on behalf of the user it serves.

The value addition of the IMS features to the multimedia service delivery is further illustrated in the example below.

3.2.1.4 IMS for Multimedia Service Delivery

The aspects that may enable IMS to be the ideal multimedia service delivery framework are the advantages offered by the SIP protocol and the IMS service architecture. The network-oriented convergence enabled by the intrinsic nature of SIP is presented in Part I and the user-centric service orientation of IMS is presented in Part II. However, it is the multimedia capabilities of the IMS service architecture that are crucial in making IMS a truly enhanced multimedia service delivery framework. It also has the ability to blend the person-to-person multimedia communication with multimedia content and application components as well as to share the service experience among many users. The ability also exists for IMS to integrate with the non-SIP and non-IMS services, thereby ensuring that the service delivered over IMS need not necessarily be developed for IMS and does not necessarily require SIP-related application components.

The operator is thereby enabled to offer those services to its subscribers that are actually provided by third party service providers located in other domains, including the Internet, without requiring a prohibitively strong technical coupling with them. IMS can also add value to the multimedia content delivered to its subscribers by third party service providers. These capabilities are discussed in the illustration depicted in Figure 3.6.

In order to experience the multimedia service, the user starts a session addressed to a public service identity (PSI) identifying the service. The PSI may be specific to the user and routed to the SIP AS according to the user's service profile (originating trigger). Alternatively, the PSI may be a shared, public one. In this case, the routeing to the SIP AS may also be based on the user's service profile (i.e. users need to be authorized to access the service so that their service profile allows routeing to the SIP AS), unless it is based on the normal resolution of the PSI to the SIP AS (i.e. all users can access the SIP AS when they issue a request addressed to it). We have already described these routeing alternatives here and also with another service use case.

The user negotiates (and possibly renegotiates during the session) the content of the service through any appropriate interface, for instance by accessing a web page supported by the SIP AS. Alternatively, this could be through a client downloaded on the terminal and communicating with the SIP AS via an application-to-application interface, like the exchange of XML documents describing the desired content of the multimedia session.

The SIP AS then performs the required actions to deliver the content (e.g. files, streaming media, web pages, applications) within the session. Depending on the type of content, the desired control of the operator over the delivery and the technical means available for each component, the SIP AS uses the appropriate mechanism for each of the components (which do not have to be co-located with the service control logic hosted by the SIP AS). This may include some of the following:

- establishing an SIP session with the component endpoint and bridging this new session with the user to the SIP AS one (typical B2BUA behaviour);

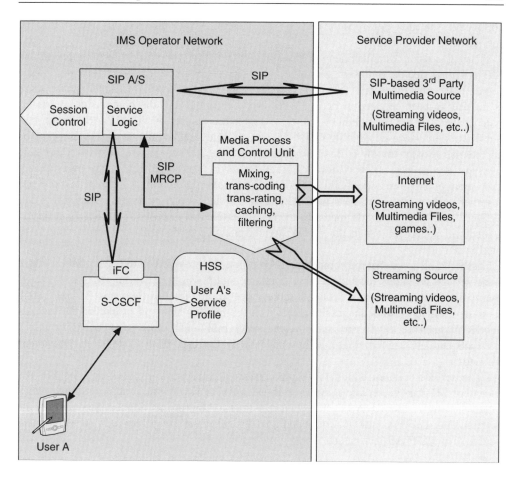

Figure 3.6 IMS service architecture in operation.

- controlling the delivery of the component via an appropriate non-SIP interface (e.g. web services, H.248 (Groves and Lin, 2007)) towards the component source and negotiating/renegotiating the SIP session accordingly towards the user;
- providing the user's terminal and/or the component source with appropriate information to establish an end-to-end connection.

In an example, the SIP AS would provide within the session a URI (e.g. HTTP URI, FTP URI, RTSP URI) via SIP content indirection or a referral, permitting the user's terminal to directly connect with the component source. In another instance, the SIP AS retrieves from the source the information required to connect to a whiteboard server and transmits it to the user's terminal, so that a whiteboard client connects to the server and interfaces with it through the relevant whiteboard protocol. In yet another instance, the SIP AS provides the component source with the information relevant to push the content to the user's terminal. In any of these instances, the SIP AS may keep a control interface towards the component source in order to terminate the delivery of the component when the service session is completed (the idea is to ensure that the component is not delivered any more after the service session is ended).

During session establishment or session renegotiation for a specific component, the SIP AS may decide to insert a number of media-level intermediaries (typically media servers) between the user's terminal and the component source(s). In such a case there is no peer-to-peer connection between the user's terminal and the component source, as both connect to the intermediary, which is under the control of the SIP AS. The potential control interface between the SIP AS and the intermediary in the network might be an alternative way for the SIP AS to synchronize the delivery of the service component with the service session (start/stop delivery). However, the usage of a media intermediary may serve other more added-value objectives, like combining/mixing different components together such as inserting text information in a video stream, caching media for better delivery quality, transcoding media to fit the capabilities of the user's terminal or inserting a localized advertisement in the media stream, possibly to decrease the service fee to be paid by the user.

It should be noted that the component sources may be provided by the operator or by third party service providers located in other domains, and possibly in the Internet. In this case, the operator acts as a service broker, adding value to individual components by integrating them in a single multimedia session, acting on the media plane and providing the level of access (no need for the user to authenticate to each individual service component provider), the QoS and the security that can be expected by the user from its telecommunications operator.

The service session, which takes place between the user's terminal and the SIP AS (usage of SIP for the service may be confined to these two entities), may terminate when either the user or the SIP AS decides that it is time to. As for individual components in the session, their delivery may be terminated through either the user's terminal, the component source or the SIP AS if it has the control means to do it.

3.2.2   IMS Development Architecture – Application Aspect   (3GPP, 2007b)

The development architecture of IMS includes the IMS application server and the procedures for supporting the ISC interface in 3GPP specification TS 24.229 (2007b), Section 5.7 provides a detailed description on the procedures at the SIP AS. However, in the context of the development architecture of IMS, this subsection presents the role of SIP application servers in the IMS architecture and procedures followed between the IMS application server and the ISC interface when the IMS application server either receives an SIP request or initiates an SIP request.

3.2.2.1   IMS Application Servers   (3GPP, 2007g, 2007k)

In IMS architecture, the transport and control layers provide an integrated and standardized network platform that enables the service providers to offer various multimedia services in the service layer. Application servers host and execute the service logic and provide the interface with the control layer using the SIP protocol. The service logic includes the way in which services are invoked, the signalling and media actions involved as well as the service interactions among the service components. Depending on its implementation, an IMS application server can host one or many IMS applications. The application server handles and interprets the SIP messages forwarded by the S-CSCF (either directly or through the OSA service capability server (SCS)) and translates end-user service logic into sequences of SIP messages, which are then forwarded to parties involved in the communication through the S-CSCF. A typical IMS application server provides the following functionalities:

- It provides an end-user service that includes communication service logic.
- It includes a software development kit (SDK) to allow service and content providers and developers to develop new services easily.
- It includes a service enabler that can be called and shared by many end-user services.
- It can interact with other ASs to generate new composite services via a service orchestration framework.

The IMS application server types defined by the 3GPP are as shown in Figure 3.7:

- SIP application servers handle pure SIP-based applications, such as the familiar presence and instant messaging. SIP application servers can act as redirect servers, proxy servers, originating user agents, terminating user agents or back-to-back user agents.

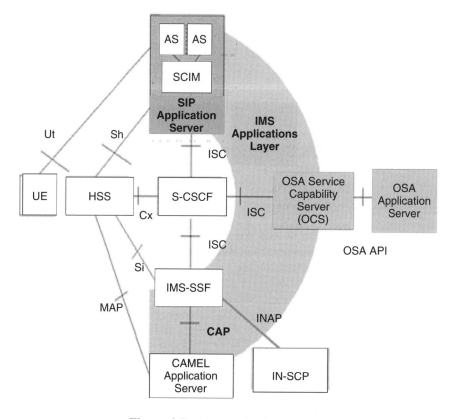

Figure 3.7 IMS application servers.

They have SIP signalling capabilities and are directly involved in the call's signalling flow. They receive SIP messages from the S-CSCF and parse them. Similarly, they generate SIP messages and send them to the S-CSCF. The application server translates an IMS services' execution in the SIP application server into a sequence of SIP procedures, such as sending invite requests and reacting to SIP responses. For services requiring media interaction, the SIP application server invokes the multimedia resource function control (MRFC) capabilities. The IMS specifications do not explain exactly how to perform this invocation, but several proposals exist for letting the SIP application server use SIP to control the MRFC (e.g. basic network media services with SIP, 8 Media Server Control Markup Language (MSCML), 9 Media Objects Markup Language and Media Sessions Markup Language).

- Open service architecture (OSA) application servers handle non-SIP-based applications, providing the gateway and the bridging between, for example, Parlay-enabled applications. OSA application servers can provide the same services as the SIP application server but have no signalling capabilities and are not directly involved in the SIP calls' signalling flow. They communicate with the S-CSCF through the OSA SCS, which maps SIP messages into invocations of the OSA API (also called Parlay) and back. From an S-CSCF perspective, the SIP application server and the OSA SCS exhibit the same behaviour. The OSA application server also has access to the HSS data, but only through the SCS. The SCS implements the diameter protocol, so it can read and update data records based on the OSA application server requests. The OSA application server mainly implements external services that could be located in a visited network or a third party platform.
- The IP multimedia service switching function interfaces with customized applications for mobile networks enhanced logic (CAMEL) application servers and IN SCPs via CAP and INAP, respectively, to provide convergence with the mobile network and the typical services are associated with SS7/INAP in the legacy fixed environment.
- Some technologies used in telephony and voice-over-IP (VoIP) application servers are also applicable to IMS application servers.

The IMS architecture lets an IMS service provider deploy multiple application servers in the same domain. Different application servers can be deployed for different application types (e.g. telephony or presence application servers) or different groups of users. The S-CSCF routes the requests to each application server based on the filter criteria (iFC) received from the HSS, as discussed earlier. The HSS stores and conveys this filter information on a per-application server basis for each user. The application server (SIP application server or the OSA SCS) uses the diameter protocol to communicate with the HSS. Diameter transports user-profile-related data, but can also transport transparent data, i.e. data for which the exact representation of information is not understood by the HSS or the protocol (e.g. service-related data or user-related information). When the HSS transfers the name and address of more than one application server, the S-CSCF must contact each application server in the priority prescribed by the iFC. The S-CSCF uses the first application server's response as input to the second application server. This concept is termed as service chaining, which shall be discussed later. During the service logic's execution, the application server can communicate with the HSS to get additional information about a subscriber or to learn about changes in the subscriber's profile.

3.2.2.2   IMS Application Server Procedures

The IMS procedures between the application server and the ISC inteface are illustrated in the following two use cases, as illustrated in Figure 3.8.

> *Case 1. The IMS application server receives an SIP request that is originated by another SIP entity. This entity could be an IMS client (e.g. phone, TV set, home PC) or an IMS application server (e.g. a messaging server, a multimedia content server). It could be located in the same operator domain as the IMS application server, or in a different one. This entity could also be a client or application server located in a non-IMS network, such as an enterprise network or the Internet. An SIP request originated from outside the IMS and routed to it can be discriminated from an IMS-issued request, and the IMS application server can adapt its behaviour accordingly. For instance, a request originated from the Internet is not authenticated by the IMS and would be processed differently based on the filter criteria.*

When an IMS application server receives a SIP request from the core network – a request that may have been initiated by an IMS client, a non-IMS client or an entity in a non-IMS network – it may support one of four different roles.

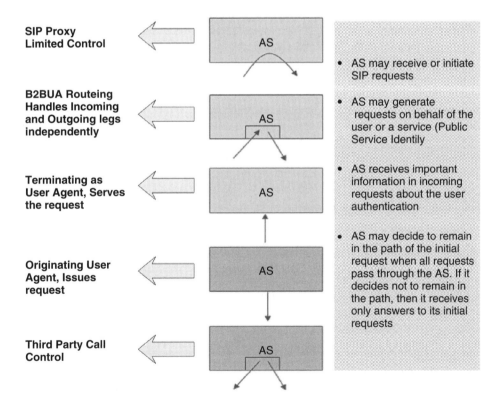

Figure 3.8 ISC-IMS application server interactions.

Acting as an SIP user agent, the IMS application behaves as an endpoint for the SIP request. The request may have been addressed to a service or service feature supported by the IMS application server (typically when the request-uri is a public service identity, or PSI), or it may have been addressed to an IMS user (the request-uri is an IMS public user identity, also known as IMPU). In this latter case, the service logic hosted by the IMS application server decides to terminate the request on behalf of the user it is addressed to. This is typically the case for presence requests terminated at a presence server (the presence request is addressed to the user whose presence is sought). Other examples include a call termination service or an IMS messaging store, which decide that the destination user cannot accept the call or the message right now.

Acting as an SIP redirect, the IMS application server will redirect the initiating party to another destination. We do not expect an IMS application server extensively to support this (feature-poor) role. The IMS application may also decide to act as an intermediary in the end-to-end SIP interaction that takes place between two SIP entities (e.g. client to client, client to other AS, other AS to other AS, other AS to client).

Acting as an SIP proxy server, the IMS application server has very little opportunity to impact the end-to-end interaction. This behaviour is adequate in cases where the AS hosts service logic that has to apply only prior to enabling the end-to-end interaction (e.g. authorization, target selection). The AS performs its logic, then lets end-to-end SIP signalling go on without any interference. It can also be used when the AS hosts service logic that monitors end-to-end SIP signalling without any intention to interfere.

When the AS hosts service logic that requires greater control on the end-to-end SIP interaction, it has to support the so-called routeing back-to-back user agent (routeing B2BUA) behaviour. As a B2BUA, the AS acts as an endpoint to both parties in the SIP interaction. In doing so, it may decide to appear explicitly as an endpoint to them (by inserting a PSI as a recipient/originator of the SIP interaction) or to remain transparent to the endpoints. Here are examples of situations where service logic needs to rely on a B2BUA, in scenarios where there is a need to modify the body of a SIP request (e.g. change a codec in the session description), a potential need to transfer a session during its course (e.g. to another party, to an announcement) and a need to insert a media plane intermediary in a multimedia session (e.g. to mix/adapt/control content, to insert in advertisement).

Service behaviour based on Direction of Requests

This is possibly the most important feature of the IMS service architecture when you want to integrate a non-IMS SIP application server with an IMS network, more especially in the context of VoIP services. The ISC interface (more especially the initial filter criteria that govern the forwarding of SIP requests over ISC) distinguishes between requests that originate from a certain IMS identity (IMPU or PSI) and those that terminate at this IMS identity. Moreover, it permits a distinction to be made between the case where the IMS identity is currently registered with IMS or not (only from 3GPP R7 for requests originating from a nonregistered identity, i.e. requests initiated on behalf of a nonregistered user by an IMS application server).

This architectural feature (shared with IN) permits different service logic to be executed depending on the direction of a request. For instance, taking classical call control services, it permits an outgoing call screening service to be executed only for originating call requests, and call forwarding services only for terminating call requests. It also permits an IMS instant message to be forwarded to a store and forward messaging AS only in the case where the recipient of the message is currently not registered with IMS, and therefore unreachable through IMS connectivity.

A clear distinction between originating and terminating parties (and their respective services) is mandatory in a multioperator environment, in which the parties might be subscribed to different operators. However, most VoIP application servers that exist in the market today were implemented for a closed, single service provider environment like the enterprise. They therefore tend to execute both originating and terminating services at once. Such a behaviour is incompatible with the IMS service architecture, and must be corrected when the AS is ported on IMS. On the other hand, the direction of SIP requests is not an issue when porting a presence server on IMS (the presence logic depends on the SIP requests more than their direction).

Note that, while the direction of a request (originating, originating unregistered, terminating, terminating unregistered) is information used to determine how SIP requests should be forwarded to IMS application servers, it is not specified how to convey this information in the SIP request that is to be forwarded. A typical approach to make the AS aware of the direction is to define a distinct AS SIP URI for each direction in the initial filter criteria (IFCs) or to add a specific parameter in the AS address. In both cases, the information about the direction is made explicit when provisioning AS addresses for iFCs in the HSS.

Case 2. The IMS application server issues an SIP request to another SIP entity. This SIP request may be the consequence of the initial reception of an SIP request. Alternatively, it may be addressed to a client or a server located in a non-IMS network, like an enterprise network or the Internet.

An IMS application server can issue SIP requests on its own, without this request being tightly related to a request the AS previously received. The generated request might be a side effect of one or several SIP requests previously received by the AS, of interactions with an end-user performed through a non-SIP interface (e.g. web page, SMS) or of interactions with a third party service (e.g. web services), but these are only examples. Anything can do, and the more intelligent the service is, the more spontaneous the generation of SIP requests can be.

SIP Roles

The IMS application server can act as an SIP user agent. It is the initiator of the request and it will support the end-to-end SIP interaction with its SIP peer, whether this is an IMS client, an IMS application server or another SIP entity outside the IMS.

The IMS application server can also act as the third party initiator of a dialogue between two other SIP endpoints, like two IMS clients or an IMS client and an IMS application server. In this case, the AS acts as an initiating back-to-back user agent (initiating B2BUA). This role can be used, for example, by a service to automatically set up a call between two users or implement a click-to-dial-back feature supported by a web page (the user clicks on a link to get called back by an operator; the service logic selects the operator and establishes the call between the two).

Note that an alternative way to implement third party control is for the application server to use SIP REFER towards one of the SIP entities it wants to involve in the dialogue (e.g. the AS asks User A to set up a session with User B). This approach is simpler to implement from an application server perspective, as it does not require the usage of an initiating B2BUA, but it also requires the support of the REFER method by the SIP client, which is not given in current SIP networks. The approach may also have an impact on how charging will be performed.

Originating IMS Identity

When it initiates a SIP request towards another IMS or non-IMS entity, the IMS application server (more precisely the service logic it hosts) has the choice between two possibilities. The AS may act on behalf of an IMS user and use an IMPU of this user as the identity initiating the request. Doing so, the AS can impersonate a user it serves and, for instance, send an instant message or subscribe to the presence of a third party just like the user would. Note that this ability of an AS to issue an SIP request on behalf of a user (which may not be registered with the network at the time the request is initiated) is the reason why 3GPP had to introduce the initiating unregistered session case in the R7 specifications of initial filter criteria.

The AS may alternatively populate the P-Asserted-Identity header with a PSI, thus endorsing an identity associated with the service logic that initiates the request. For instance, an application server may initiate a request as a specific conference, voice mail account or chat room. Because it has a secure connection with the IMS core network and is part of the IMS trust domain, the AS can directly set up the P-Asserted-Identity header with the IMPU of the user on whose behalf it acts on or a PSI, ensuring that the IMS request was duly authenticated by the IMS network.

Routing of the Request

3GPP initially tended to be very restrictive about how the routeing towards the IMS core network of an SIP request initiated by an AS could be performed. Fortunately, the latest specifications permit room for variants.

When the request is sent on behalf of a user, the AS may have to route the request to an S-CSCF serving the IMPU of the user, in order for originating services associated with the user to be executed (in this case the AS has to insert the 'orig' parameter to indicate that this is an originating request). The routeing of the request to this S-CSCF might be direct, which implies that the AS knows the address of an S-CSCF serving the IMPU (possibly through a previously received request or by retrieving it from the HSS via Sh) or through an entry point to the network serving the IMPU (which is less efficient but requires less knowledge from the AS).

Alternatively, and if the operator policy allows it, the AS may directly route the request to the network serving the destination address, thus bypassing potential originating services and charging procedures associated with the IMPU. This flexibility was not part of initial 3GPP ideas, but the authors always supported it as a simpler approach placing fewer constraints on the AS, and which is certainly adequate for some services.

The procedure for the case where the request is initiated by a PSI is similar, except that when the request is routed to an S-CSCF serving the PSI in order to apply originating procedures and execute originating services, the address of this S-CSCF is a priori static and can be configured in the AS (but it can also be retrieved from the HSS as well). Note that the case where the AS has to route the request to an S-CSCF serving the IMPU or PSI requires an IMS-specific behaviour that will impact non-IMS SIP application servers when they are ported on IMS.

3.2.3   Service Composition

Service composition in IMS is based on integration of multiple applications acting on a single communication session (call instance). The use of functional applications as building blocks allows service providers to mix and match different application offerings without the need additionally to customize their interworking principles. It typically involves the integration of two or more services or service enablers. These are combined according to some service logic and functions to create a hybrid application. The concept of service chaining is an important concept in service composition.

3.2.3.1   IMS Service Chaining

The IMS service framework has the capability to chain different application servers (and their services) in the signalling path between two or more service endpoints. Service chaining extends the SDP concept by providing an environment to facilitate the rapid development of value-added applications that are structured as a sequence of web service invocations. The service chaining environment includes service creation tools to specify and validate the application logic (i.e. to define and test the sequence of web service invocations, prior to deployment), as well as a service chaining server that manages the execution of the application logic; i.e. it manages the execution of a service chain. A complete service chaining architecture must address multiple requirements:

- It must provide a framework to combine multiple individual network services to deliver a richer, multimodal user experience.
- It must enable the seamless transition from one network service to another, maintaining both user sessions and session information.
- It must work across multiple devices and multiple networks.
- It must provide an integrated user interface, supporting sequential and parallel invocation of services within a single user session.
- It should use web technology to allow easy customization of the service chaining logic.

3.2.3.2   Service Chaining Architecture

As shown in Figure 3.9 the implementation of the service chaining architecture relies on three key technology components hosted within the service delivery platform:

1. The service registry holds information on the underlying network functions that are available and can be invoked via service interfaces.

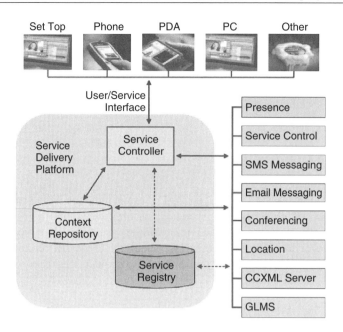

Figure 3.9 IMS service chaining architecture.

2. The context repository holds context information for a given user or group session, so that this context information can be maintained for the duration of the service chain.
3. The service controller executes the service chaining application logic that invokes network functions (via their services interface) in response to user requests and other events.

As each service in the chain corresponds to the invocation of an independent network function, one of the main challenges for the service chaining architecture is to ensure a seamless transition from one web service to the next. To achieve this, network session information, session state information, user information and other contextual information is maintained in the context repository independently of each given service invocation. For example, session-specific contact information that is extracted from the buddy list function (GLMS) can be stored in the context repository and then delivered as input to other network functions as they are invoked by the service controller in response to user requests; i.e. once the user has chosen the list of friends with whom he/she wants to communicate, that subset is stored in the context repository. When a user requests the messaging or conferencing functions, the service chaining application retrieves that context information (in this case the buddy list) and passes the information on to the requested services. Other session state information can also be shared in this way between the elements of the service chain.

As an example, in a mid-term voice-centric usage of IMS it is possible to chain application servers as follows (see Figure 3.10):

1. A voice features server supports typical call control services for the end-user, such as call forwarding or call screening.

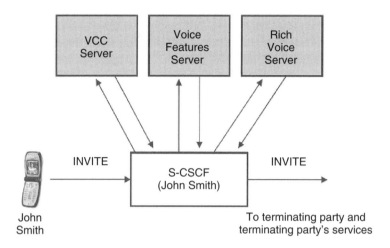

Figure 3.10 Service chaining example.

2. A voice call continuity (VCC) server supports the domain selection and seamless handover for handsets supporting both WiFi over IMS and GSM/UMTS over circuit-switched calls (e.g. a call is handed over from WiFi to GSM when the handset leaves WiFi coverage).
3. A server supports one or several rich voice services, i.e. services combining an IMS data session (e.g. best effort video, whiteboard) with an existing voice session (CS or IMS, depending on the VCC).

The combination of these three servers together (in the right sequence) according to the service profile of the end-user stored in the HSS allows rich voice services for a voice call supported either in the IMS or the circuit-switched network, with the possibility to switch between the two networks depending on the access being used by the handset, and this together with standard call control services. This composition is controlled by service profile information stored in the HSS and does not require any of the application servers to be aware of the others. The mechanism allows some IMS services to be used as building blocks, which can be combined in different ways in various products and for customizable user experiences. Obviously, it only applies to cases (as in the example) where no service interaction issue exists between the services supported by the different application servers or where these issues are solved by chaining the application servers in the right sequence.

Service orchestration is an important task in service composition and IMS enables the following means for service orchestration:

- Terminal-based orchestration
  - Combined service logic resides in client on end-user equipment
- Interaction manager-based orchestration:
  - Controlled by operator in the IMS domain.
  - Operator provides the developers with access to IMS application servers (AS).

- Applications are typically SIP containers with an exposed SIP servlet API.
- Hybrid application can appear as a single AS to IMS network.

- Parlay X-based service orchestration:
  - Parlay X web services gateway can provide interface to external entities.
  - Application logic exists outside the operator network.
  - Requests for telecom services are made using Parlay X.
  - Web services gateway provides:
    - Protocol translation,
    - Policy enforcement,
    - SLA enforcement.

The service chaining procedure described previously has the following issues:

1. It may lead to heavy processing by the S-CSCF of SIP interactions with application servers, in addition to the other tasks it already has, i.e. virtually doing everything in the IMS core network (e.g. user registration and authentication, end-to-end SIP routeing, generation of charging information, etc).
2. Multiple SIP-based interactions between the S-CSCF and application servers may lead to bad end-to-end latency for service delivery. These latency issues may potentially lead the operator to limit service chaining for the sake of service usability.
3. Chaining actually concerns application servers, and not individual services hosted by each of these application servers.

Issues 1 and 2 should be addressed via service co-location. It should be possible for an operator to co-locate on the same server, for a given user, services that are likely to be chained together or that generate other SIP interactions (e.g. presence based service accesses presence). In this context, out of the above-listed service orchestration methods, the service composition and interaction manager-based (SCIM-based) orchestration is of interest. Therefore the SCIM performs in the application server, and at the service level, what the S-CSCF does in the network and at the application server level. Its role is therefore to select and chain the services that need to be executed when an SIP standalone or initial request reaches the application server.

The SCIM exploits and complements the IMS capability to have a personalized chaining (composition) of services on the signalling path associated to an SIP request. Of course, SIP requests may relate to a large variety of services and are not limited to telephony or even to session control. The SIP chaining mechanism of the SCIM is very similar to the one supported by the S-CSCF. The user 'service profile' used by the SCIM might have similarities and differences with the initial filter criteria of the service profile used by the S-CSCF. All potential enrichments desired for chaining decisions can be implemented at the SCIM level. The SCIM is one of the orchestration mechanisms supported by an IMS application server, which is SIP centric. Web services-oriented service orchestration is to be supported by a complementary component of the platform. The platform is typically implemented as a JAIN SLEE or Java EE (J2EE) server supporting SIP servlets. In the latter case, the SCIM can be implemented as an application router in JSR 289 (JCP, 2008).

Such a SCIM aims to enrich the user-oriented nature of the IMS service architecture and to provide a highly customizable and differentiated service experience between IMS

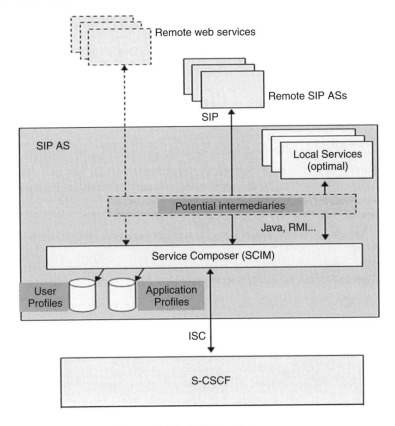

Figure 3.11 SCIM architecture.

users. At the same time, it can fulfil many typical service interaction management use cases that are targeted by service broker standardization. Distinct features of SCIM are illustrated through the following examples, as illustrated in Figure 3.11.

An IMS application server needs to follow some rules when interacting with the IMS core network. For instance, when a service initiates an SIP request out of the blue, it needs to include relevant information such as an original ID for the SIP dialogue/transaction or a charging vector, and it may be mandated to route the request to a specific S-CSCF when this request is initiated on behalf of a user. This task may be handled by the SCIM, if it is not handled by another component or by the application itself.

As part of its service selection and chaining task, the SCIM may forward an SIP message to remote application servers. By doing so it partly takes over the responsibility of the S-CSCF and permits the IMS core network to be offloaded from a part of the application layer-related SIP interactions.

A remote application server may not comply with IMS extensions to SIP, and may therefore not be able to exploit them (for instance a non-IMS application server may not be able to extract the asserted identity of the request initiator from the 3GPP specific P-Asserted-Identity header). In this case the SCIM may adapt SIP messages between IMS and the non-IMS application server.

A remote application server may belong to a third party service provider. The SCIM or a complementary component may support policies for interactions with this third party. The SCIM may possibly translate SIP requests into web service requests or other application-related interfaces.

3.2.4   IMS Service Creation – Generic Service Platform Aspects

IMS service creation may involve device-side development and/or server-side development. The application logic can be device-centric or server-centric. Server-side considerations are:

- Applications can be hosted by operator, enterprises or third parties.
- Applications can potentially use the session initiation protocol (SIP), Parlay or Parlay X web service interfaces.
- Third parties may provide their own application frameworks.
- Within operator network, operators can use SIP servlet APIs such as the Java specification request (JSR) 116 and JSR 289 (JCP, 2003, 2008).

Device-side framework are:

- JSR 180 (JCP, 2009a) addresses SIP stacks for J2ME environments.
- Handsets will increasingly include SIP stacks.
- JSR 281 (JCP, 2009b) will standardize a comprehensive IMS framework.
- JSR 164 (JCP, 2005a) defines the SIMPLE (SIP for instant messaging and presence leveraging extensions) presence.
- JSR 165 (JCP, 2005b) defines SIMPLE instant messaging.
- Device vendors may define additional device-side application frameworks.

The service creation environment (SCE) is typically a collection of service creation tools and resources to register, test and deploy native applications as well as third party applications. The IMS specifications do not define the application server's internal architecture. In order to understand how to build IMS applications and deploy end-user services, it is essential to examine the technologies that are available for service creation. Currently, the technologies for building IMS applications are varied; e.g. some are used to implement a complete application server, others can be used to implement just a layer of the application server. Some of them are SIP-dependent, so they can build only an SIP-based IMS service and SIP application servers. Others are protocol independent, so they can be used to build both SIP and OSA application servers. These application server technologies are classified into three families: SIP programming techniques, APIs and service execution environments. SIP programming techniques let application programmers access basic SIP functionalities to program the SIP applications. The two techniques in this category are the SIP common gateway interface (SIP CGI) and the SIP servlet. These techniques are SIP-dependent, so they can only be used for SIP application servers.

3.2.4.1   SIP CGI   (Khlifi and Grégoire, 2008)

SIP CGI was inspired by HTTP CGI, a tool for creating dynamic content for the web. When a server receives a SIP request, it invokes an SIP CGI script. The server passes the message body to the script through its standard input and sets environmental variables containing

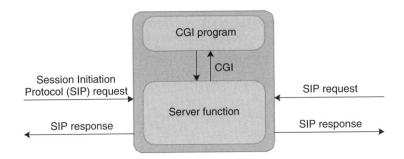

Figure 3.12 CGI script-based SIP application server (from Khlifi, 2008).

information about the message headers, user information and server configuration. The script performs some processing and generates signalling instructions for the server to execute. The SIP CGI script can instruct the server to generate an SIP response, proxy a request, create a new request, or change a request's headers. Figure 3.12 shows the functional model of a CGI-based SIP application server. In a CGI-based IMS SIP application server, IMS end-user services are written as CGI scripts. The server calls a CGI script for each incoming call. The script runs the service logic and returns to the server a list of actions to perform on the SIP request.

3.2.4.2 SIP Servlet Programming (Khlifi and Grégoire, 2008)

The SIP servlet API is a Java extension API for SIP servers, inspired by the HTTP servlet concept. A SIP servlet is a Java-based application component that is managed by a container and performs SIP signalling. These containers, sometimes called servlet engines, are server extensions that provide servlet functionality. Servlets interact with SIP clients by exchanging request and response messages through the servlet container. The container passes objects representing SIP messages to the servlet. The servlet has access to all the SIP message headers through those objects and, with this information, decides how to respond to a message. Servlets can answer or proxy requests, create or forward responses and initiate new SIP transactions. The container provides many services that the SIP servlet can exploit, such as automatic retries, message dispatching and queuing, forking and merging, and state management. The servlet itself only manages high-level message handling and service logic. Figure 3.13 shows the relationship between the servlets and the container as well as the different methods of the servlet interface. An IMS servlet-based SIP application server is an SIP servlet container. IMS applications are SIP servlets (or a group of SIP servlets). When the server receives a new request, it applies some preconfigured rules to select the servlet (the application) to process the request. The servlet executes the service logic and invokes the container capabilities to send and receive SIP messages.

Several APIs for building communication application servers can be used as part of an IMS application server. These APIs wrap up network and protocol functionalities into an easy-to-use abstract software component. The APIs provide high-level object-oriented interfaces that let programmers implement communication applications. Figure 3.14 shows an API-based IMS application server. When the API and application reside on different machines, a remote-procedure call mechanism can allow interaction between them. Parlay

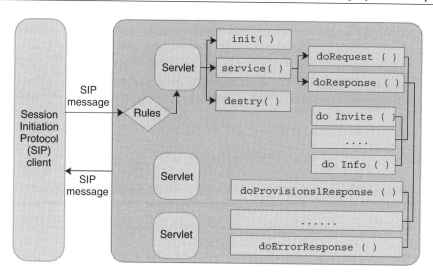

Figure 3.13 SIP servlet-based SIP application server (from Khlifi and Grégoire, 2008).

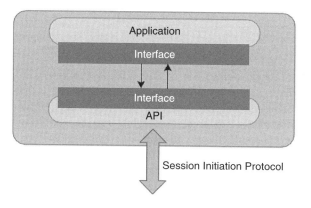

Figure 3.14 API-based application server (from Khlifi and Grégoire, 2008).

(Parlay, 2008) is a set of API specifications for managing network services such as call control, messaging and content-based charging. The Parlay group first proposed it for telephone networks and the 3GPP later adopted it as the OSA API to give the OSA application server access to the IMS network functionalities. The Parlay API supports all call-control functionalities that previous telephony APIs provided and offers some new features such as mobility, presence and data session control. Moreover, Parlay is the only API that includes VoIP and SIP systems in its specifications. Two versions of the OSA API exist:

- the standard version, a simple API specification that can be programmed using any object-oriented programming language;
- the web service-based version, or Parlay X.

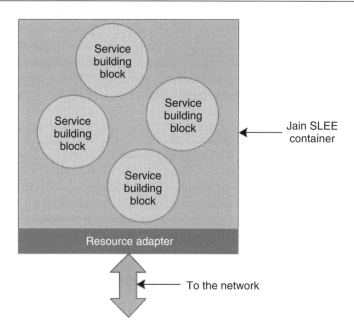

Figure 3.15 A Jain SLEE application service architecture.

Existing service execution environments are based on the JAIN-SLEE architecture or the OSA-PARLAY-X service execution architecture.

A service logic execution environment (SLEE) is a high throughput, low latency event-processing application environment designed for communication applications that can be used to build IMS application servers. Jain SLEE (http://jainslee.org) is the Java specification of the SLEE concept. It specifies the runtime execution environment, called a container, and communication services' internal architecture. A Jain SLEE service is a collection of reusable object-oriented components – service building blocks (SBBs) – running inside the container. Figure 3.15 shows the architecture of a Jain SLEE-based application server.

An SBB is a software component that sends and receives events and performs computational logic. An external resource, such as the communications protocol stack, the SLEE container or another SBB event, can generate events. A Jain SLEE application server interacts with external resources, such as network protocols and TAPIs, through resource adapters, which adapt resources to SLEE requirements. An IMS SLEE-based application server is a Jain SLEE container, and IMS applications are SBBs. Jain SLEE is not related to any signalling protocol. It can be used for both IMS SIP application servers and OSA application servers. For an SIP application server, it needs a resource adapter for SIP and for an OSA application server, it needs a resource adapter for the OSA API.

3.2.4.3 IMS Device Implementation

Device side development can be at a low level or a high level. Low-level development requires a more complex code development. Higher level interfaces such as Java JSR 281 (JCP, 2009b) will expose service standardized capabilities such as group list management, push-to-talk over cellular (PoC) and messaging. SIP traffic will be compressed and encrypted between the device and the network. Mobile device OS platforms supporting IMS include Mobile Linux, Symbian, Windows Mobile.

3.3   IMS Services – Possibilities

The technical advantages offered by the IMS architecture has been discussed in various literatures published so far. Chapter 2 presents the core network capabilities of the IMS, while Section 3.1 presents the service capabilities offered by SIP and IMS, such as service control and service provisioning features. However, the true success of IMS depends on the translation of these technical and service capabilities offered by IMS into innovative services with enhanced end-user experience. In other words, IMS should not only radically improve the end-user service experience, but also enable a wide variety of new services that end-users will be willing to use and to pay for. The positive signs of an IMS services take-off are dependent on the following:

- Business system readiness – IMS vendors release IMS-ready products and equipments.
- Technology maturity – mature standards enable mature technology.
- Interworking between carriers – cooperation between various standard bodies such as 3GPP, 3GPP2, TISPAN, and PacketCable facilitates the interworking.
- End-user acceptance – last but not the least is end-user acceptance, which relies on new services as well as enhanced service experience.

The mobile industry is in a transition phase from traditional voice and short message service-centric business to a variety of new and exciting multimedia services and applications. Telephony and messaging is about to be complemented by the next generation of person-to-person applications, shared applications such as two-way radio sessions (push-to-talk), sharing a view, sharing files, shared whiteboards and multiplayer game experiences, etc. It also brings the ability to combine existing services in exciting ways, e.g. when adding a multiplayer game during a push-to-talk session and adding location capability to push-to-talk sessions and so on. The IP multimedia subsystem (IMS) will also enable new services between mobile and fixed devices. The potential of IMS services is enormous. Its inherent capabilities for distributed integration of services, reusability of services, global service transparency and reach guarantee the quick development of ubiquitous, personalized, multimedia communications and entertainment services.

Taxonomy of IMS Service

Several classifications of IMS services are possible. For instance, IMS services can be categorized based on the utility they provide. They can provide a richer user experience by seamlessly integrating various media components such as messaging, video, streaming media and others. Examples of these services include push-to-talk over cellular (PoC) and video share. These services are designed to enrich the person-to-person communication experience.

Some services provide access to content and entertainment. These are classified as person-to-machine services. In this case, IMS becomes a secure vehicle for distributing premium content. Examples of these types of service include mobile music, streaming content and news, and others. Other sets of services include machine-to-machine or collaborative services developed to address context-sensitive and/or community-based applications such as enterprise services.

Beyond all the above ways of classifying the IMS services, a more generic way of categorizing services follows.

3.3.1 Blended Service

An alternate way of categorizing IMS services would be based on the service building blocks that are used to build the IMS service. The IMS operators and service providers have certain service-building functionalities that are at their disposal to churn out new and innovative services. Blended services are built combining these service building blocks, which offer enhanced service building capabilities. They are designed to be deployed once and be reused across multiple applications. The blended services can further be divided into three types based on the services that are blended.

Services that use the inherent service capabilities provided by the IMS architecture as a service building block are termed the IMS-based blended service. The service capabilities of IMS referred here are the session control, generic messaging framework, the event subscription framework, etc. An example IMS-based service would be voice call continuity (VCC), which makes use of the session control capabilities, event subscription services, push-to-talk over cellular (PoC), which makes use of the inherent service features offered by IMS, and the XDM enabler, which makes use of the user-specific service-related data that resides in the IMS core network.

The second type of blended service is those services that require additional functionality other than the core IMS functionalities such as an application server and can be termed the service enabler-based blended service. Examples of such a service are Presence, Instant Messaging, Location, Gaming, etc.

The third type of blended services are services other than a typical telecom service that blends or integrates with the IMS to offer innovative end-user experience, such as IPTV and IMS-based web services. Key to IMS and SIP architecture is the ability to integrate with other non-IMS or non-SIP applications. After all, SIP is modelled on the HTTP and Internet paradigm and therefore IMS/SIP applications can integrate with other non SIP-based services, such as RTSP-based media streaming, SOAP/XML web-based services, etc. Blending of IMS and the web service is dealt in detail in Chapter 5 and hence is out of the scope in this section.

Non-IMS-based blended services are composite services created by combining simpler, more basic components from various non-IMS services. For example, a blended service providing TV viewers with on-screen notification of incoming telephone calls and the opportunity to handle those calls through remote control commands can be created by combining existing telephone call components with TV viewing components. Since component services interact with one another as part of blended services, service blending is a powerful way to create many interesting and distinctive new services. In particular, service blending is a much more powerful means of service creation than service bundling, which is simply the association of one service with another. The potential of a blended service is demonstrated with the IMS-IPTV blended service as an example.

The service enablers discussed in this subsection all have a common quality. They are reusable components within the core service network and can be leveraged by other applications for the purpose of obtaining specific information or invoking some well-defined function. Because they are reusable, they need to be implemented only once and thereafter are available to any service authorized to invoke them. Reusable components have the advantage of yielding consistent functionality irrespective of which entity invokes them, assuming that entity is authorized to use the enabler. The enablers then act as stable building blocks for more complex, higher-order services that may incorporate the basic functionality of one or more enablers. The availability of service enablers lowers the cost of implementing the more complex end user services by avoiding the need to recreate a given functionality. As a result, the cost of implementing the end user service is reduced in terms of both capital and operational expense.

3.3.1.1 IMS-Based Blended Service

Voice Call Continuity   (3GPP, 2007n)

Voice call continuity is a home IMS application that provides capabilities to transfer voice calls between the CS domain and the IMS. VCC provides functions for voice call originations, voice call terminations and domain transfers between the CS domain and the IMS and vice versa. Figure 3.16 shows the VCC architecture defined in 3GPP TS 23.206 (2007n).

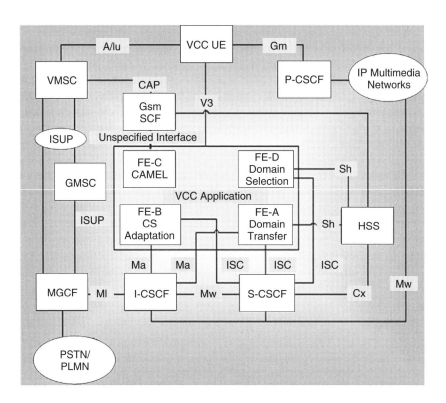

Figure 3.16 Voice call continuity architecture.

In the context of rapid developments in VoIP and the Internet, 3GPP defines the relevant VCC specifications based on IMS in its R6 and R7 to meet both carrier and subscriber requirements on service continuity in different access modes.

The VCC application typically comprises of a set of functions required for a VCC UE to establish voice calls and control the switching of the VCC UE's access leg between the CS domain and IMS while maintaining the active session. The domain transfer function (DTF) uses the ISC reference point towards the SCSCF for execution of the domain transfer functions. The domain selection function provides selection of the domain to be used for delivery of a VCC UE incoming call, as specified by operator policy or user preferences. However, operator policy shall have precedence to user preferences. The CS adaptation function (CSAF) acts as the VCC UE proxy into IMS for CS originated calls and CS legs established for domain transfer to CS. The CAMEL service function operates either independently or in collaboration with the CSAF. The VCC UE is a VCC capable user equipment with an active VCC subscription. It supports voice over both IMS and CS domains.

The key component of the VCC solution is a VCC application server in the IMS network. The VCC application server offers the control to the VCC service subscriber session and thus realizes a WiFi and GSM bidirectional handover. The gsmSCF (gsm service control function) can be performed by the SCP (service control point) of the existing network. In cooperation with the handover AS, it performs the anchor procedure in the IMS domain when the VCC subscriber handover from the GSM network to the WiFi network thereby ensures that the VMSC (visited mobile switching centre) routes the call to the appropriate HOAS. Figure 3.17 illustrates that the handover function in the service layer is controlled and implemented by the handover AS (VCC application server).

HOAS offers the following functionality:

- The HOAS performs a similar function to B2BUA. It is the first point to be triggered when a VCC subscriber makes a call, and the last to be triggered when a

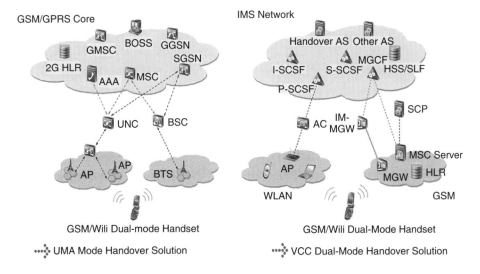

Figure 3.17 Voice call continuity architecture.

VCC subscriber receives a call, thus ensuring that the handover latency is effectively reduced.

• HOAS also solves the CS/IMS domain selection problem and is responsible for controlling the origin of a new session to replace the one to be handed off.
• It provides interface interconnection with other NEs (network elements) and, by interacting with subscriber information via the iFC from the S-CSCF, facilitates redirection when the UE (user equipment) goes unexpectedly offline from the WiFi network.

For a detailed explanation of the Interfaces and the precise message flow, refer to the 3GPP TS 23.206 (2007n) Specification for more details.

Currently, there is also an alternative solution for the dual-mode handover, the UMA (unlicensed mobile access) (3GPP, 2007a). Subscribers receive different experiences through the two modes. The UMA solution realizes RAN and the WiFi RAN handover in GSM access networks, and expands the CS domain access network. This gives subscribers the same experiences as delivered by CS domain services. Although the UMA solution is simple and easy to launch, it does not represent a long-term solution due to its inherent QoS, capacity and non-SIP disadvantages. The VCC solution accomplishes CS-IMS domain handover. As an application of IMS domain, VCC subscribers are able to enjoy the same experiences as other IMS services. IMS architecture is also access independent; thus the IMS-bound VCC solution is capable of realizing seamless voice and multimedia session handover for various terminals such as mobile phones, telephone sets and PCs. It is also applicable to a range of networks, such as CDMA, GSM, UMTS, PSTN, Fixed BB, WiFi and WiMAX, and can be implemented by fixed, mobile and integrated carriers. Therefore, it is the best choice to implement the dual-mode handover service.

Push-to-Talk over Cellular (PoC)   (Garcia-Martin, 2006)

The push-to-talk service over a cellular public network emerged in the mid-1990s first in the USA. Since then effort has been going on among different cellular network operators and vendors around the world to standardize this service. Ericsson, Motorola, Nokia and Siemens submitted the technical standard specifications for PoC to the Open Mobile Alliance (OMA) standards body in August 2003. Two specifications, PoC Release 1.0 and PoC Release 2.0, have been completed. Both were passed as input to the OMA for standardization. The specification is based on the Third Generation Partnership Project's (3GPP) IMS architecture. The OMA specified PoC service is based on the voice over Internet protocol (VoIP) and relies on the IP multimedia subsystem (IMS) as a service enabler.

The OMA PoC specifications outline the following features or connection capabilities of PoC:

• One-2-one communication. This is the basic capability for setting up a voice call between two users. The called subscriber can receive the call automatically or manually.
• One-2-many communication. This feature enables a subscriber to establish a voice call towards a group of people. One of the group members is the speaker and others are listeners. A participant gets the right to speak through the floor control mechanism. There are three modes of group communication supported by the mentioned version of PoC specifications:

1. Ad hoc. Ad hoc group communication is formed when a subscriber selects more than one other subscriber and invites them to join in the talk session. The difference between ad hoc group talk and prearranged group talk is that for the latter case a unique identifier has been previously allocated for the instant group.

2. Prearranged. Participants in the session are defined beforehand. The server invites (INVITE) the rest of the participants when a certain member of the existing group invites others in the group session. A prearranged PoC group has a permanent URI by which all the group members can be contacted. An example is sip:swimming_team@serviceprovider.com.

3. Chat. In a chat group talk, users join and leave an existing chat group as time goes by. The chat group session is established as soon as the first subscriber joins in.

Functional Benefits of the PoC Service

The service presents a superb opportunity for mobile operators to boost their existing average revenue per user (ARPU), as well as win new customers, by providing a popular two-way radio service over attractive and advanced cellular phones. Here are the benefits:

- Differentiates with cost-effective voice service capable of generating incremental revenue because it is not a substitute for existing cellular services.
- Anchors multimedia applications, providing operators with a future-proof platform for always-on voice communication services.
- Based on IP technology, the PoC solution efficiently uses network resources, reserving them only for the brief duration of voice transmission rather than for an entire call session.
- Allows service providers to differentiate the service to correspond to the most diverse requirements of the various user segments and to implement it across access technologies.
- Users are always connected to their user community and can hear whenever there is any talk in any of their communities. For example, a user may be connected to several groups such as 'My Family', 'My Friends', 'My Gaming Team', etc. The user will hear when there is any talk burst in any of those groups and may reply instantly.
- The direct connection promotes efficient, spontaneous and occasional communication, especially when there is a need to be in contact with certain groups of people frequently over a longer period, such as a working day.
- PoC is also a key service enriching IP multimedia communication sessions, such as interactive gaming.
- Subscribers can make PoC calls over nationwide networks and across regional borders.
- PoC-based applications can be personalized based on customer communication needs and behaviour (e.g. for small businesses and corporate users, families, teenagers and leisure groups).

XML Document Manager Enabler (XDM)

The XML document manager server (XDMS) is an enabler that makes user-specific, service-related data available to other services and service enablers within the network. The type of data that is most often associated with the XDM is the 'list' Typically composed of universal resource identifiers (URIs), these lists represent friends, family members, co-workers or other groups with whom an individual may want to communicate.

The XML document management server is a service-independent building block that can be used and reused across any number of innovative person-to-person or group communication applications. The expense of implementing and maintaining new services that make use of lists of friends or co-workers, e.g. video conferencing, is reduced because the list management functionality need not be designed into the conferencing application. It can be accessed through a well-defined and open interface to the common network enabler.

The ability to create and manage lists of friends, family, co-workers or any group of people is well understood today as a result of the readily available IM or email applications on the market. The XDMS extends this useful capability well beyond the bounds of a single application in the network and can make the lists available to other service applications. A user could create a list of friends through an IM application on a mobile handset and access the same list through an audio or video conferencing application.

3.3.1.2 Service Enabler-Based Blended Service

IMS offers a framework for IP communication applications, across multiple networks and devices, with a well-defined standard control layer that ensures interoperability and roaming, end-to-end session management and a coherent framework for service charging. IMS also promises a clean separation of the application layer from the core network. The 3GPP IMS standards, as per their current definition, propose a detailed architecture for the transport and control layers of the next-generation telecommunication networks, but do not specify how the application layer should be structured. In order to deliver multimedia applications in an efficient and a cost-effective manner it is essential to have a horizontally layered service enabler plane that makes use of the session control, management and charging functionalities offered by the IMS in the control plane. From the illustration in Figure 3.18, an ideal service enabler plane should host a set of application-independent building blocks that offer generic functionality to support a diverse range of multimedia applications. It should act as a point of convergence and consolidation in a heterogeneous environment, while offering network-agnostic and device-agnostic APIs to application developers. These service enablers should be designed to be deployed once and be reused across multiple applications (O'Connell, 2007).

This subsection discuses a few such service enablers.

1. The *presence enabler* (3GPP, 2007m) is one of the most important service enablers. It provides a focal point within the network for the receipt of user-related presence updates from various elements throughout the network. It subsequently brokers that information to other entities (e.g. presence-based applications) who register to receive notification of changes to a user's presence and availability state. The various states would include conditions such as 'Present on the Network', 'Not Present on the Network' or 'Present But Do Not Disturb', among others.

   The Presence service is a network service that accepts stores and distributes presence information. The presence service may be implemented as a single server or have an internal structure involving multiple servers and proxies. There may be complex patterns of redirection and proxying while retaining logical connectivity to a

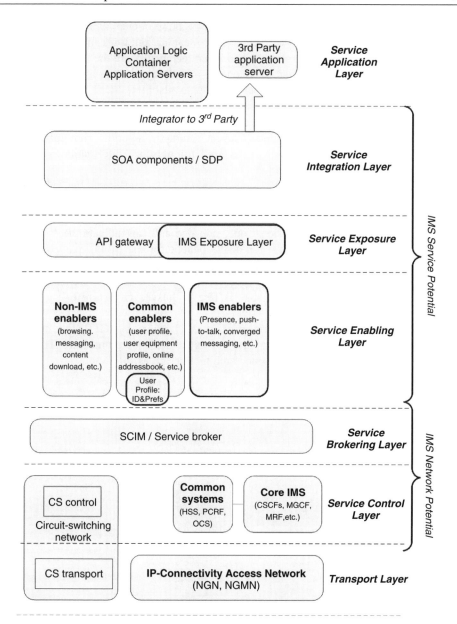

Figure 3.18 Service-centric horizontally layered representation of IMS.

single presence service. Also the presence service may be implemented as a direct communication among presentity and watchers, i.e. the server is not required.

A user client may publish a presence state to indicate its current communication status. This published state informs others who wish to contact the user of his availability and willingness to communicate. The most common use of presence today is to display an indicator icon on instant messaging clients, typically from a choice of graphic symbols with an easy-to-convey meaning and a list of corresponding text descriptions of each of the states. Even when technically not the same, the 'on-hook' or 'off-hook' state of a called telephone is an analogy, as long as the caller receives a distinctive tone indicating unavailability or availability.

Common states on the user's availability are 'free for chat', 'busy', 'away', 'do not disturb' and 'out to lunch'. Such states exist in many variations across different modern instant messaging clients. Current standards support a rich choice of additional presence attributes that can be used for presence information, such as user mood, location or free text status.

The analogy with a free/busy tone on the PSTN is inexact, as the 'on-hook' telephone status really shows the availability of network connectivity for reaching the called number, not really with the availability of the device or its user. However, when we compare the scenario of only one line and can indicate the availability or unavailability status (free/busy tone), the benefit for the calling party is very similar. They can decide if they want to start communication in a different way (try using another communication method or postpone the call). Presence goes further as it allows us to know if the state is free or busy even before trying to begin a conversation.

The architecture of the IMS presence service is defined by 3GPP in the specification 3GPP TS 23.141 (2007m). Figure 3.19 shows the presence architecture. The functions of network elements in the presence architecture are as follows:

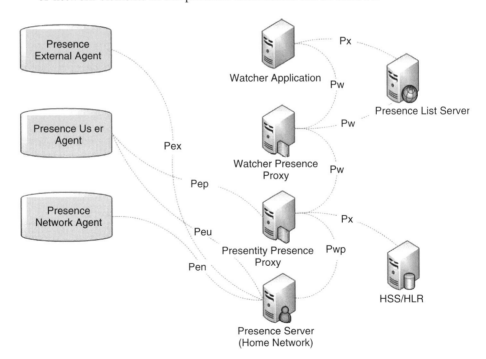

Figure 3.19 Presence architecture.

Presence User Agent provides the following functionality:

- Collects presence information associated with a presentity representing a principal.
- Assembles the presence information in the format defined for the Peu and Pep reference points.
- Sends the presence information to the presence server element either via the presentity presence proxy over the Pep reference point or over the Peu reference point.
- Manages the subscription authorization policies.
- Handles any necessary interworking required to support terminals that do not support the Peu and Pep reference points.
- indentifies itself uniquely (among the presence user agents of the presentity) when publishing presence information.

Presence network agent provides the following functionality:

- Receives presence information from network elements.
- Is able to send requests to the HSS/HLR to cause other network elements to send (or stop sending) presence information to the presence network agent. Note that this only applies where the other network element has presence information subscriptions managed via the HSS/HLR.
- Associates presence information with the appropriate subscriber/presentity combination.
- Converts the presence information into the format standardized for the Pen interface.
- Publishes the presence information to the presence server across the Pen reference point.

Presence Externel Agent provides the following functionality:

- Supplies presence information from external networks.
- Sends the presence information across the Pex reference point according to the format standardized for the Pex reference point.
- Handles the interworking and security issues involved in interfacing to external networks.
- Provides functionality to resolve the location of the presence server associated with the presentity.

Watcher Presence Proxy provides the following functionality:

- Address resolution and identification of target networks associated with a presentity.
- Authentication of watchers.
- Interworking between presence protocols for watcher requests.
- Generation of accounting information for watcher requests.

Presentity Presence Proxy provides the following functionality:

- Determination of the identity of the presence server associated with a particular presentity.
- Authentication of the watcher presence proxy.

- Authentication of the presence user agent.
- Generation of accounting information for updates to the presence information.

Service and functional benefits of the presence enabler

The presence enabler gives the service provider an entity that can 'watch' and maintain comprehensive state information about individual users. The presence or absence of a user, based on their registration state with the network, can be augmented with other information such as 'Do Not Disturb', some of which may be user configurable or obtained from multiple network sources. This rich, globally coherent user information can be provided to network components, including presence-enabled clients on the user device, either on demand or via subscription.

The availability of a common presence enabler in the network allows it to project a coherent image of an end-user's presence state, irrespective of the service application or client requesting the information. In the absence of a commonly specified presence facility, earlier network architecture with self-contained or 'siloed' services could easily render different views of a user's presence, depending on where a request originated. Coherent presence information enhances user satisfaction because the network is able to render more consistent and accurate information.

2. *Messaging*. The phenomenal growth in SMS text messaging illustrates the popularity and revenue potential of mobile text communication. Instant messaging and group chat enablement enhances text messaging by allowing two or more users to engage in a real-time text messaging 'session', and accommodating differences in the devices or clients they may use. The ability to have mobile (and fixed) group chat sessions in real-time integrated with other IMS-enabling services such as presence and location services will provide further growth for network operators. IMS additionally enhances the service by enabling easy integration with other services, such as gaming, to further increase the service's attractiveness.

There are two modes of instant messaging: pager mode instant messaging and session-based instant messaging (Figure 3.20). Pager mode instant messaging uses the SIP MESSAGE method (Aboba *et al.*, 2005). In the session-based instant messaging mode the SIP INVITE method is being used. The media is sent via RTP (real-time transport protocol). The actual message from the user will be sent over the message session relay protocol (MSRP) once a session is established.

Service and Functional Benefits of a Messaging Enabler

The use of messaging enablers in the network allows an operator to utilize a common infrastructure to support a variety of messaging technologies, including IM, SMS, MMS and E-mail. This results in a simpler service environment that is easier to build and less expensive to maintain and enhance. Along with offering them to end-users as freestanding applications, the IM and group chat services can be integrated into other mass-market applications such as mobile games or other entertainment applications. Because the common infrastructure facilitates greater interoperability among disparate messaging systems, there is an increased opportunity to generate additional revenue. Messaging services are presently able to command fairly high service fees relative to the bandwidth they consume when compared to bandwidth-intensive services like

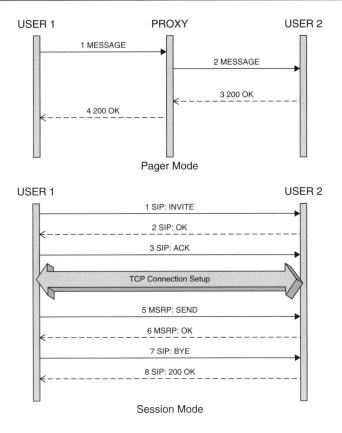

Figure 3.20 Pager mode and session mode instant messaging.

video streaming or video sharing. It follows that an enabler that simplifies the service topology of the network so that it is reusable across a variety of messaging services normalizes message handling across network boundaries and enriches the messaging experience of the end user will have a positive impact on an operator's business. Finally, this enabler will allow the operator to drive and brand their own IM-based community services.

Instant messaging is a concept that is commonly available and well understood by users in either the desktop or mobile environment. Making the various forms of instant messaging (SMS, two-way paging, IM, E-email, content sharing) interoperable with one another while hiding the complexity from the user will have a positive impact on customer satisfaction.

3. A *location enabler* makes information about a network user's physical location available to requesting entities according to security and privacy rules enforced by the network. The information can be made available on request or it can be captured at specified intervals. The enabler gives an application the ability to obtain and use location information without the need to interact with the complexities of the low-level elements in the network that provide the data (e.g. positioning servers).

An application can request alert notifications from the location enabler when a particular distance has been travelled or whenever a geographical boundary has been crossed by the user. The enabler will automatically perform periodic 'checks' of the entity's location. When the specified conditions are met, the requested application is provided along with the location information of the user. The applications are not required to conduct the polling of the positioning servers and then interpret the results.

Functional and Service Benefits of a Location Service Enabler

A Location enabler hides the complexities associated with obtaining the geographic coordinates (x, y) of a user. While the positioning function itself can be viewed as a discrete element considered for use in other top-line applications, the function of an enabler is something more intelligent and contributory to overall end-to-end call processing. Providing a reusable resource for location information also helps to ensure that authorization and a privacy rule are consistently enforced and simplifies the provisioning of these rules. The preservation of privacy and safety are particularly important when making location information available about an individual using the network. For example, alert conditions can be set within the enabler, as discussed above. The requesting application will only be triggered when those conditions are met, reducing message traffic within the network. Importantly, it is in the ability of an enabler to track and notify changes in location that highlight its value.

Users benefit in many ways from the availability of location-enabled services. The possibility of locating places of interest, friends or family (when authorized), or determining their own location on demand has obvious benefits for users and can be applied in any number of applications.

3.3.1.3 Non-IMS-Based Blended Service – IPTV   (Beck and Ensor, 2007)

Figure 3.21 illustrates a typical IPTV use case. Alice and Bob are connected to an IMS through telephone access networks. When a call/session control request is made, signals and associated media move from the access network to an IMS through signalling and media gateways. Within an IMS, these signals are transmitted as SIP messages. Call/session control signals go from signalling gateways to call/session control elements. These control elements examine the call request and determine whether any applications should be executed in association with the call processing. If so, the session control element invokes application servers to execute the appropriate applications. Once these applications have been executed, the call request may be passed through gateways and then through an access network to the called party's device or to call/session control elements associated with the called party (in the same or in another IMS), which can also invoke application servers. Multiple application servers can be associated with a single session. Call/session controllers can invoke these servers directly, but have only limited means of controlling interactions among the invoked applications. The IMS specifications include an SCIM (service capability interaction manager) function, which can support more robust control of application interactions. An IMS SCIM behaves as a specialized application server. It uses SIP to communicate with an IMS call/session controller and SIP-based application servers to coordinate applications within

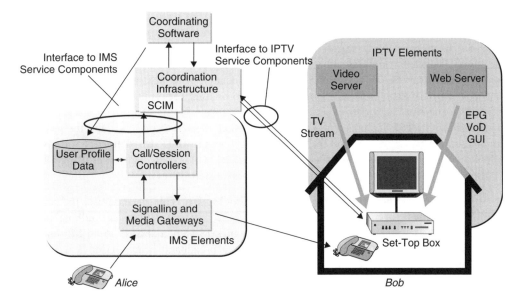

Figure 3.21 IPTV architecture (from Beck and Ensor, 2007).

the context of IMS sessions. In Figure 3.21, the SCIM is shown as part of a coordination infrastructure.

Typically, an IPTV service is based on two fundamental parts content delivery and service management. Content delivery involves moving TV programming material from producers to consumers. These data transmissions generally involve multiple media servers (which are connected through one or more networks) and consumer STBs. Live content is usually streamed to consumer STBs for immediate display on the consumer TVs. Video-on-demand content also may be streamed, but it is more often transferred asynchronously as files, which are stored on consumer STBs for later display. Service management has two aspects: consumer control of content delivery and management of service subscription records. Consumer controls include displays of available content, e.g. electronic program guides (EPGs), program (channel) selection, etc. Management of subscription records includes billing and payment, feature subscription updates, etc. Figure 3.21 presents a simplified view of IPTV service elements, representing each part as a single server connected to the consumer STB. Various parts of an IPTV service may be used to create a blended service. For example, an IMS-IPTV blended service might use IPTV content delivery functions, an EPG and IPTV billing and payment functions, but not use IPTV feature update functions.

Use of RTSP to Deliver an IPTV Blended Service

Traditionally RTSP (Schulzrinne, Rao and Hanphier, 1998) has been the protocol of choice for streaming media applications and the current VoD over IP. An obvious approach then is to extend RTSP as needed by the new blended services. However, RTSP does not address the new requirements of converged services where multiple heterogenenous streams merge.

An alternative approach is to consider the possibility of using SIP features. While currently the SIP is used for voice communication services, the protocol was historically designed to signal a wide variety of media services. Driven by the requirements associated with IP-based communication services, the SIP has been extended over the years to address many similar requirements currently associated with next-generation multimedia services. The SIP enables the IMS infrastructure to be used for session setup and streaming control to enhance the video experience across platforms. Some of the key advantages of using SIP with RTSP includes access to service enablers such as presence and the possibility for session pause, restart and transfer. Recent works in IETF has defined solutions that allow interoperability between SIP and RTSP (Arkko *et al*., 2006; Marjou *et al*., 2008).

Some of the other auxillary blended services are listed below:

1. *Video sharing between a mobile phone and set-top box.* This use case uses the concept called *video forking*, illustrated in Figure 3.22. Two subscribers are talking on the phone. User A is also watching IPTV via an IMS signed session to an SIP-enabled set-top box; the TV content is provided by a broadband bearer. User B is using a mobile phone but is interested in the video that is being described by User A and would like to watch the video on the mobile device via a broadband interface such as WiMax or Wi-Fi, available on the multimode mobile device.

    (a) User A's set-top box sends an invite to User B's SIP-enabled mobile phone. This invite advertises the URL of the video that User A is watching.
    (b) User B's mobile was previously registered with the network and was found to be authorized for IP video and registered with the IP video application server. User B's video profile is in the HSS.
    (c) User B's mobile device is able to send a request for the video to the IPTV AS, which obtains necessary information from the HSS. The video is delivered to User B's mobile device.

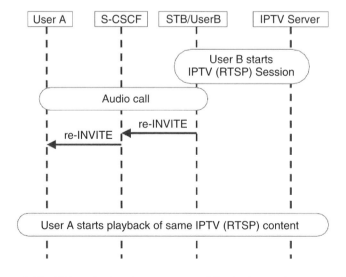

Figure 3.22 Video sharing between a mobile device and a set-top box.

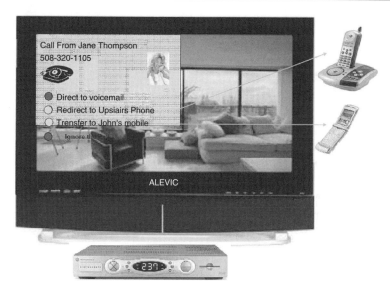

Figure 3.23 Caller ID on IPTV (from Montpetit and Ganesan, 2007).

2. *Caller ID on IPTV* (Montpetit and Ganesan, 2007). In this use case (see Figure 3.23), the application server enables an SIP-based STB to receive call information such as calling number, calling name, etc., from the incomming call. In addition, the STB is also capable of call control such as redirection of the call to a voicemail server or another phone.

(a) The user's STB was previously registered with the network and was found to be authorized for blended services; it is registered with the blended service AS. The user's profile is obtained from HSS.

(b) A call is sent to the user's home phone via IMS; because of the IMS common identity the AS is aware of the video session on the user's TV.

(c) The AS sends a reinvite to the user's SIP-enabled STB. This re-invite advertises the information about the incoming call, such as actual information or link to the information.

(d) The client application in the STB displays the information of the call and, based on the action, the respose to the AS chooses to end the call or route it to the phone or send it to another location, etc.

(e) In addition the user preferences could use preconditions such as 'Do Not Disturb' while the VoD is on or send a call from a family member in priority, etc.

3.3.2 End-User Services

These are the services that are directly consumed by the end users. They could include the base telecom services that are enhanced and may become part of a more complex service, such as multimedia telephony, conferencing, video sharing, etc. They also consist of social networking applications that offer a common instant messaging (IM) platform

across PC, wireless devices and TV that interface with the IM features of leading social
networking sites.

3.3.2.1 Video Conferencing

The IP multimedia subsystem (IMS) video-conferencing service extends the point-to-point
video call to a multipoint service. Video-conferencing requires an IMS conference bridge
service, which links the multiple point-to-point video calls together and implements the
associated service logic. The video telephony connections are made point-to-point from
the terminals to the conference bridge, which takes care of joining the point-to-point connections into a conference. The conference bridge is not concerned about the underlying
infrastructure and client devices and assumes that audio and video connections are provided
by the appropriate standard and that these connections are delivered over the IP network.

3.3.2.2 Video Share

A real-time video sharing service is a peer-to-peer, multimedia streaming service that can
be offered entirely as a packet-switched service or as a 'combinational' service, combining
the capabilities of the circuit-switched and IMS packet-switched domains. In a combinational scenario, the service enriches the user experience during a circuit-switched telephony
call by, for example, exchanging pictures, video clips or live video over a simultaneous
IMS packet-switched connection. This enables operators to leverage upon circuit-switched
infrastructure, telephony performance, user behaviour and experience already in place. Even
following the evolution to all-IP networks, CS telephony enrichment will be maintained for
the gradually diminishing CS domain. In both combinational and full packet-switched scenarios, the media is delivered and consumed almost in real-time. There is only a marginal
delay, thus providing the experience of being there and 'sharing the moment'. The spirit is
always live, even when sharing a stored video clip, since there is the possibility that users
can have an ongoing voice conversation at the same time.

The introduction of person-to-person video share during a voice call provides a key
differentiating service. Video share takes place in real-time during a normal voice call. One
or another of the users sends a live camera view or video clip from the device. Both users
see the same video and can discuss it while they watch, and then remove the unidirectional
video element spontaneously as needed.

Functional Benefits of the Video Share Service

Video share is a new compelling service that provides new business opportunities for service
providers. For example, video share:

- Produces a new stream of user-generated data traffic, which will increase the packet
  data revenues generated during voice calls.
- Extends voice calls, by providing users with additional topics to discuss.
- Is more cost-effective via a packet-switched network than via a circuit-switched
  network.
- Allows convergence between the mobile and PC domains, greatly increasing the potential user base.

- IP-based video services have a clear evolution path for improved video quality and efficiency.
- The ability to add and remove video share during the call is a fundamental characteristic enabled by the IMS architecture. As illustrated in the use case above, video sharing can be a natural expansion of voice calls where important or interesting events can be communicated visually and while a voice call continues. Video sharing has an element of spontaneity that can increase as the video capture devices continue to improve in quality. The availability of a UMTS or HSPA network and IMS service ecosystem that can make the experience increasingly gratifying are likely to increase subscriptions to the service, as well as frequency of use.
- An easy-to-use video-sharing application can have utility for both consumers and businesses that want or need to share a combination of video and audio information. Instances where such capability would be an aid to law enforcement, journalists or health care professionals would be as numerous as personal usage, such as a teen calling a friend or family member to share a video and get an opinion of a purchase about to be made.

3.3.2.3 Multimedia Telephony   (3GPP, 2007c)

IMS-enabled voice and video calls are carried over a packet core network (VoIP). Video telephony is seen as a critical end-user service in mobile networks. The session initiation protocol (SIP) enables voice and video telephony person-to-person and multiparty sessions over an IP network. The issues around VoIP and video telephony calls are around quality of service (QoS) in packet core networks, as well as on interoperability with the PSTN and legacy phones, and interworking with existing domains for video telephony such as H.323 (Levin, 2003) and H.324M (ITU-T, 2007). The bandwidth requirements are driven by the coding schemes in the terminals and the video quality requirements of the users. Thus, with similar coding schemes, packet-switched video telephony has similar bandwidth requirements as circuit-switched video telephony. However, packet switching gives more freedom in balancing the bandwidth and video quality requirements. The bandwidth and quality of service requirements for the connection are requested from the network by the terminal at the connection setup phase (PDP context activation).

In accordance with the service definition and requirements in 3GPP TS 22.173 (2007c), the IMS multimedia telephony communication service specified herein allows multimedia conversational communication between two or more endpoints. An endpoint is typically located in a UE, but can also be located in a network entity. As for traditional circuit-switched telephony, the protocols for the IMS multimedia telephony communication service allow a user to connect to any other user, regardless of operator and access technology. The IMS multimedia telephony communication service consists of two principal parts: a basic communication part and an optional supplementary services part.

4

IMS Deployment

This chapter is dedicated to perform a reality check on the actual deployments of NGN and IMS technologies. A discussion on some of the crucial deployment concerns are presented and the most important hurdle is identified as 'interoperability'. The efforts and results of the interoperability trials conducted by the MSF, OMA, GSMA and the SIP forum organizations are presented. The chapter further discusses the actual deployment approach adopted by the fixed network operators, the cable and the mobile operators. Building a real-life IMS test network and the role of test networks in the deployment of new technologies is investigated. Open-source initiatives that make a valuable contribution to building test networks and test deployments of the technology are discussed. Various open-source activities as well as the tools and resources available within the open-source community are highlighted in this chapter.

4.1 Deployment Concerns

Despite the vast technological and business advantages IMS promises to offer, it is yet to witness large-scale commercial deployment. IMS is a disruptive technology which aims to change the layout of the existing communications world from being vertically integrated to horizontally integrated. It is a highly complex technology that is continually evolving and hence it is natural to face many hurdles on its road to deployment.

Complex and Immature Technology

One of the main deployment concerns from the network operators and service provider's perspective is the complexity of IMS technology and the complexity of the standardization. There are a lot of standard bodies involved in the standardization of IMS and associated technologies such as FMC and NGN. Most standards are still not mature and are constantly evolving. There currently exists a large array of specifications from various standard bodies such as the 3GPP, TISPAN, PacketCable, MSF, OMA, GSMA, etc. Each of these standards

IMS: A Development and Deployment Perspective   Khalid Al-Begain, Chitra Balakrishna, Luis Angel Galindo and David Moro
© 2009 John Wiley & Sons, Ltd

are in different stages of development, giving rise to uncertainity among the IMS players. There are certain gaps in the standards as the initial phases of the standards focused on architectural aspects and only recently has the focus shifted to real-world deployment issues, such as QoS, security, reliability, charging, interoperability and other management functions.

Apart from the standardization chaos, it is a challenging task for the operators and the service providers to understand the IMS technology, which is complex in itself. Other issues that need to be addressed for a successful commercial deployment are reliability and reachability as IMS-based service uptime is critical to network operator revenue. Security, billing and roaming (visited networks) are certain key areas that still need to be addressed. This is naturally making many carriers and other IMS players in the industry cautious about rolling-out IMS technology.

Business Case

Recent years have witnessed advancement in radio access technologies, such as 3G, HSDPA and LTE, capable of a higher data rate, low latency and highly suited for multimedia service delivery. The all-IPfication of telecom core networks using the pre-IMS softswitches and phasing out of IMS in a piecemeal fashion offer various alternative options to the operators for multimedia service delivery. Hence, big investments such as IMS require a stronger business case. Given that the future of multimedia services is massively influenced by regulation, competition and global economics, it is not surprising that IMS business cases are not always easy to make. Furthermore, it becomes critical to choose the right strategy for doing business in a multicarrier–vendor environment.

Service Development

Currently there is a lack of available software tools to allow third party application developers readily to design, develop and deploy new IMS applications. One way of filling this gap is by enabling greater cooperation among operators and developers to strengthen the service development environment. As IMS technology is deployed, a lot of network capabilities would be exposed to the third parties. There are software extensions such as the Java extensions or APIs to make the developers' job easy. However, there is lack of an open development environment that bundles these extensions and offers a service creation enviroment to third party developers. As depicted in Figure 1.11 of the Section 1.5.1, the native telecom environment offers protocol-based development. Hence the complexity of development is high and the number of skilled developers is low. This greatly limits the innovation or rather the number of services developed and increases the cost of service creation. However, in the Internet, the scenario is just the opposite, where the development environment is web-based using higly abstracted languages and scripts. Hence, the number of developers is very high as the complexity of development is low and there are higher chances of service innovation.

Another concern of the industry is the service delivery platforms (SDPs). IMS can be viewed as one of the components that allows network capabilities to be exposed to the SDP, and the SDP in turn can expose the capabilities to the actual application developers (refer to Sections 1.2.5 and 3.2 for a detailed description of the IMS and service delivery frameworks). However, carriers clearly need control over the release of IMS capabilities to third parties. Therefore there may have to be controlled release of IMS capabilities, either via the SIP, or via web services or via Java.

Other Challenges

- Handset compatibility with IMS services and the lack of high-quality dual-mode capable handsets that could be used in multiple networks is a great deterrent in IMS deployment. The revenue management, billing and rating of calls in a dual-mode environment is a 'new frontier'.
- Battery life of the handsets is a concern given that multimedia streaming applications could drastically drain the battery resource.
- With differences due to operator implementation of early IMS and IMS being delivered in releases, the chances are that different vendors may conform to different releases. This may result in incompatibility and conformance issues.
- Authentication mechanisms and subscriber-centric policy management issues need to be addressed. Offering access-agnostic IP applications trigger the need for subscriber-centric or user-centric policies and approach.
- Roll-out of IMS may create additional traffic. Hence, there is need to manage and avoid severe congestion problems.
- Challenges still lie ahead for session and event-based charging.

4.1.1 IP Challenges

With IMS enabling the fixed mobile convergence gives rise to a common IP bearer network which will enable large-scale communication networks to be constructed. This implies that IMS architecture demands a larger address space, which cannot be provided by the IPv4 addressing scheme. Early deployments of IMS are being implemented on the currently existing IPv4 network, while the long-term deployments of IMS require a large address space provided by the IPv6 addressing scheme. IP-related challenges faced by the operators deploying IMS can be summarized as:

1. To support both IPv4 and IPv6 versions of addressing.
2. To plan and manage a migration path from IPv4 to IPv6 as well as interoperability issues arising out of the coexistence of the two networks.
3. IPv4 networks typically consist of two IP domains: a globally routeable public domain and a locally routeable private domain. Issues introduced by network address translation (NAT) place an additional burden on the interaction between the access networks and IMS core network.

Adoption of IPv6 is viewed as the key to deploying next-generation fixed and mobile services. IPv6 is the next-generation Internet protocol that succeeds IPv4. The evolution from IPv4 to IPv6 is primarily intended to support the peer-to-peer model of Internet connectivity and network efficiency and to encourage innovation. IPv6 supports these by incorporating the following enhancements to IPv4:

- expanded IP addressing capabilities;
- header format simplification;

- stateless address auto-configuration;
- improved support for extension headers;
- flow labelling capability;
- any cast capability;
- inherent support for IPSec and Mobile IP (MIP) (Koodli, 2005);
- simplified header processing;
- less fragmentation of the address space, leading to smaller routeing tables;
- support for route optimization in Mobile IP;
- support for address auto-configuration;
- reduced dependency on translation devices;
- more sophisticated flow identification, potentially leading to improvements.

4.1.1.1   IPv6 Transition Issues

The worldwide transition from IPv4 to IPv6 has begun. It is likely that this transition will occur at different rates in different regions of the world. Some regions are already moving aggressively to support IPv6. The progress of transition is dependent upon the required urgency to expand the address space to support next-generation applications. In the interim, it is expected that there would be coexistence of IPv4 and IPv6 networks. Therefore, both the access networks and the IMS core network will be required to be dual stack, capable until the transition to IPv6 is complete. In general, the transition mechanisms can be grouped into following mechanisms:

- dual stack (enables IPv4 and IPv6 to coexist in the same devices/networks);
- tunnelling (includes configured and automatic tunnels, encapsulating IPv6 packets in IPv4 packets and vice versa);
- protocol translation (enables an IPv6-only device to communicate with an IPv4-only device).

Dual Stack   (3GAmericas, 2008)

This refers to an IP-capable device (e.g. a host computer) supporting simultaneously both IPv4 and IPv6 (Figure 4.1). This enables applications to communicate across either an IPv4

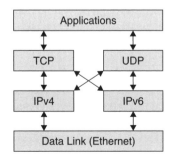

Figure 4.1 IPv4–IPv6 dual stack.

or IPv6 network. Typically, the preferred network is based on name lookup and application preference. Dual-stack routers support both IPv4 and IPv6 routeing protocols, and are able to forward both IPv4 and IPv6 packets. Many of the routers available today support this dual-stack capability. Dual-stack routers in conjunction with dual-stack application servers enable a gradual transition to IPv6, in which legacy IPv4 applications and devices can coexist with newly transitioned IPv6 applications on the same dual-stack network. Requiring all new IP-capable devices to support dual-stack routers is desirable in providing a flexible operational environment for transitioning to IPv6. Ideally, the IPv6 transition process would consist of two phases:

- Replace all IPv4-only devices with dual-stack devices.
- Once all devices support both IPv4 and IPv6, introduce IPv6-only devices.

However, it should be noted that dual-stack devices require IPv4 addresses and therefore do not mitigate the IPV4 address exhaustion problem. It is therefore likely that service providers will have to start deploying IPv6-only devices before all IPv4-only devices have been converted to dual-stack support. Therefore, besides dual stack, other transition mechanisms will be required.

Tunnelling   (3GAmericas, 2008)

This refers to a technique to encapsulate one version of IP in another so that the packets can be sent over across a network that does not support the encapsulated IP version. This technique allows two IPv6 islands to communicate across an IPv4 network, or vice versa. Tunnelling enables different parts of the network to transition to IPv6 at different times. There are two main categories of tunnelling: configured and automatic tunnels. Configured tunnels refer to a manually configured tunnel within the endpoint routers at each end of the tunnel. The endpoints are statically configured; hence, any changes to network numbering would require modification of tunnel endpoints. In contrast, automatic tunnels refer to a device dynamically creating its own tunnel to dual-stacked routers for sending IP packets within IP. The IPv6 Tunnel Broker (Durand *et al.*, 2001), 6to4 (Carpenter and Moore, 2001), Teredo (tunnelling IPv6 over UDP through NATs) and ISATAP (intrasite automatic tunnel addressing Protocol) send IPv6 packets encapsulated in IPv4 and can be referenced as IPv6-over-IPv4 mechanisms while the dual-stack transition mechanism sends IPv4 packets within IPv6 and can be referenced as an IPv4-over-IPv6 mechanism. Security risks associated with automatic tunnelling should be considered, prior to introducing automatic tunnelling. For example, automatic tunnelling could enable user nodes to establish tunnels that bypass an enterprise's security mechanisms (e.g. firewalls, intrusion detection), reducing the depth of security as well as allowing unsolicited traffic. Even though automatic tunnelling provides an extremely flexible transition mechanism to move to IPv6, the risks associated with each transition mechanism should be analysed.

Translation   (Choi and Fischer, 2006)

Early IMS deployments were introduced using IPv4 and NAT traversal mechanisms. Implicit support for private addressing is required in IPv4 because of the limitation of IPv4 globally accessible address space. All packets with private addresses must traverse a NAT at the edge of the private IP domain in order to route in the public domain. The end user conveys signalling information to IMS servers in SIP and sets up a media path using the session description protocol (SDP). SIP/SDP are upper-layer protocols that contain IP-related fields. Because NAT functions at the network and transport layers translate only the IP and TCP or UDP headers, a solution is required to solve the problem of NAT traversal in SIP. The problem of NAT traversal in SIP has received significant investigation and a number of mechanisms have been presented in the IETF to resolve this problem (Rosenberg and Schulzrinne, 2003; Rosenberg *et al*., 2002). However, these mechanisms are not yet widely available. One approach, the implementation of symmetric response to record the IP port in combination with the received address in the SIP headers, assists in resolving IP-related values for an end user behind a NAT, but leaves unspecified the media path IP address and the ports in the SDP. The IMS core network does not modify the contents of the SDP. There are several potential solutions for NAT traversal with respect to IMS:

- Insert an IMS ALG on the signalling path to translate the IP-related fields in the SIP headers (for signalling) and in the SDP (for media) for users behind an NAT.
- Include unilateral self-address fixing mechanisms at the client (e.g. simple traversal of UDP through NATs (STUN) (Rosenberg *et al*., 2003), traversal using relay NAT (Audet and Jennings, 2007) and interactive connectivity establishment (ICE) (Rosenberg, 2007, 2008)).
- Implement a middlebox communications (MIDCOM) (Swale *et al*., 2002) server solution to control IP addresses and ports. The IMS/ALG function is deployed at the edge of the network. The IMS/ALG function may be supported by a session border controller (S/BC), because an S/BC typically sits at the edge of the network and is designed to fill specific technical gaps when delivering multimedia services over the IPv4 packet network. In general, the S/BC provides the following capabilities:
  - management of NAT translation functions at both the IP and the application layers (e.g. SIP/SDP/real-time transport protocol (RTP));
  - management of data firewall pinholes in subscriber premises for voice-over-IP (VoIP) traffic (e.g. signalling and bearer);
  - management of VoIP traffic from/to the overlapping private IP address spaces across multiple enterprise networks;
  - protection of an operator's internal networks: topology hiding and denial of service (DoS)/distributed denial of service (DDoS) attack protection (e.g. call gapping and malformed packet filtering);
  - call admission control (CAC) as a way to ensure that existing communication paths are maintained at a contracted service level by bandwidth rate limiting; and
  - SIP/H323/MGCP ALG (i.e. an H.323 to SIP protocol conversion function) (Levin, 2003).

However, the S/BC deployment as an IMS ALG creates complications with IPSec, SIP signalling compression, quality of service (QoS) support, and even proxy discovery as currently defined in the 3GPP/3GPP2 IMS architecture. Accordingly, in order to support these mechanisms (e.g. in UMTS), it is not recommended that NAT be placed between the end user and the P-CSCF.

In UMTS deployment scenarios, the end-user and the P-CSCF are placed in the same IP domain. End-users are allocated IP addresses through PDP context activation. These IP addresses may be in the private domain, but since the end-user and the P-CSCF are in the same IP domain, the end-user signalling traffic is routeable to the P-CSCF. For wireless networks, the NAT function is located after the IMS call session control functions (CSCFs). Locating the NAT function after the IMS CSCFs can expose the CSCFs to overlapping private IPv4 address domains. This also must be addressed in configuring the IMS network. Unilateral self-addressing fixing (UNSAF) mechanisms allow the end-user to determine its IP address outside the NAT. These mechanisms require additional functionality in end-user equipment and specialized servers. Early IMS deployments may include the S/BC to solve the issues of NAT traversal; however, the self-address fixing mechanisms provide a more efficient, long-term solution. As mechanisms such as ICE and MIDCOM become more commonly supported in client equipment, the need for an IMS ALG may be supplanted.

Figure 4.2 depicts the IPv4-to-IPv6 transition scenarios.

4.1.2 Devices Challenge

From an end-user perspective, the terminal is the most important piece that actually enables the service to be experienced. It greatly determines the user perception or the QoE of any service. However, IMS-related standards have addressed the terminal as a blackbox since the terminal internals do not affect interoperability with the network. The standards have specified only the interaction with the network, leaving all the internal architecture and interactions as an implementation option. Figure 4.3 summarizes the device-related challenges.

This has resulted in the need to install a specific client software for each service provided by the operator, such as an enabler client, application client, etc. There is a need for dynamic installation and noninstallation of these software. With several services running on the same terminal, it is essential for the terminal to offer the required QoE, ensure efficient use of resources and orchestration of internal operations. This triggers the need for a horizontal layered internal organization for terminals, but currently there is a lack of specifications of APIs between the layers. The problems that exist in the terminals today are summarized in the figure. Application clients must be specifically developed for each target terminal model. Such a terminal-dependent development of applications results in an inconsistent QoE across different terminal types and development requires more time, resulting in an increase in the time-to-market intervals. Portability of applications across different terminal types becomes difficult, resulting in users not being able to share applications.

One possible solution is to achieve homogeneity across terminal platforms and technologies. Java presents the potential of bringing in the homogeneity since it is platform independent. However, there are differences in implementation of the JSRs, leading to compromise in portability and compatibility.

Open Mobile Terminal Platforms (OMTP, 2008b) have recently published a first version of IMS functional requirements, addressing the basic capabilities (basic profile), integrated behaviour for the IMS framework and COMMON high level APIs for IMS core and main enablers. These recommendations, along with the OMA enablers and the recent 3GPP specifications, such as ICSI, IARI (3GPP, 2007g), etc., would help to harmonize the operation of IMS terminals. 3GPP TS 23.228 R7 (2007g) introduced the ICSI/IARI identifiers as a mechanism for UEs to provide a hint to the network on the AS they wish to be linked into the signalling path.

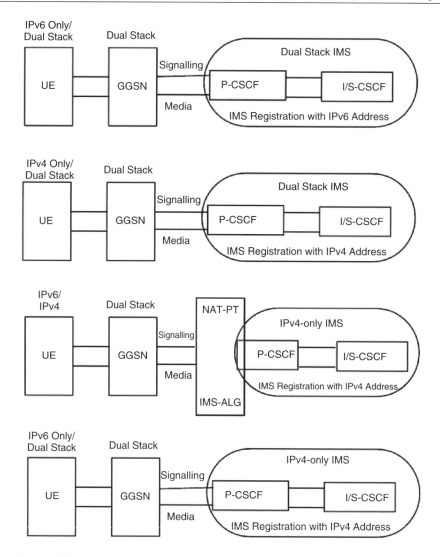

Figure 4.2 IPv4-to-IPv6 transition scenarios (from Choi and Fischer, 2006).

Agreements between operators, terminal vendors and middleware providers may lead to the inclusion of a common or, at least, similar middleware in IMS terminals. Open SW developer communities may also help to build a common set of modular resources to be used on top of heterogeneous IMS frameworks.

4.1.3 Operations and Management Issues

4.1.3.1 Staffing

The IMS is a complex technology and it will be necessary for mobile operators to upgrade the IP skills of the staff to ensure that the content management and network resource

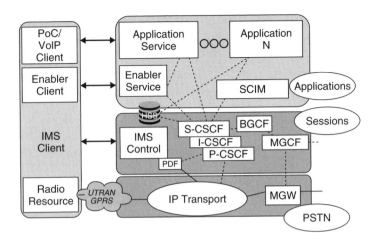

Figure 4.3 IMS device challenge.

management is seamless and, furthermore, that the customer needs of quality of service and security are met. To meet exceptional customer service requirements, highly skilled personnel are required, particularly in the initial phases of deployment, as the success of IMS depends on customer feedback and quality of experience.

4.1.3.2 Charging

The content management and service provision on IMS is complex and a billing scheme is still unclear. The increasing demand for content-intensive services provides operators with great revenue-generating opportunities and has prompted the need for dynamic solutions with interactive pricing capabilities. For real-time multimedia services, the billing system should be able to manage the real-time information provided by the IMS network entities. The IMS services depend a great deal on third party content providers. The billing system must, therefore, possess the flexibility to support any basis of charging that service providers may demand. The IMS pricing will be based on creative communication bundles and value-added offers and must be scalable enough to handle any surge in transaction volumes. However, it must be ensured that the services are charged at a cost acceptable to the users and the expected revenues are achieved.

Another issue that is hindering the operators is the operation and management of the IMS, including the associated processes for fulfillment, activation, billing and questions related to the integration of BSS systems into the IMS network infrastructure. Some operators pointed out that there is currently no centralized entity defined for IMS policy and resource management. This is extremely crucial in order to take advantage of some of the efficiency potential within the architecture shared resources.

The IMS policy charging rules function (PCRF) is the key element that is supposed to pull the policy and online charging functions together and handle such issues. However, with the slow progress of IMS policy standardization, there is a risk that operators will implement separate systems to bridge the gap. Therefore there could be one handling basic policy decisions, another that deals with traffic filtering and another that decides on bandwidth.

The result would begin to resemble the vertical service stovepipes that IMS is supposed to help eliminate.

4.1.3.3 QoS

QoS becomes a major concern for operators when calls are routed through other service provider networks, where they cannot really control the QoS. On an all-IP world, carriers cannot clearly define or evaluate QoS data (traffic measurement of calls or Erlangs, etc.) the way they can in the circuit-switched TDM world. This introduces unnecessary delay in transcoding in the network. The voice quality and the KPIs must be improved (compared to the circuit, dramatically). The upshot of this could be complaints for poor voice quality, one-way RTP, CODEC negotiations and dropped calls.

End-to-end quality of service is a major concern of operators presently and they are looking into finding ways to ensure quality of service over heterogeneous networks. Different access networks provide local QoS mechanisms but they cannot be applicable to NGN networks because of greatly resource-consuming multimedia services. Due to the converged nature of IMS, it is supposed to provide user services in all access networks at all times. Therefore, high service availability, resource management and quality are essential. To provide QoS over the core transport network IMS uses procedures such PDP and RSVP for admission control and resource reservation.

While some vendors are pointing to increases in signalling traffic, operators are also becoming increasingly concerned with the signalling delay, as call setup time, handover time, etc., are increasing as there is a lot of extra interaction, given the extra network elements involved. Furthermore, there is more latency involved in the IP network compared to the rugged SS7 network, so service providers believe the latency is a key area on which to focus on their testing. This latency is also raising new issues, which these operators never faced in the TDM world, including RTP streams coming from multiple resources for an endpoint, which creates garbled voice.

4.1.4 Security

While IMS is an attempt to simplify the network core design, it has made the management and monitoring of that network more complex than in the past. Much of the security built into IMS so far is pretty basic and is in an early stage. One additional problem that IMS has had to solve for fixed (not mobile) networks is the use of non-SIM-based handsets, as this makes fixed networks inherently less secure than mobile ones. It also introduces another, very practical issue, in that IMS assumes that every end-user has the ability to register with the network as intelligent SIM-based handsets can. IMS infrastructure requires default real-time security monitoring. It requires a full set of security at its access, control and media planes. An access plane requires authentication and encryption for securing its users (including support for IMS over UMA links), as well as global protection against denial of service attacks on the IMS core elements. The control plane requires full deep packet inspection capability on SIP control sessions with the capability to protect against targeted SIP DOS attacks, network address translation traversal capability with topology hiding and rich QoS capabilities. The media plane is the most challenging security plane, as high performance is required across multiple applications, including RTP verification,

IP denial of service protection, per flow traffic mirroring to ensure compliance with lawful intercept and per flow quality of service policing and marking to ensure prioritization of real-time media based upon the codec negotiated by the SIP control plane.

As a result, there are a lot of information security considerations that need to be taken into account in an IMS network, including the usual suspects (identity theft, caller ID spoofing, potential SS7 network breaches, security and integrity of the data streams, DoS/DDoS attacks, convert channels and SPAM/SPIM/SPIT, among others). The key concern for most carriers is to ensure identity, integrity and enforcement of strong authentication on an SIP-based session coming from a terminal that is not trusted outside the wireless domain. Most SIP devices are not currently equipped with a strong security mechanism such as the SIM card on a mobile handset.

4.2   Interoperability

Interoperability is always crucial to the wide-scale deployment of a new technology by operators, but for IMS it is still work in progress inevitably, the standards are still evolving and the products are still relatively new. Although the core IMS elements, such as the CSCFs, media gateways and controllers, are reasonably solid, there are other aspects of the IMS network, such as application servers, service interactions, QoS, security, etc., in a more evolutionary stage. Hence, while announcing the IMS-ready products, it is essential that they are interoperable and future-proof. IMS would also have to accommodate the existing plethora of non-IMS devices. Standards work is already under way, for example, on the issues involved with delivering IMS services to a simple mobile handset (OMTP, 2008b). The other overhead is keeping track of the crucial migratory issues, interoperability and interconnectivity with existing network infrastructure. IMS is not the only way of delivering IP-based services. Some operators may feel they could manage without IMS, at least until there is longer-term experience available. This makes the interoperability issue particularly more critical for IMS. The key to deploying IMS-based solutions while IMS standards are still evolving is interoperability. Interoperability issues are currently being addressed through testing in public forums, such as the IMS Forum Plugfest, IMTC, ETSI and in real-time interoperability tests on service providers's premises.

4.2.1   Interoperator Interoperability

Interworking of IMS services across different IMS networks is typically acheived by roaming implementations as prescribed by the standards. Two possibilities exist for interconnection between IMS services providers. The first possibility is achieved by establishing a direct connection between two service providers using leased lines and, secondly, by following the roaming implementations such as the GRX, IPX and the Diff Serv as defined by the GSMA R.34.

The GSM Association (GSMA) recognizes the importance of SIP/IMS as the platform for next-generation mobile multimedia using the evolving GRX (general packet radio system (GPRS) roaming exchange) and the subsequent IPX (IP exchange) to interwork the service providers. In order to promote the technical and commercial interoperability of mobile multimedia services, GSMA brought together the world's leading mobile operators, vendors and GRX carriers in October 2004 to conduct SIP trials (GSMA, 2007).

The GSMA's comprehensive programme of SIP trials addresses all of the requirements to assist mobile operators deliver a full service portfolio. The result of this industry-coordinating initiative is a series of international multioperator, multivendor, multicarrier SIP trials designed to enable the global interoperability of SIP services. This will ensure that services will work smoothly right from launch and discourage the formation of isolated islands of connectivity.

Enabling global interoperability of SIP services requires the exchange of IP traffic between mobile operators and other service providers in order to provide an end-to-end service to the consumer. This exchange process is referred to as IP internetworking (IPI). The GSMA SIP trials were developed as part of the GSMA IP program, which has proposed a solution to the challenge of enabling global IP interworking and interoperability based on four key concepts:

- Openness. The solution is entirely open to all IP service providers to provide ubiquitous services.
- Quality. End-to-end service level agreements (SLA) will ensure reliable delivery of IP services for consumers, as well as billing transparency and security.
- Cascading payments. Cascading value and obligations through the chain of delivery ensures quality and fair distribution of value from the customer through to the providers.
- Efficient connectivity. Efficient routeing of IP services starts with one connection from the service provider and no more than two interconnections to the other service provider.

However, it is not necessary to develop entirely new network technologies for this purpose. The open standard GPRS roaming exchange (GRX) platform is IP-based and well placed to develop into what the GSMA terms the IP Packet eXchange (IPX).

4.2.1.1 GPRS Roaming Exchange (A Framework for IMS Interworking Networks with a Quality of Service Guarantee 2008)

The GPRS roaming exchange (GRX) was created as a platform for enabling mobile data roaming. GRX networks have evolved purely from the exchange of 2.5G and 3G packet-switched roaming services additionally to facilitate interworking of MMS. In the GRX model the operator can replace the need for multiple connections to other operators with a single (or few) logical connections to a GRX. GRX networks interconnect with other GRX networks to form a global roaming network for mobile data roaming. The roaming model for GPRS is to route traffic from the visited SGSN (serving GPRS support node) to the home GGSN (gateway GPRS support node) across the GRX network. Therefore, a subscriber will egress his or her service provider GGSN when roaming. This backhauling mechanism across the GRX can introduce latency and jitter.

4.2.1.2 IP Exchange

In order to facilitate IMS (IP multimedia subsystem) roaming, the GRX has evolved into the IP eXchange (IPX) (see Figure 4.4). IPX would support secure control plane traffic via SIP/SDP (session initiation protocol/service delivery platform) as well as the QoS-enabled

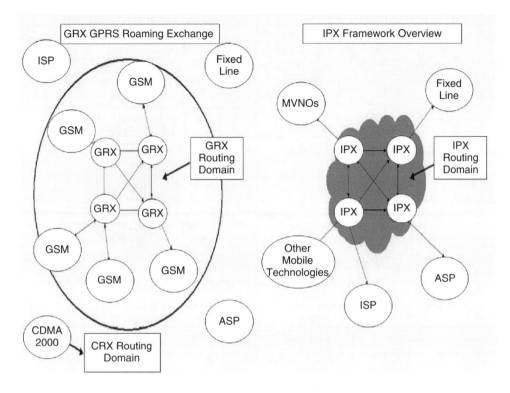

GRX GPRS Roaming Exchange

IPX Framework Overview

Figure 4.4 Evolution to IPX from GRX (from GSMA, 2007).

user plane via the RTP (real-time transport protocol). The IPX provides interconnection between different service providers, i.e. mobile and fixed operators, other service providers such as ISPs (Internet service providers), ASPs (application service providers) and later possibly content providers and other stakeholders, in a scalable and secure way with special focus on guaranteed QoS.

IPX providers manage the exchange of traffic and associated basic control information. The IPX provider can offer both technical and commercial interconnect capability via a single agreement with the service provider. One mechanism is to use SIP to convey QoS information. One can also leverage the subscriber profile for QoS information exchange. The goal would be to use the policy management mechanism to exchange QoS and other service policy information, as suggested in this chapter. The detailed mechanisms will require further studies and possible standardizations and are currently beyond the scope of this book.

Figure 4.5 shows the multilateral IPX model, which is GSMA's preferred solution to interconnect IMS service providers.

An IPX proxy acts as a hub between IMS networks and facilitates interworking and inter-operability from both technical and commercial perspectives. An IPX proxy is an SIP proxy with additional functionality to meet mobile operator requirements, including key functions such as accounting, brokering, security, routeing, protocol conversion and the capability of transporting and mediating, where necessary, control plane and user plane packets between different IP multimedia networks. IPX proxies are independent of applications or services.

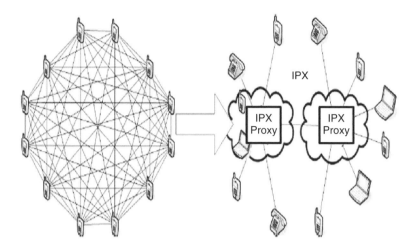

Figure 4.5 Multilateral IPX model (from GSMA, 2007).

It is possible to run any SIP-based service via IPX, including new peer-to-peer (P2P) applications regardless of the protocol. Using an IPX proxy does not automatically mean using a commercial broker model. It is entirely possible to use an IPX proxy just for routeing traffic and make agreements bilaterally or via contact brokering.

4.2.1.3   GSMA SIP Trials   (GSMA, 2007, 2008b)

The ongoing SIP trials were launched in October 2004, as part of the IP interworking initiative, and were led by the GSMA and sponsored by TeliaSonera, who originally proposed the concept. The trials consist of a number of test campaigns designed to demonstrate practical and operational deployment of SIP/IMS, as well as the interoperability of terminals and IMS platforms. The test campaigns focus on end-to-end service delivery and service interoperability using the IMS and an IPX proxy. They are tests in a real service environment and test the interworking of IMS systems at the network-to-network level and end-to-end interoperability at the service level. They validate the actual user experience.

SIP/IMS is a key enabler for mobile multimedia services and the manner of its deployment is critical. End-users expect ubiquitous service availability. To assist in achieving this, the goals for the SIP trials are to:

- Demonstrate interoperability of SIP-based services across networks, platforms and devices as an essential, practical step towards ubiquitous service delivery.
- Confirm services can be delivered to end-users with a low and predictable latency and with the appropriate QoS.
- Verify that necessary accounting information can be captured at the IPX level, which is an important objective as the standards process often neglects such key commercial considerations.

The trials have focused on the technical and operational aspects of interworking and basic SIP interoperability testing, including the ability to obtain necessary billing and internetwork accounting data. The main SIP-based services and applications tested to date are peer-to-peer IMS services such as instant messaging, Video Share and gaming and client server services such as push-to-talk over cellular (PoC).

Three main stakeholder groups are involved in delivering interoperability across networks, platforms and devices:

- mobile network operators (service providers);
- GRX/IPX carriers (the interoperator backbone);
- vendors (infrastructure, platform and terminal suppliers).

Every test campaign is managed by a lead operator and involves a set of interworking teams, each representing a pair of operators together with their respective vendors and GRX carriers. Campaign results are summarized in a document based on reports from each participating company. The results are available to all GSMA members. The substantial practical benefits generated by the trials remain with the participants.

4.2.1.4 SIP Trial Outcomes (GSMA, 2008b)

- First European SIP trial. Particular emphasis was placed on testing the IMS network-to-network interface (NNI) as the GSMA focuses on the interoperator domain. This first European campaign used standardized IMS core systems from multiple vendors interconnected through the GRX routeing domain. Three different peer-to-peer SIP applications from Nokia, 'voice instant messaging, video sharing and gaming', were tested over both 2G and 3G access networks.
- Second European campaign. This built on the interworking scenario introduced in the first campaign by demonstrating and testing several technical extensions of the IPX model:
  - connection between IPX proxies;
  - IPv6 and IPv4/6 conversion;
  - basic performance-related measurements;
  - accounting data collection by IPX proxies.
- First Asia-Pacific campaign. This was based on the first European campaign and included IPX proxy connectivity (hub-to-hub connectivity) and tested several IMS applications from both Nokia and Siemens. Nearly 40 test cases of real-time Video Share, peer-to-peer gaming and multimedia instant messaging were successfully tested among various IMS platforms. Successful session establishment and overall user experience were obtained for all applications and the IPX proxy-to-IPX proxy connection was verified. This campaign established interoperability between applications from different vendors.
- Second Asia-Pacific campaign. The second set of trials in Asia included additional trial participants from the Asia-Pacific region and had a similar scope to the second European campaign:
  - connection between IPX proxies;
  - IPv6 and IPv4/6 conversion;
  - basic performance-related measurements;
  - accounting data collection by IPX proxies.

- First Tel-URI campaign. The primary goal of this campaign was to leverage the interworking scenario by demonstrating and testing infrastructure telephone number mapping (ENUM) architecture feasibility for Tel URI/SIP URI flows. The trial also investigated the generation of accounting information from session data by the IPX proxy. A Video Share application from Ericsson and a gaming application from Siemens that support ENUM were used for the tests.
- First Video Share campaign. An important initial step in the first campaign was the definition of a specification for an interoperable Video Share service. The resulting GSMA Video Share specification does not require specific application servers in the network. Video Share interoperability enabled by IPX proxies was successfully demonstrated and it was verified that the ENUM architecture supports the Video Share service.
- Second Video Share campaign. The second Video Share campaign enhanced the Video Share service definition based on the results from the first campaign. It focused on Video Share charging principles and included specific IPX proxy test cases on accounting, security and performance. Both Tel URI and SIP URI end-user addressing schemes were tested.

4.2.2 Interoperability of Services and Service Enablers

4.2.2.1 GSMA Rich Communication Suite   (GSMA, 2008a)

The rich communication suite (RCS) initiative is an effort of a group of like-minded industry players for the rapid adoption of mobile applications and services providing an interoperable, convergent, rich communication experience. The RCS initiative includes network operators, and network and device vendors. The RCS initiative uses an iterative, agile methodology to deliver a consistent feature set, implementation guidelines, example use cases as well as demonstrations around interoperable reference implementations based on profiling of existing standards and specifications. The deliverables of the RCS initiative are intended to be provided as input to industry associations for consideration as reference specifications. RCS is an evolution of legacy services focusing on user experience.

The RCS initiative work is divided into a sequence of phased efforts. The RCS Phase 1 effort of the RCS initiative focuses on a core feature set with the intention of realizing practical interoperability demonstrations between different devices and network infrastructures by Mobile World Congress, Barcelona, in February 2008. The following areas have been identified as being part of the RCS for the initial phases (see Figure 4.6):

- enriched call;
- enhanced messaging;
- enhanced phonebook.

The enriched call experience initially provides the capability to share multimedia content during a call. The forms of multimedia sharing available at a given time between the communicating parties are shown to the call participants to eliminate unpleasant errors when either party cannot share chosen multimedia (e.g. when not within 3G coverage).

The enhanced phonebook allows guaranteed communication through capability enhanced contacts. In addition, enhanced phonebook means that communication can be initiated from the phonebook by selecting a communication type (e.g. calling or messaging). The enhanced

Figure 4.6 Rich communication suite (from Ericsson, 2008).

messaging allows the possibility to view and trigger all communication (including calls, SMS, MMS, instant messaging) in a conversational view, where the user can see the communications history. The conversational view is similar to chat history in instant messaging services. Users gain value from the simplified communications experience and from the availability of the richest possible messaging services for continuing communications dialogue.

RCS provides implementation guidelines for IP communication solutions including voice, instant and multimedia messaging, video and presence. Capability-enhanced phonebook, media sharing, group messaging and secure authentication are seen as important aspects too. NSN and Nokia are contributing to the RCS initiative by planning interoperable services between mobile devices and PC terminals, across the different access networks, providing also a selected set of supplementary services and backwards compatibility with legacy services such as SMS and voice.

4.2.2.2 Interoperability Challenges   (Ericsson, 2008)

The current focus of the RCS initiative is to overcome some of the existing obstacles to reaching these future services. Currently, standards are being developed in different fora, IETF, 3GPP and ETSI/TISPAN, while some IMS-based services are standardized in OMA. However, these standards cover a very broad range of services and with great latitude in what functions need to be included in each, which will lead to different combinations of services and functionality. By allowing a common, well-defined set of features, based on profiling of the available standards, which is agreed upon by device manufacturers, infrastructure vendors and operators, a common denominator of interoperability can be achieved with a common feature set, providing an engaging solution for end-users and thus driving service uptake. Additionally, the feature set allows for seamless interworking with existing services but also provides a bridge to allow users to upgrade to richer services and forms of communication.

RCS services are built using a set of standardized features that use the IMS architecture to enable service integration that provides a smooth user experience and interoperability between service providers. RCS services can be used both in mobile and fixed network environments and are interoperable between these two network environments. RCS gives the possibility to create a trusted connection between users and network environments. The presence service (OMA SIMPLE presence) plays a key role in RCS. The discovery of the rich communication capability is the key enabler to increase the usage of these new services. Publication of the end-user current communication capabilities guarantees that service completion can be realized for these new communication services. In the RCS Phase 1 feature set, presence information is used to communicate not only the communication capabilities but also personalized contact features including a photo, availability and status text. Presence can be conveniently shown in several ways: in the phonebook and in places where services are launched. User feedback on presence services has been very positive. In addition to presence services, RCS includes a set of common industry standard communication services that, taken together, form a comprehensive and innovative user experience. The additional communications services comprise Multimedia Messaging, Chat, File Transfer, Video Share and Image Share.

RCS also includes several other services such as the feature to centrally back up (and restore) the contacts in the local phonebook to a safe network repository, minimizing the impact of lost, broken or stolen devices. There is also the feature to access network-based directories and search and retrieve contacts to the local phonebook on the device. After a user has established a trusted communication connection with another user, the presence service enables them to see what services (communication capabilities) each has available. The 3G Video Calls, Multimedia Messaging, Chat and File Transfer services are examples of possible communication capabilities for a given user. Service attributes for Video Share and Image Share are included as part of the voice call setup. The service profiles enable integration of services that give users a smooth and seamless service experience. Existing services, including video, voice, MMS and SMS, as well as a variety of future services, can be integrated into RCS. In addition, as new services are added or when users upgrade their devices, the new service capabilities can be published to users via the presence service. The RCS enhanced phonebook allows a number of useful capabilities to aid communication. Contact information is no longer restricted to just the ones stored on the device. The user can also search and retrieve contacts from the mobile network directory or other fixed network directories (e.g. white pages or yellow pages), or even a service-specific directory (e.g. a gaming service). In addition, the RCS enhanced phonebook can provide a quick summary of communication with each of your contacts. The detailed view of an individual contact in the phonebook will reveal the complete profile of that contact, including their presence and most recent communications capabilities. The communications capabilities information provides an important clue as to how that contact can be reached. Communication with any contact is as easy as identifying them in the RCS enhanced phonebook and then selecting the wanted communication type.

4.2.2.3 Business Rationale of RCS   (Ericsson, 2008)

From the operator perspective, the business rationale for the RCS is based on how it provides a common feature for addressing multiple customer requirements and how it enables the operator to attract new subscriptions and increased usage of communication services.

- The RCS increases the attractiveness of the operator communication service offering by integrating presence and capability information and other new communication services. Capability information increases the usage of communication services by informing the user of the recipient's supported communication methods. Exposing service capabilities also gives the operators the flexibility to gradually deploy communication services at their own pace and means to advertise the new services to the customers.
- Today the user needs to know which messaging platform their friends use (MSN, Skype, etc.) and connect to them with the user credentials (e.g. name email) used by that application. The RCS initiative enables messaging just based on phone number information, which is already stored on the users phonebook and makes connecting to friends easy no matter which operators these friends have a subscription with.
- The RCS strengthens the existing operator business model by answering the customer demand for personalization and self-expression through new means of communication. This enables the operator to maintain ARPU levels and subscriber loyalty by introducing a better service experience for the same price.
- Mobile advertising in instant messaging services and with conversational messaging user experience could possibly become a mechanism for a new revenue source for operators.
- The RCS allows sustaining messaging pricing plans. If all messaging goes to Internet-based services (e.g. email or IM such as Skype, MSN, Google Talk), all messaging revenues are generated by external service providers. This would not allow for messaging-specific pricing, but all messaging would be part of a general data plan.
- The RCS further differentiates the operator on multimedia services compared to, for example, Internet service providers and so leverages and democratizes usage of IM and multimedia sharing.
- The RCS also improves the messaging user experience over GSM/3G through increased interoperability between operators. The increased availability of rich messaging services has not happened as widely and as quickly as for SMS.
- The RCS gives the possibility to monetize the usage of PC communication by integration to the mobile network environment. The scope of the RCS initiative includes mobile-to-mobile and mobile-to-fixed network connections.

4.2.2.4 OMA Initiatives (OMA, 2008)

To acheive true interoperability, it is essential to minimize the market fragmentation and enable seamless interoperability. Individual organizations often address standardization issues within one specific scope. By linking the activities of a number of organizations, the Open Mobile Alliance aims to address areas that previously fell outside the scope of any existing organizations, as well as streamlining work that may have been previously duplicated by multiple organizations.

Most significantly, the Open Mobile Alliance is designed to be the centre of all mobile application standardization work, enabling the creation of mobile services designed to meet the needs of the end-user. To grow the mobile market, the companies supporting the Open Mobile Alliance will work towards stimulating the fast and wide adoption of a variety of new, enhanced mobile information, communication and entertainment services. The implementation of OMA specifications results in benefits to the end-users, providing easy-to-use mobile services that are interoperable across countries, operators and mobile terminals.

The objectives of the OMA could be summarized as:

- to enable consumer access to interoperable and easy-to-use mobile services across geographies, operators and mobile terminals;
- to define an open standards-based framework for permitting services to be built, deployed and managed efficiently and reliably in a multivendor environment;
- to establish a standards forum (the OMA) for the mobile industry to function as the driving force for creating service-level interoperability; and
- to drive the implementation of open services and interface standards, using a user-centric approach that guarantees rapid and broad adoption of mobile services.

The OMA interoperability process (IOP) defines interoperability testing events designed to help achieve a number of goals for OMA and its members. These goals are:

- ensuring the quality of OMA specifications and
- enabling vendors to verify and test the interoperability of their product implementations.

The OMA interoperability process (IOP) has been established around the concept of regularly held testing events, which are commonly known as 'OMA TestFests'. These are hosted by OMA in connection with an operator or vendor, in a confidential and secure testing environment, where member companies can bring their implementation(s) to test in multiple multivendor combinations. It has extensive programmes for service enablers that provide architectures and interfaces that are independent of the underlying wireless networks and platforms and are intended to work across devices, service providers, operators, networks and geographies. The current organization of OMA consists of the following work groups as depicted in Figure 4.7:

- *Requirements*
  - Specify use-cases and identify interoperability and usability requirements.
  - From use-cases provide requirements to work groups.
  - Coordinate the requirements work and verify the consistency of requirements in work groups.
- *Architecture*
  - Define of overall OMA architecture.
  - Enable specification work in work groups.
  - Assure adherence of specification work to OMA architecture.
- *Security*
  - Secure communication protocols between mobile clients and servers at transport and application layers.
  - Security and trust services such as authentication, confidentiality and integrity provided by and to mobile clients and servers.
  - Secure interactions with entities using secure hardware tokens.

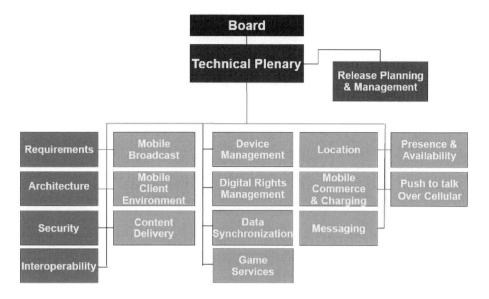

Figure 4.7 OMA organization.

- *Interoperability*
  - Identify, specify and maintain processes policies and test programmes to ensure interoperability for OMA enablers and end-to-end services.
  - Consolidate affiliate IOP groups into a common understanding of IOP activities.
  - Study and understand best practices.
- *Instant messaging and presence service*
  - Identify, specify and maintain requirements, architecture, technical specifications and test events.
  - Accelerate adoption of ubiquitous instant messaging and presence services.
- *Mobile web services*
  - Define application of web services with OMA arhiteture.
  - Provide application of web services converged with the work of external activities.
  - Generate recommendations, specifications and best practices to apply web services.
  - Leverage existing standards where applicable.
- *Device management*
  - Protocols and mechanisms for management of devices.
  - Setting of initial configuration data in devices.
  - Updates of persistent data in devices.
  - Retrieval of management data from devices.
  - Processing events and alarms generated by devices.
- *Data synchronization*
  - Specifications for data synhronization.
  - Develop similar specifications, including but not limited to SynML technology.
  - Provide conformance specifications and a set of best practices.

- *Location*
  - Specifications to ensure interoperability of mobile location services on an end-to-end basis.
  - Enable mobile location services.
  - End-to-end architectural framework with relevant application and contents interfaces, privacy, security, charging and billing and roaming.
- *Games services*
  - Interoperability specifications, APIs and protocols for network-enabled gaming.
  - Enable game developers to develop/deploy mobile games to interoperate efficiently with OMA platforms.
  - Enable cost reduction for game developers, game platform owners and service providers.
- *Mobile commerce*
  - Bring industry players together to coordinate the effort on m-commerce.
  - Provide an overall view on m-commerce.
  - Provide secure transactions, engage wireless network operators and meet financial industry requirements.

4.2.2.5 OMA Mobile Broadcast

Mobile broadcast has significant business potential as it enables cost-efficient delivery of media content to a large audience. The media industry has content that interests mobile users while the mobile industry has infrastructure to manage distribution of media content to mobile users as well as to collect revenues from mobile consumers for the media content. However, technology fragmentation could be a huge deterrent in realizing the business potential of mobile broadcast services.

The OMA BCAST working group was created and chartered to offer the following functionalities:

- Define the requirements for the enabler for mobile broadcast services.
- Specify the set of functions needed to enable the mobile broadcast, including but not limited to service guide and service/content protection.
- Package the functions as a service enabler, which facilitates global interoperability and is bearer independent in order to be useful for a diverse and heterogeneous infrastructure.

Figure 4.8 depicts a typical horizontally layered OMA broadcast architecture. The OMA BCAST enabler will support mobile broadcast services delivered over different systems, DVB-H, 3GPP MBMS and 3GPP2 BCMCS, but is not limited to these. It would be available on any broadcast bearer technology. These systems are called BDSs (broadcast distribution systems). For each of these BDSs, the OMA BCAST specification will contain an 'adaptation specification' document. Over these transmission technologies lies the OMA broadcast enablers, based on globally interoperable service enablers. The actual mobile broadcast services make reuse of these enablers to deliver the broadcast services.

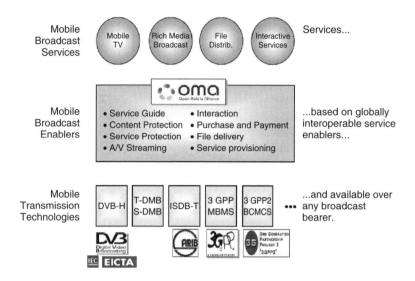

Figure 4.8 OMA mobile broadast.

4.2.2.6 Instant Messaging and Presence

IMPS (instant messaging and presence service) is a set of specifications striving to ensure interoperability in mobile instant messaging. The initiative was started in 2001 by three leading telecommunication companies, namely Ericsson, Motorola and Nokia. The initiative was originally known as the Wireless Village but has since then been consolidated into the Open Mobile Alliance (OMA), which is an industry forum for developing market-driven, interoperable mobile service enablers. In order to minimize interoperability issues, the initiative is open to participation from industry supporters interested in providing input regarding ongoing specification work. The aim of IMPS is not only to provide an exchange of messages and presence information in mobile networks but also to facilitate the creation of gateways to other instant messaging services.

Finally, the plethora of existing proprietary solutions effectively prevents service providers from marketing, providing and charging for their services in an efficient and economic way. By contrast, the OMA web service technology makes numerous service enablers accessible to any service provider. Service providers benefit from simplified, standardized mechanisms for providing and managing services. The confidence of content providers (if not the same as the service provider) will be further strengthened through the introduction of interoperable mechanisms for downloading rich media content in a secure and manageable fashion, e.g. using OMA solutions for downloading content and managing digital rights.

4.2.3 SIP Interoperability

Although IMS largely uses the SIP as defined by the IETF for signalling purposes, there are many IMS-specific extensions to SIP protocol. These extensions lead to interoperability issues with other SIP networks. IETF SIP-3GPP SIP interoperability is critical for IMS-based SIP applications are interoperable with the non-IMS SIP networks such as the Internet.

There is a lot of activity going on in the area of SIP interoperability. The most important of them all is the SIPit event hosted by the SIPForum twice-yearly, week-long test fests; in July 2004 the Forum released its first SIP Forum test framework (SFTF) v.1.0, an open-source test suite for SIP developers (see SIP Forum Releases Test Framework (SIPForum, 2008)). This is intended to allow SIP device vendors to test for common protocol errors and thus improve the interoperability of devices.

Equally, all the new SIP and related standards being introduced create new areas for interoperability issues. The mid-2005 SIPit saw testing of security aspects (SIP over TLS and SRTP) and RFC 3236 (Baker and Stark, 2002) for DNS location of SIP servers, for example. The SIPForum also runs a parallel series of test fests for the emerging presence and instant messaging standards SIMPLEt.

In short, SIP interoperability, while steadily progressing, is inevitably going to be an ongoing issue as the technology continues to develop. Further, being only one protocol among the many involved in IP-based networks, it is inevitably intertwined in commercial products with much wider issues of equipment interoperability. It is no secret in the VoIP world, for example, that it is still a big issue whether a softswitch from vendor X will work with a media gateway from vendor Y.

The MSF is developing and testing interoperability agreements that address such practical issues. For example, it has developed a scenario to examine the interactions among protocols such as SIP, MGCP and H.248 (Groves and Lin, 2007), while providing basic calling features such as call forwarding, call waiting and three-way calling.

As shown in Figure 4.9, from an SIP interoperability point of view, the essential interfaces are:

- Gm-interface (UE-IMS);
- ISC-interface (IMS-AS);
- Mm-interface (IMS-other SIP networks).

This implies that SIP interoperability issues are three dimensional and SIP interoperability activities should resolve the following issues:

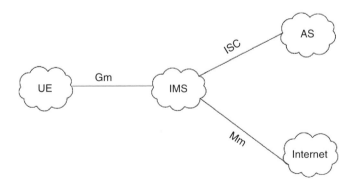

Figure 4.9 SIP interoperability.

- The existing SIP clients should be able to access IMS services.
- The existing SIP servers should be used by the IMS architecture as the IMS-AS.
- IMS networks should interwork with other non-IMS SIP networks such as the Internet.

Currently, the SIP client and SIP server need modification to be able to interwork with the IMS network. Similarly, for the IMS network to interoperate with non-IMS SIP networks, appropriate gateways are essential. However, the Mm interface that connects IMS with external networks is still evolving in the 3GPP standards specifications.

4.2.3.1 Multiservice Forum (MSF)

The MSF framework is designed to achieve a viable architecture based on IAs that facilitate complete interoperability between system suppliers and carrier networks. Within the MSF framework, members share a common agreement on architecture, interfaces and protocol implementations focused on commercially viable solutions. Conceptually, the MSF framework 'occupies' the space between technology exploration and product development, where vendor and carrier issues are resolved and implementations agreed upon. MSF IAs specify mandatory and optional functions, ensuring interoperability between devices from different vendors. The Global Multiservice Forum interoperability testing is a key component of the MSF process, also known as the GMI events. Figure 4.10 depicts the scope of MSF activity.

GMI 2002

The MSF took its design process and created an architecture, implementation agreements, and test plans focused on an event to be held at the end of 2002 as GMI 2002. This Global

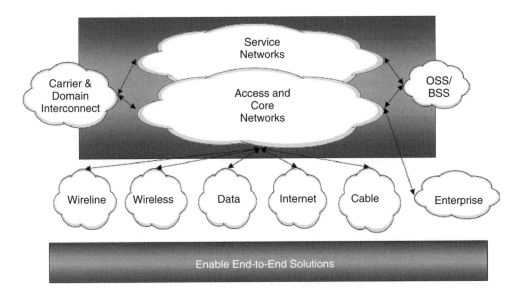

Figure 4.10 Scope of the MSF activity.

MSF Interoperability Event was to be held at three international locations to prove true global interoperability. It was to demonstrate MSF's Release 1 Architecture for Multiservice Switching Systems utilizing the following signalling and control protocols:

- MEGACO/H.248 (Groves and Lin, 2007);
- bearer independent call control (BICC);
- session initiation protocol (SIP);
- SIP for telephones (SIP-T).

GMI 2004

The MSF GMI 2004 Event provided an excellent foundation for which the MSF can build IMS realizations. GMI 2004 demonstrated the use of media gateway controllers, session border controllers, service brokers, SIP application servers, Parlay and Parlay-X gateways and applications, and Media Servers. With diligence, the MSF can build upon its previous work and utilize the experience and lessons learned to make enhancements to support full IMS implementations. By focusing on physical implementations and interoperability, the MSF will continue to leverage its core competence and deliver back to its member companies and the industry at large.

GMI 2006

Global MSF interoperability 2006 (GMI 2006) represents the first international, multivendor validation of IMS (IP-based multimedia subsystem), the underlying framework that will enable the long-heralded, long-awaited next-generation network (NGN) as well as the array of advanced services that these converged networks will deliver.

More specifically, unrestricted roaming allows subscribers and devices to access the network from anywhere, eliminating the boundaries that once separated fixed, mobile and wireless networks. This is the essence of fixed mobile convergence (FMC). Subscribers can initiate a session on, for example, a wireless LAN (WLAN) at work and receive the same service set they would receive across the cellular infrastructure or from their home office. Equally important, IMS ensures that this unfettered movement neither disrupts nor degrades the quality of service (QoS) of whatever value-added service a subscriber is using. Finally, IMS supports true multimedia services, which can range from voice over IP (VoIP) and text messaging through streaming audio and video-conferencing, to name a few possibilities. In addition to evaluating this essential aspect of IMS, GMI 2006 also validated the Multiservice Forum (MSF) Architecture Release 3 as a full peer network to a pure IMS implementation. This demonstrates the ability of vendors whose products comply with all relevant MSF implementation agreements (IAs) to interoperate in this environment. GMI 2006 also demonstrated:

- effective and enforceable QoS using a session border controller (SBC) and bandwidth manager (BM);
- IP carrier interconnect/interworking;
- security interoperability in an FMC environment;
- third party applications and service brokering;
- interworking of priority calling between PSTN and NGN.

GMI 2008

GMI 2008 is the fourth in a series of biannual events designed to test standards compliance in network scenarios of interest to major carriers. Building on the success of previous events, GMI 2008 provided the first real network multivendor trial of innovative new services deployed over a converged Internet protocol (IP) multimedia subsystem (IMS) infrastructure. Incorporating an IMS core compatible with the Third Generation Partnership Project (3GPP) Release 7 (R7) standards, the MSF Release 4 (R4) architecture introduces the concept of access tiles to support a range of both fixed and wireless access technologies. The IMS core provides the basis for introducing a range of service capabilities, including IMS-based IP television (IPTV), service oriented architecture (SOA) gateways and location services. Voice over IP (VoIP) softswitches are also supported to maintain backward compatibility with previous MSF architecture releases and to reflect real-life deployments in carrier networks.

4.2.3.2   IMS Forum

The IMS/NGN Forum is a global nonprofit industry association dedicated to the advancement of IP multimedia subsystem applications and services interoperability. They are a proactive group of industry leaders, experts and visionaries focused on revenue-generating services and best practices for IMS applications convergence. The IMS Forum is the creator and organizer of IMS Plugfest, NGN Plugfest and IMS/NGN Certified, the industry's only event focused on IMS and NGN services interoperability verification and certification.

The IMS/NGN Forum's mission is to accelerate the interoperability of IMS applications and M-Play NGN services, enabling enterprise and residential consumers to benefit quickly from the delivery of quadruple play voice, video, Internet and mobile services over broadband via wireless, cable and fixed networks. IMS applications and services comprise enhanced VoIP, IP Centrex/IP PBX and business unified communications including fixed mobile converged services, femtocell, LTE, video-conferencing and web collaboration, entertainment including IPTV and gaming, NGN/IMS networks including GSM, UMTS, WiFi, WiMAX, DSL, Cable and Optical. The IMS/NGN Forum includes OSS/BSS, billing/charging and security in the working group activities.

The IMS/NGN Forum technical working groups are involved in the creation of test plans, the execution of the IMS Plugfest and NGN Plugfest, authoring whitepapers and developing IMS/NGN best practices and guidelines. The marketing working group plans and participates in the IMS/NGN Forum's aggressive outbound marketing activities, consisting of analyst briefings, industry events, press activities, speaking engagements, bylined articles, case studies and more.

The IMS Plugfest programme is on an aggressive ramp to intensify the testing and interoperability of these critical areas as the IMS ecosystem matures. The IMS/NGN Forum in cooperation with UNH-IOL and its membership will continue to be aggressive in its mission to help the IMS industry develop to its fullest potential. IMS as a technology is maturing and more work needs to be done as standards continue to evolve, but the good news is that it is a demonstrably real, standards-based technology with real vendors delivering real products and services today. The Plugfests, held every 3–4 months, bring together industry-leading IMS vendors from around the world, all of whom build and test real IMS networks. It is the IMS Forum's vision that results from the Plugfests will serve as proof points for removing barriers to the adoption of IMS, and that the Plugfests will add industry-recognized certification for IMS applications and services inter-operability.

Scope of the Plugfests:

- Phase I: completed
  - Define and establish the 'reference' IMS test network.
  - Basic interoperability between control and application layers (HSS, x-CSCF, MGF, AS, UE).
  - Device registration.
  - Subscriber database interactions.
  - Basic call flows and charging for them.
  - IP-to-TDM interactions (media/signalling gateways).
  - Examine the interaction of north/south interfaces to application servers.
- Phase II: in progress
  - Enhanced interoperability and interaction east/west (AS to AS).
  - Nomadic services (moving profiles between multiple IMS core networks).
  - Presence services.
  - VoIP, instant messaging.
  - Video and multimedia.
  - Downtime prevention/service assurance: reliability, availability and security.
  - User and application profile handling.
  - Testing and certification of the 'gm interface'.
  - Basic roaming (visited networks).
  - Billing applications – online and offline charging.
- Phase III: future testing
  - IPTV.
  - FMC.
  - Forward migration, moving from 2G to '3G IMS' (inclusive of cable, mobile and fixed).
  - Service level traffic variations to maintain high quality services.
  - Prevent downtime and customer churn.
  - More sophisticated online charging scenarios.
  - Additional services for businesses and unified communications using a common IMS architecture.
  - IMS clients.

4.3 IMS Deployment Strategies

Based on the findings from the study of the existing IMS deployments, the commercial requirements have been derived. For the sake of clarity they are classified according to the end-user, fixed operator, mobile operator and fixed mobile convergent operator.

4.3.1 End-User Requirements   (Bang, Jorstad and Jonvik, 2007)

End-users will have a variety of end-devices (STB, PC, handheld portable devices, home equipment, MP3 players, etc.) with access to several networks. Installation, operation and maintenance will become complex. Prices for voice are eroding and voice is seen as a feature instead of a service. Users will request from their operators:

- Valuable service. The user is interested and willing to pay only for services that are useful and valuable for him or her. The user is not interested in the network infrastructure but only on the services that are useful and usable. IMS as such is not interesting for the user but the IMS services. Services that are fancy and funny but not really useful for the user will be abandoned in the long term.
- User-friendly service. The service must be easy to use and the user should intuitively figure out how to use it. For difficult features, there should be instructions that are accessible to the user in a simple way. The service must be implemented for nontechnical people.
- Working first time. The service must preferably work for the user at a first try because it is very unlikely that the user will try again. The service must therefore be deployed and tested carefully before being offered to the users.
- Service reliability. The user must able to rely on the knowledge that the service is working when he/she needs it. This means that the service must be working most of the time, preferably 99.99% of the time.
- Service robustness. The service must be robust, i.e. it must continue to work no matter how serious a mistake the user has made.
- Service ubiquity. The service must be operative at any time and no matter where the user is and what device is being used.
- Uniform user experience. There must be the same service everywhere on any device and on any network. Nowadays, each user may have several devices of different types and the service must appear more or less the same to the user regardless of the device. The service must also be delivered continuously when the user is moving and the handset is changing networks. The user must have a uniform user experience.

4.3.2 Fixed Operator Requirements   (Bang, Jorstad and Jonvik, 2007)

Operators are moving towards next-generation networks (NGNs), which should consolidate their traditional voice networks and other service delivery networks into a single converged IP-based network. NGNs in Europe are based on IMS and SIP as the session initiation protocol for voice services.

- Ability to offer IP telephony both on fixed and mobile networks. Although IMS was intended originally for mobile networks it should be usable for both fixed and mobile networks. Indeed, fixed operators view IMS as a technology that enables them to expand their activities into the mobile domain. By being an MVNO offering mobile subscriptions that use mostly wireless LAN hotspots and the mobile networks only when it is necessary, fixed operators can win back the amount of traffic lost to mobile operators in the 1990s.

- Incremental service deployment. At the first stage, the focus is on a few successful existing services such as voice, but it must be possible gradually to introduce new advanced services with time.
- Reduction of OPEX/CAPEX. The introduction of IMS is considered by fixed operators as a step to convert their legacy core infrastructure to IP and should require investment other than traditional telecom equipment. In addition, the operation and management costs should also be smaller due to the openness of IP equipment.
- Integration with legacy systems. It must possible to integrate IMS with legacy infrastructure seamlessly and to maintain the same pre-IMS quality of service.
- Separate charging for connection and service usage. Since fixed operators want to offer their services on the mobile networks a separate charging for connection and service usage is required. They must also have the possibility to distinguish the services and charge differently.

4.3.3 Mobile Operator Requirements   (Bang, Jorstad and Jonvik, 2007)

Mobile markets in Europe are saturated with mobile telephony and prices are eroding. Data services that require much more bandwidth have not been taken up in the market.

- Transparent IMS. As observed in the existing IMS commercial deployment, the mobile operator's promotion should be focused on innovative services such as video sharing, presence, messaging, etc., but not on IMS. On the contrary, IMS should be in the background as a technology enabler.
- Richer communication. To be successful, it is not sufficient to promote IMS only as a VoIP service. IMS should be promoted as richer communication that is provided on multiple communication media (voice, video, text, digital content, etc.).
- Incremental service deployment. At the first stage, the focus is on a few new services such as VoIP, PoC and video sharing, but it must be possible to introduce gradually new advanced services with time.
- Integration with legacy systems. It must possible to integrate IMS with legacy infrastructure seamlessly and to maintain the same pre-IMS quality of service.
- Integration with non-SIP-based applications. For some mobile operators, it is crucial that IMS could be easily integrated with non-SIP-based applications. This means that proper interfaces must be provided.
- Shorter time-to-market with IMS. The development of new combinatorial applications making use of multiple communication channels such as voice, video, text, content, etc., must be simpler and the time-to-market must be reduced considerably.
- Openness. The IMS platform must be open and allow the integration of third party services.
- Service roaming. For mobile operators, it is crucial that service roaming is supported and that the user can have access to the IMS services no matter whether he/she is located independently of the network used.
- Customization/segmentation. It should be possible for mobile operators to customize and package the IMS services in various ways to target different target segments. A service profile management must be provided to allow easy modification of the service. The management of the user interface and application flows must be flexible.

It should be possible to perform a remote update of the user interface to change the branding and themes.

- Uniform user experience across terminals. It must be possible to ensure a uniform user experience across all terminals. Since mobile terminals come with different form factors and capabilities, transcoding and adaptation capabilities must be provided. Both terminal and vendor independency must be supported.

4.3.4 Fixed Mobile Convergent Operator Requirements (Bang, Jorstad and Jonvik, 2007)

For a fixed and mobile operator IMS could be the catalyst for service convergence across wireless and wireline networks. The main requirement is profitability, i.e. to keep the costs down and revenue up in a very competitive market. Profitability is closely linked to the number of subscribers and usage of services.

- Same service both in fixed and mobile networks. For the fixed mobile convergent operator, it is crucial to be able to offer the same services in both the fixed and mobile networks. The goal is to promote the usage of the operator's network as much as possible. Mobile phones with an IMS client must be able to operate in both the fixed and mobile networks.
- Richer communication. To be successful, it is not sufficient to promote IMS only as a VoIP service. IMS should be promoted as richer communication, which is provided on multiple communication media (voice, video, text, digital content, etc.).
- Advanced applications involving both fixed and mobile devices. For a convergent operator, it is crucial to be able to offer convergent applications that make use of both mobile terminals and fixed devices such as stationary PC, IPTV, set-top box, home gateway, etc. It is therefore necessary with an IMS client for stationary devices.
- Service continuity. The services offered must be able to flow seamlessly over heterogeneous devices without interruption. The user must not have to worry about the change of networks or devices.
- Interoperability. Communication is universal and users from one operator must be able to interact with users subscribed to other operators. There must therefore be interoperability:
  - between operator's services (presence, community with other operators);
  - between client frameworks;
  - between devices and device classes;
  - between fixed and mobile devices;

Generally, the Tier 1 carriers and service providers that are implementing IMS are doing so as part of a much larger programme of next-generation network and service architecture development. However, for the end-users to experience the use of IMS, it can be done through introduction of new services. Small providers that are deploying IMS are doing so to obtain a service advantage in a specific area. As a result, the recent surveys (Lightreading, 2008) indicate the emergence of a small group of initial key but trivial services such as residential and business voice over IP, voice fixed mobile convergence, IP Centrix, etc.

A growing number of operators of different types worldwide are offering a few key services based on IMS, even if this is offered with a fairly minimal configuration of the IMS core and a few application servers. This section looks at a few examples of IMS deployments from the different types of carriers and service providers perspective (Lightreading, 2008).

Mobile Operators

Far EasTone Telecommunications Co. Ltd claims to be the first Taiwanese operator to offer services that use IMS. The mobile operator has deployed IMS and HSPA to support voice FMC, and in August 2007 launched a voice service using VoIP over WiFi over HSPA via a fixed wireless terminal. Customers can use a WiFi phone, PC client and mobile phone on a single number, and it is also possible to access the service via public WiFi hotspots. A strategic rationale for the service is to target its existing mobile customers who take their broadband service from other providers. Service features include personalized ringtones and intelligent voice mail, which provides different answering messages depending on the status of the phone, such as 'no answer', 'busy', or 'turned off'.

The Danish incumbent, TDC A/S (Copenhagen: TDC – message board), is using IMS to support IP Centrex services targeted at SMEs as a replacement for PBXs, which TDC no longer supplies. The customer buys a basic package and can then customize the solution via a website.

Telefnica SA (NYSE: TEF – message board) says it has been running commercial service traffic in Spain with IMS since mid-2005 as part of its strategy to offer IP-based multimedia services. Services include residential telephony, teleworking services and enterprise voice services, such as IP Centrex. New services are to be added; a video call service was launched in June 2007, allowing video calls between fixed videophones, 3G mobile phones and TVs (via a set-top box).

For a major operator, such as a large incumbent, IMS is likely to be only a part of a much larger network and service restructuring for an IP-oriented world. AT&T provides a good example of IMS's role in such a transformation in a US context, illustrating the need for this type of provider for something that provides IMS-type functionality. Years ago, the company decided that it needed to integrate and consolidate all of its networks into a single global MPLS-enabled IP network. In turn, this meant creating a layer on top of that IP/MPLS network to support both existing and future services of all types – voice, video and data.

AT&T concluded that there was then no currently available architecture for a carrier of its size that could handle the services that it was then offering and was also planning to offer. Therefore the company started a programme to design an architecture for that layer, which it called real-time services over IP (RSOIP). However, the subsequent arrival of IMS standards and their extension from mobile to fixed networks persuaded AT&T to modify RSOIP and adopt IMS into it (see AT&T Defines Service Creation Platform in Lightreading, 2008).

BT is also using IMS-enabled services as part of a massive current migration of its entire UK network to a converged IP-based NGN in its 21st century network (21CN) programme. This went live in a limited pilot area of the UK (in South Wales) in November 2006 and is scheduled to be completely rolled out across the UK by 2010. Among other things, the 21CN will involve delivering the standing PSTN service via call servers and a converged multiservice access.

The initial service focus is expected to be:

- Consumers: VOIP (BT's Broadband Talk), PSTN simulation, convergence and advanced voice services.
- Enterprise: SME Centrex/hosted PBX, enhanced voice services, major corporate solutions, corporate fusion.
- Carrier: CP-to-CP transit services.

BT sees a positive business case for IMS from the cost reductions that it will generate in terms of platform consolidation, reductions in operational expenditures and training costs, and so forth. The customer experience should be improved (e.g. through convergent and blended services and QoS management) and, of course, both the company and customers should gain from the potential for new services and their faster time-to-market. IMS, however, is only part of a much bigger scheme that BT envisages for 21CN, embracing support for a universal SDP that integrates both telco-type services/applications and Web 2.0 approaches. IMS would form part of the basic functional enablers, but above it would be a layer of larger 'reusable capabilities', such as home environment, authentication, customer profile and preferences, application-driven QOS, session control, presence, instant messaging, directories and the like. In turn, above this, would be a layer of BT service components that would be potentially reusable for all customer segments and would include an end-to-end series of high-level components:

- incoming communication channels;
- communication relationship;
- scheduling;
- communication routeing;
- queuing;
- message storage and retrieval;
- media transform;
- delivery to device.

Above this would be a layer of BT services based on these components. Finally, above the BT services layer and using published BT APIs would be a layer of third party and end-user applications and services created as mash-ups. BT describes this approach as Web21C service execution. Essentially, the architecture is based on IMS, but is extended with SoA and reusable capabilities. Service execution will deliver consumer, SME, major corporate solutions and wholesale carrier-to-carrier interconnect. 21CN and Web21C are an open architecture, intended to work with industry and developers.

4.3.5 Summary of Real Deployments

For fixed and broadband operators, the key elements of IMS are as follows:

- a cost-effective means to deliver and charge for 'Internet style' applications and services;
- a means to convert its legacy core infrastructure to IP;
- a means to offer a fixed mobile convergent telephony service;
- a means to penetrate the mobile world.

For mobile operators, the key elements of IMS are as follows:

- a means to provide voice, video, data and multimedia services that can be accessed over different types of devices;
- a means to launch advanced services like video sharing.

For total operators, the key elements of IMS are as follows:

- a means to offer an enhanced communications experience and a range of applications;
- a cost-effective means to deliver and charge for multiple applications and services;
- a means to offer a fixed mobile convergent telephony service;
- a means of integrating both SIP and non-SIP applications;
- a means to provide end-to-end quality of service;
- a means to provide end-to-end security.

4.4   IMS Test Networks

4.4.1   The Need for IMS Test Networks

Next-generation networking architecture proposes to merge technologies from telecommunications IN networks, Internet, web services, broadcasting and entertainment. The proposed architecture is highly complex in terms of signalling and interworking. The complexity coupled with the claims of NGNs of providing ubiquitous and seamless multimedia services across heterogeneous networking environment and end-devices proves to be a huge challenge on the road to its deployment. Any new technology triggers the need to measure its performance, capabilities, applicability and suitability prior to its deployment. Similarly, research based on next-generation networks necessitated a complete and realistic testing ground for verification and validation of the proposed technology. On the one hand network testing grounds are traditionally restricted to laboratories where the setup is highly controlled and isolated from real-life scenarios; on the other hand testing on commercial networks is typically forbidden at the risk of disrupting the live services. Thus, the majority of the existing communication testbeds lack the scale, the extendibility and the infrastructure to represent and emulate a true next-generation networking environment. In order to implement, evaluate and test the feasibility of the proposed mechanism, a test network was mandatory. Implementation and evaluation done on a live test network provides valuable measurements that could be used by network operators, service providers and application developers in their effort towards NGN deployment.

The need to build a live test network was triggered by the following requirements:

1. To emulate a next-generation networking architecture compliant to the 3GPP standards at all layers such as the connectivity plane, control plane, service enabler plane and application plane.
2. To benchmark and evaluate the performance of the next-generation multimedia services from an end-user perspective (quality of experience) on a commercially deployed WWAN network. It can be noted that WWAN links are severely plagued with problems like high round trip times (RTT), fluctuating and relatively low

bandwidths, frequent link outages and burst losses; hence it is imperative to measure the perceived quality of experience of the end-user on a real-life WWAN setup.
3. Proof-of-concept implementations.
4. To create a foundation for:

- next-generation application development and validation;
- prototyping service delivery platforms;
- infrastructure component testing;
- networking technologies integration and interoperation.

4.4.2 Building a Holistic IMS Test Facility

NGN is designed to offer a common service delivery platform that spans multiple access network technologies at the edge. In order to achieve access independence and smooth inter-operation between the circuit-switched and packet-switched networking technologies, the 3GPP's IMS, which forms the core network of NGNs, conforms to the IETF standards. The test network design is based on the 3GPP's reference architecture for the IP multi-media subsystem (IMS) (3GPP, 2007g). The bottom-most layer in the architecture is the *transport layer*, which provides the bearer services and hence is also termed the connectivity plane. This is where the SIP signalling is initiated and terminated when a session is established with end-users. In order to represent truly a real-life connectivity plane, the testbed comprises all of the commonly used access networking technologies, such as Ethernet (Popular LAN), wireless (WLAN), mobile access (GSM, GPRS, UMTS) and the PSTN network.

The layer above the transport layer is the *control layer*. It forms the core of the NGN architecture, which handles the session control and management, charging and other support functions.

The layer above the control plane is the *service enabler plane*. It consist of components that provide the service logic. The service enabler layer makes use of the underlying session control layer for the provision of advanced seamless mobility-based services. This layer provides abstraction from the underlying networking technologies and session control in the form of sophisticated and reusable application programming interfaces (APIs), thus creating an efficient environment to create services.

The layer above the service enabler layer is the *application layer*, which consists of application servers, service development tools and third party services. It provides the base for validating the underlying entities.

Across all of the above-mentioned layers, the network design provides engineering, integration and conformance. In order to emulate a next-generation network, it was essential to have the components of the session control plane, service enabler plane and application plane to reside in an operator's core network, which are then accessible to end-users connected via heterogeneous access networks. Since it was not practical for us to install our equipment inside a commercial operator network, the test network was designed to include a VPN connectivity with the operator network, thus virtually extending the operator network to the IMS lab setup at Glamorgan University. The network is designed to be accessible by a host of access technologies including the mobile and wireless access. Figure 4.11 shows the global view of the test network.

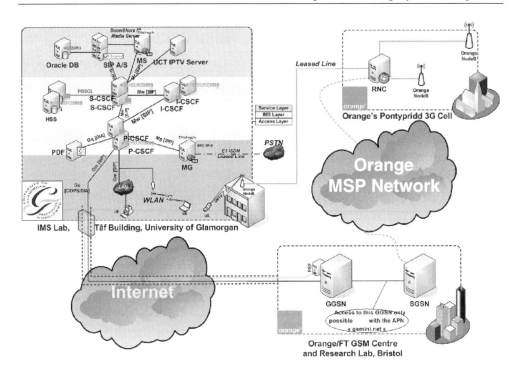

Figure 4.11 Global view of the IMS test network and service architecture.

4.4.2.1 Access Networks

Next-generation networking is all about mobility and hence it was imperative to equip the test network with mobile access technology (GPRS, UMTS, HSDPA). Orange, the UK's leading mobile operator, supported the test network by installing an indoor node B equipment in the lab and receiver antennas all across the building. This provided special UMTS/HSDPA coverage to users in the lab and created a unique research scenario wherein there was UMTS/3G coverage within the building and on stepping out there was an automatic handover to Orange's public GPRS network. The indoor node B within the Taf Building at the University of Glamorgan is linked to the RNC (radio network controller) in Pontypridd, the closest 3G cell from Orange. As discussed in the previous section, it was required to connect the test network with the operator's network to emulate a real-life scenario. This is provisioned in the network design via an IPSec VPN connectivity between the Orange Bristol Lab, where Orange's 3G network node is located, and the University of Glamorgan. This facilitated to backhaul HSDPA/UMTS/GPRS traffic to the test network directly from the cellular provider's network. This also ensured that the path between the test network and the cellular provider's network was never the bottleneck, thus enabling the measurements made on the test network to be close to the real-life measurements and of great value to the operators. A restricted set of SIM cards were released that had access permissions to connect to a specific APN (access point name) in the GPRS networks, called gemini6.net, which in turn permitted access to the GGSN at the Orange network at Bristol. Subsequently, the mobile devices had access to the IMS network at the University via the VPN. Wireless access points were installed in the lab terminating on the backbone switch, one hop away from the test network.

ISDN 30 E1 termination is installed in the lab by British Telecom to provide PSTN access to the test network. This link is connected to the media gateway (Cantatas IMG1010 appliance), which converts the SS7 signals to IP, enabling the standard telephony network to access the IP-based core network in the test network.

An interworking element part of the media gateways and the test network are equipped with the Cantata Technology IMG 1010. It is an integrated media and signalling VoIP gateway that provides any-to-any voice network connectivity, enabling the delivery of SIP services into legacy PRI, CAS and SS7 networks, as well as IP-to-IP trans-coding for network peering applications. The IMG 1010 simultaneously supports PRI and SS7 signalling, plus SIP and H.323 (Levin, 2003).

Session Control Plane

The session or service control plane is the heart of the test network. Figure 4.11 shows the network elements consitituting the core network. The test network has two implementations of the IMS core network: one is a proprietary implementation developed as an in-house project at the France Telecom R&D labs and the other is the open-source implementation by FOKUS (2006) called the openIMS core. The two implementation are deployed on two different IP subnets and emulate two difference operator networks. This equips the test network with the additional feature to test interoperator roaming functionalities. Both implementations are based on the 3GPP standards, although of different releases.

Service Enabler Plane

The key feature of IMS is its ability to integrate with other non-SIP-based services such as RTSP-based media streaming, SOAP/XML web-based services, etc. For this purpose the 3GPP defined reusable components within the core service network that can be leveraged by other applications for the purpose of obtaining specific information or invoking some well-defined IMS functionality. Being reusable components, they are implemented and deployed once, thereafter invoked by any authorized service yielding in consistent functionality irrespective of which entity invokes them. The enablers further act as building blocks for more complex services that may incorporate the basic functionality of one or more enablers. Thus, availability of service enablers lowers the cost of implementing the more complex end-user services by avoiding the need to recreate a given functionality.

- Presence is one of the basic service enablers that could be used by any service to incorporate the user presentity feature. Presence is a basic service derived from the instant messaging that involves making a user's status available to others and the statuses of others available to the user. Presence information may include: person and terminal availability; communication preferences; terminal capabilities; current activity; location and currently available services. A typical example of a presence-specific application will be a phonebook embedded with presence information, making it dynamic. Dynamic presence will be the initial information the user sees before establishing communication; hence it might affect the choice of communication, method and timing. An OpenSERV SIP application server runs a presence service in the test network. It runs on a virtual public address on the box that is running the P-CSCF on port 5060 and has a domain name presence.glam-oims.net. The presence server was integrated with the OpenIMS by defining it as one of the AS in the service profile of the OpenIMS HSS.

- With media servers becoming more capable and intelligent, the traditional telephony services on IP are augmented with an intelligent interactive voice response (IVR). The IVR responds to a DTMF tone detected by a pre-recorded or dynamically generated audio to direct the call further and helps in handling large call volumes. IVR has been highly criticized for its poor design, user-unfriendliness delays and for its unethical business usage, such as enforcing advertising material, etc., on the callers. In order to counter these criticisms IVR is supplemented with visual menu and multimedia files, which are termed the interactive voice video response (IVVR). We have implemented a similar interactive response service on the test network, wherein the response could be in text, audio or video; hence we termed the service interactive media response (IMR). The IMR service is implemented as a service enabler on the SIP application server. IMR is deployed as a SIP application archive (SAR) on the application service. An SAR is simply a Java archive (JAR) with additional requirements for the directory structure and content. The IMR SAR is invoked by requests that are forwarded by the SCSCF. The IMR SAR initiates the request for media from the media server and sends out an early media announcement to the end-user while it is being connected to the streaming source. This enables the IMS operators to personalize the third party streaming services to the user's preferences by providing the required quality of service. It also enables the operators to generate revenue by pushing personalized advertising media to the user.

Application Plane

The SIP application server and a media server constitute the application plane. For our implementation, we have used the ubiquity SIP application server, which is a commercial carrier-grade server that uses the Jain SIP implementation of the SIP/RTP stack. The National Institute of Standards and Technology (NIST) provides a public domain implementation of the Java standard for SIP called Jain SIP (2008), which serves as a reference implementation for the standard and is compliant with RFC 3261 (Rosenberg et al., 2002). The SIP servlet API specification defines an environment for execution of network-based SIP applications. It is implemented on application servers that support SIP and optionally also HTTP and the J2EE platform. It builds on the HTTP servlet API and, like the HTTP servlet, defines an API format for application packaging. The API aims to allow applications complete control over SIP signalling while at the same time hiding much of the nonessential complexity of SIP, which is not relevant to application developers. The media control entities are the components that provide the media processing, mixing, transcoding functionalities in order to deliver highly adaptive and personalized multimedia content to end-users according to their current networking and service context. The test network is equipped with two media servers, one from Cantata Technologies and the other a Vision XML server from NMS communications. Media servers have two functional units, the controller (MRFC) and processor (MRFP), thus separating the functions of media control from media processing. SIP AS controls the media content by interacting with the media controller and the media processor, processes the media on behalf of the SIP AS, provides transcoding to the end-user and, in our implementation, also provides media mixing features as in a conference. The media server permits the SIP parameters to be configured and decides how it shall be controlled by the SIP application server. Apart from the media servers, we have an open-source transcoder FFMPEG (2008) installed on a ubuntu flavour of Linux. FFMPEG not only performs real-time transcoding of the multimedia content but it also enables transrating, i.e changes the sampling rate at which the media is encoded without changing the codec format. It offers various options to change the multimedia resolution, frame rate, bit rate, frame size, aspect ratio, etc.

4.4.3   IMS Open-Source Initiatives

This subsection aims to present a thorough investigation on open-source tools, software and projects available to facilitate IMS-based research and the study of next-generation networking technologies. There have been open-source initiatives spanning across various elements of IMS and NGN technology starting from end-user equipment to core network components, to auxiliary components such as media servers and media gateways. There are benchmarking and testing tools available too. These open-source tools facilitate proof-of-concept implementations of IMS technology and enable prototyping and innovative IMS-based services (see Figure 4.12).

4.4.3.1   IMS Architecture

Open IMS Core

The open IMS core is designed and developed by the FOKUS Fraunhoufer Insititute (FOKUS, 2006). It is an implementation of IMS call session control functions (CSCFs) and a lightweight home subscriber server (HSS), which together form the core elements of all IMS/NGN architectures as specified today within 3GPP, 3GPP2, ETSI TISPAN and the PacketCable initiative. The four components are all based upon open-source software (e.g. the SIP express router (SER) or MySQL). The open IMS core has formed the heart of the larger test environment called the open IMS playground (FOKUS, 2006).

The open IMS core is available with all the key components and interfaces as an open-source package. It permits interconnection with operator IMS test networks and performs interoperability on IMS components S-CSCFs, media gateways, SIP application servers and others. It provides a suitable environment for development of new multimedia applications, application platform extensions and facilitate research on IMS mobility, QoS and security aspects.

The FOKUS open IMS playground (FOKUS, 2006) acts as an open technology test field with the target to prototype and validate existing and emerging IMS standards and to extend the IMS appropriately on top of new access networks as well as to provide new seamless multimedia applications. All major IMS core components, i.e. x-CSCF, HSS, MG, MRF, application servers, application server simulators, service creation toolkits and demo applications, are integrated into one single environment and can be used and extended for R&D activities by academic and industrial partners. All these components can be used

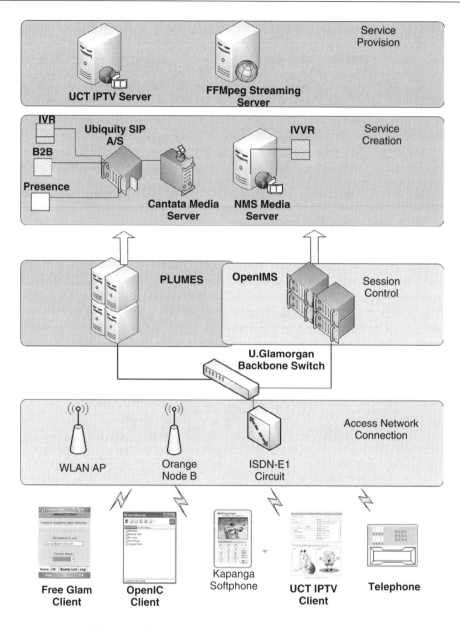

Figure 4.12 Horizontal-layered service architecture.

locally on top of all available access technologies or over IP tunnels remotely. Users of the open IMS playground can test their components by performing interoperability tests. The SIP express router (SER) (SIPS, 2007), one of the fastest existing SIP proxies, can be used as a reference implementation and to proof interoperability with other SIP components. The conformance of the SIP protocol to the standard can be tested by usage of the SIP conformance test suite. However, the major focal point of the IMS playground is put on the application server side. A variety of platforms enable rapid development of innovative services. The different platform options, each with its strengths and weaknesses, can be selected and used according to the customer needs.

Joint work with the industry has led to the development of IMS components in line with industry standards and a comprehensive interoperability framework for component testing. Network operators using the open IMS playground can perform interoperability and benchmarking tests on their components while developers creating IMS applications can gain proof-of-concept experience through the numerous available programming platforms. With the release of the open-source IMS core the possibilities for joint collaboration have escalated; via the BerliOS platform industry and academia alike, the source code for the IMS core components can be downloaded and their own localized IMS platforms created. This open-source approach increases the pool of developers involved in the project and stimulates academia and industry to explore and enhance NGN/IMS technology. The eventual interconnection of these emerging NGN testbeds would further provide skills exchange and would address the globalization of markets.

The open IMS playground includes the following components:

- The open-source IMS core (OSIMS) is the heart of the open IMS playground. It is open source and includes:
  - home subscriber server (HSS) FHoSS;
  - call session control functions (CSCFs).
- The open IMS client (OpenIC) runs on a mobile or fixed terminal. It is available in Java and .NET.
- The open IMS SIP AS (SIPSEE) is an SIP application server providing a converged SIP and HTTP service environment for service creation.
- Open service enablers (OpenSE) implement Parlay X web services, which provide a widely accepted, standardized way to access telecommunication functionality in next-generation networks (e.g. IMS) via web services.
- The OSIMS management console (OMACO) provides the means for testing, monitoring and controlling the vital IMS core network components.
- IMS evaluation and testing is used to check functionality, interoperability and conformance of IMS components.
- The XML document management server (XDMS) makes user-specific service-related information accessible to service enablers.
- The presence server provides a standard compliant presence enabler service for many applications like instant messaging, push-to-talk over cellular, video calls/conferences and others.
- The generic bootstrapping architecture (GBA) enables strong authentication and single sign-on for HTTP services offered by IMS environments and the Internet.
- The policy and charging control architecture (PoCCA) constitutes the link between the access network and the IMSCore to provide access, QoS and resources control in the IMS.
- The converged open messaging server (COMS) implements an OMA compliant service enabler and offers common reusable instant messaging capabilities.

Mobicents

While the open IMS core provides the core network capabilities and network-centric features to enable research on the IMS and NGN technologies, mobicents is the other open-source IMS architecture that provides service creation and service execution features.

Mobicents is the only open-source VoIP platform certified for JSLEE 1.0 and SIP servlets 1.1 compliance. Mobicents brings to telecom applications a robust component model and execution environment. It compliments J2EE to enable convergence of voice, video and data in next-generation intelligent applications. Mobicents fits in as a high-performance core engine for service delivery platforms (SDP) and IP multimedia subsystems (IMS). It enables the composition of service building blocks (SBBs) such as call control, billing, user provisioning, administration and presence-sensitive features. The Jain SLEE specification allows popular protocol stacks such as SIP to be plugged in as resource adapters. The SLEE service building blocks – SBBs have many similarities to EJBs. The extensible standard architecture naturally accommodates integration points with enterprise applications such as Web, CRM or SoA endpoints.

Monitoring and management of mobicents components comes out of the box via SLEE standard-based JMX and SNMP interfaces. This makes mobicents servers an easy choice for telecom operations support systems (OSS) and network management systems (NMS).

4.4.3.2 Application Servers

UCT IPTV Server (UCT, 2007; Waiting *et al.*, 2008)

The UCT IPTV server is an SIP application server that streams up to three channels to multiple destinations. The server is built on top of the eXosip library for SIP signalling and the gstreamer library for media delivery, and is released free on the Internet under the GPLv3 licence. The server is compatible with any client that is capable of receiving and decoding the H.263 1998 video standard (Bormann *et al.*, 1998) and MPEG1 audio standard. The server should not be confused with a video-on-demand server, which serves a different media stream to each client. Its primary goal is to serve a limited number of packet-based media streams to as many clients as possible, similar to regular digital terrestrial and satellite television broadcasts. The server is designed to fit tightly within the IMS architecture; therefore, unlike other IPTV solutions it uses SIP exclusively for signalling. The benefit of this design is that the preexisting IMS QoS, service provisioning and charging mechanisms can all be reused for this service, hence leveraging the service delivery platform that IMS provides. Furthermore, this reduces the complexity of client software as the existing SIP stack can be reused. The IPTV server is provisioned in the HSS as an application server. This enables the network operator to add and remove users from this service by adjusting their initial filter criteria. Users that have the IPTV service enabled in their user profile are able to join a channel by submitting an 'Invite' request to the application server that includes their preferred media IP address and port numbers for both the audio and video components of the video stream.

OpenSIPs – Open SIP Server (SIPS, 2007)

This is a mature open-source implementation of an SIP server. OpenSIPS is more than an SIP proxy/router as it includes application-level functionalities. OpenSIPS, as an SIP server, is the core component of any SIP-based VoIP solution. With a very flexible and customizable routing engine, OpenSIPS unifies voice, video, IM and presence services in a highly efficient way, thanks to its scalable (modular) design. The OpenSIPS project is a continuation of the OpenSER project.

Scenarios where OpenSIPS fits are:

- VoIP service providers (residential);
- SIP trunking;
- SIP load-balancing;
- SIP front end (for SIP termination);
- white-label solutions;
- enterprise services;
- SIP router (LCR for multi-GWs).

OpenSIPS can be used as:

- SIP registrar server;
- SIP router/proxy (LCR, dynamic routeing, dialplan features);
- SIP redirect server;
- SIP presence agent;
- SIP IM server (chat and end-to-end IM);
- SIP to SMS gateway (bidirectional);
- SIP to XMPP gateway for presence and IM (bidirectional);
- SIP load-balancer or dispatcher;
- SIP front end for gateways/Asterisk;
- SIP NAT traversal unit;
- SIP application server.

SailFin   (SailFin, 2008)

Project SailFin is based on robust and scalable SIP servlets technology on top of a deployment quality, Java EE-based GlassFish. We are working towards achieving JSR 289 (2008b) compatibility, adding high availability and clustering features, and integrating with existing GlassFish services.

JSR 289 updates the SIP Servlets API and defines a standard application programming model to mix SIP Servlets and Java EE components. Java EE services, such as web services, persistence, security and transactions, enable faster development of smarter communications-enabled applications.

Other open-source SIP application servers are Asterisk, MysipSwitch, FreeSwitch and SipX.

Asterisk   (Asterisk, 2008)

Asterisk is an open-source software implementation of a private branch exchange (PBX) that runs on OpenBSD, FreeBSD, Mac OS X and Sun Solaris. It supports voice over IP protocols, such as SIP, H.323, IAX (inter-Asterisk eXchange) and MGCP (media gateway control protocol). Thus, Asterisk can interoperate with many SIP telephones and

Asterisk PBXs. Asterisk contains many features including voice mail, conference calling, interactive voice response and automatic call distribution. All these made Asterisk a very popular software implementation of PBX. The current release version is Asterisk 1.4.1 as of March 2007.

4.4.3.3 Auxillary Components

UCT Policy Control Framework   (UCT, 2007; Waiting *et al.*, 2008)

Policy frameworks allow a network operator to have strategic control over every aspect of the network architecture, ranging from resource reservation to end-user experience, using pre-defined policies. While policy control already exists at various layers in existing networks the policies are generally static and have little interaction with applications. As networks converge and the number of services increase and they become more personalized, end-to-end policy coordination will become critical. Standards bodies have recognized the benefits of such a management framework to control resources and admissions in the transport layer; most notably the 3GPP have defined the policy control and charging architecture (PCC) and ETSI TISPAN have defined the resource and admission control subsystem. These architectures specify only logical architectures and not physical implementations; furthermore, the systems are highly complex and will require significant investment to deploy in a commercial environment. Operators will find it difficult to commit the necessary capital unless the systems have been tried and tested. The UCT policy control framework is a 3GPP-compliant, open-source and freely distributed architecture specifically designed for QoS policy control in a single IMS domain. The framework works in coordination with the FOKUS open-source IMS core. It is largely based on the 3GPP PCC specification and defines a policy decision function (PDF), policy enforcement point (PEP), policy repository and web management interface. The framework provides a flexible environment for the easy creation and execution of policies encouraging further research and innovation in the field.

UCT PEP   (UCT, 2007; Waiting *et al.*, 2008)

This is a subelement of the policy and charging enforcement point that enforces dynamic policy rules in the transport layer and monitors flow usage. The element is Java based and implements the diameter Gx interface to the PDF. This interface is used to receive policy rules and to send transport-layer events to the PDF. Upon receipt of the policy rules the UCT PEP translates the policy to transport-layer specific configuration information. Linux traffic control is used to create a DiffServ router with several different service classes; this router acts as a gateway allowing only authorized IP flows, and depending on the definition of the policy rule the packets for each IP flow are marked and queued accordingly.

SEMS   (SEMS, 2008)

SEMS is a media and application server for SIP-based VoIP services. It shows good performance doing basic services like announcements and conference for combination with external application servers, and, thanks to its easy-to-use and flexible application development framework and back-to-back user agent support, application logic and media serving can be combined in the same process. It can provide a full VoIP system together with the SIP express router (SER). Basic applications like announcement, pre-call announcement, RBT, conference, voicemail, mailbox, etc., and lots of example applications are available. Scripting is in Python and a simple state machine description language. Support is given for all commonly used free codecs (g711, gsm, iLBCi, speex, adpcm, 116, etc.), wideband, ZRTP encryption, SIP registrar client, XMLRPC server/client, DIAMETERi client and more.

Open-Source OSS   (OSS, 2008)

Members of the Telemanagement Forum have realized that while the management of next-generation networks and services poses significant technical challenges, the present supply chain, market configuration and business practices of the OSS community are an obstacle to rapid innovation and adoption of NGOSS. The present processes undertaken to test, develop and realize Telemanagement Forum standards utilize only a small portion of the Telemanagement Forum community, which can cause lengthy lead times. Moreover, the validation, testing and development of Telemanagement Forum standards by implementation are primarily performed within short-term catalyst projects that discontinue upon completion or are performed internally within companies where the results are not widely visible to the community. Forums for open development could potentially provide a medium to shorten this supply chain for the deployment of workable systems. By delivering an open OSS testbed, this project hopes to provide a new and complementary route to expedite the testing, development and realization of Telemanagement Forum standards. The software in the testbed will be made permanently available for use by vendors, integrators and service providers, and software will evolve using an open-source approach under a licence or accreditation scheme to be agreed with the Telemanagement Forum technical council. It is not the intention to create a set of open-source OSS at carrier grade standard, but to provide Telemanagement Forum members with free access to a permanent and evolvable set of NGOSS reference software that can be extended as the industry chooses. It is envisaged that this open-source approach in the TMF context will help:

- validate and provide feedback on standards through development and testing of open OSS software based on open-source or free software components;
- expedite realization of NGOSS by service providers through development and sharing of NGOSS-compliant adaptors to commercial and open-source OSS products;
- promote standards development by providing a reusable pool of zero-cost, working NGOSS reference software, infrastructure and tools that can be used in further development projects.

4.4.3.4   IMS Client Software

1. *The IMS Communicator* (IMS-Communicator, 2004) is an open-source IMS client emulater built on the Jain SIP reference implementation (RI) and the Java media framework API by the researchers at PT Inovacao Portugal. The IMS communicator extends the SIP communicator project with new IMS-specific features, such as the implementation of the AKAv1 authentication algorithm, use of the IMS public user

identity (IMPI) during registration, subscription to the registration event package, support for the precondition mechanism and early media capabilities. In order to implement these features the project contributed to the extension of the Jain SIP RI that previously did not support the 3GPP extensions to SIP.

2. *UCT IMS Client* (UCT, 2007; Waiting *et al.*, 2008)

This was the first IMS client released to the open source community. The aim of the project is to provide true IMS signalling, proof-of-concept implementation of several rich services and a mechanism with which to test other IMS network components. The aim of the project is not to create a stable consumer grade product but rather a platform from which new IMS enablers can be implemented, evaluated and refined. The implementation of the client is achieved by the use of several free open-source libraries: the oSIP and eXosip libraries for SIP signalling; the gstreamer framework for media coding, decoding and transport; the libcurl library for HTTP support; the libxml library for XML parsing; and the GTK library for the graphical user interface. The client itself is released under the GNU Public License Version 3 (GPLv3), which allows users the freedom to modify and redistribute the software for their own purposes. This is particularly useful for research projects that aim to produce innovative services that require modifications to the client software, some of which are discussed in this chapter. We now evaluate the features of the UCT IMS client that have allowed for the evaluation of several interesting performance metrics.

3. UCT IMS client is designed to be used in conjunction with the Fraunhofer FOKUS open IMS core. The client has been developed by the Communications Research Group at the University of Cape Town, South Africa. At present the client is still in active development and there are several known bugs.

The client supports AKA authentication and emulates IMS signalling as far as possible. The current version supports voice and video calls (numerous codecs), pager mode and session-based instant messaging, presence, an IPTV viewer and an XCAP client.

There are many open-source implementations of the SIP stack, out of which PJSIP is a prominent one. Other resources to the developer, such as testing tools, are available in the open-source community.

SIPP   (SIPP, n.d.)

SIPP is the open-source test tool used for performance testing of the SIP protocol. SIPP is one of the best tools to use for the performance testing because it can establish and release multiple calls with the INVITE and BYE methods. Besides, it can also read XML scenario files describing any performance testing configuration. It features the dynamic display of statistics about running tests, periodic CSV statistics dumps, TCP and UDP over multiple sockets or multiplexed with retransmission management, regular expressions and variables in scenario files, and dynamically adjustable call rates.

SIPP can be used to test many real SIP equipments like SIP proxies, B2BUAs, SIP media servers, SIP/x gateways, SIP PBX, etc. It is also very useful in emulating thousands of user agents calling the SIP system.

FFMPEG   (FFMPEG, 2008)

This is a computer program that can record, convert and stream digital audio and video in numerous formats. FFMPEG is a command line tool that is composed of a collection of free software/open-source libraries. It includes libavcodec, an audio/video codec library used by several other projects, and libavformat, an audio/video container mux and demux library.

FFMPEG can also grab from a live audio/video source. The command line interface is designed to be intuitive, in the sense that FFMPEG tries to figure out all parameters that can possibly be derived automatically. You usually only have to specify the target bit rate you want. It can also convert from any sample rate to any other and resize video on-the-fly with a high quality polyphase filter.

Part III

Convergence – The Road Ahead

5

WIMS 2.0: Convergence of Telcos with Web 2.0 Facilitated by IMS

The telecom world has changed dramatically in the last few years, especially mobile communications. The use of intelligent networks has brought a whole new set of ground-breaking services. However, the developers needed to master the infrastructure of the operator in order to set up these services, which limited the possibility of opening the market to other service providers. Any operator's portfolio comprises a long list of services, but only 5% of them are used by 90% of the customers, showing that customer's demands are not aligned with operator's services. The user, who is not aware of the network technology, is not satisfied with the applications that are available. In order to overcome these limitations, the mobile world has introduced a new control architecture based on IP technology: IMS. IMS, as described in this book, is much more than a mere architecture; it enables new ways of fulfilling the user's needs, splitting up the market, reducing the time-to-market and increasing the creativity for producing new services. At the same time, we have seen the arrival of new, highly social services in the Internet world, the so-called Web 2.0 (O'Reilly, 2005) services. Now, the tendency is to offer these Internet services (Internet-based services (IBS)) in the mobile world and this causes a conflict with IMS applications, as IBS is already under exploitation, whereas IMS is still a promising future rather than a reality. IMS and IBS seem to be similar in concept, but they differ in relevant points, such as the business models, the need to worry about the underlying network (e.g. QoS), the strict control over the terminal or other O&M issues. Basically, IMS is still a 'walled garden' while IBS is an 'open garden'. Having spent a good deal of resources in the setup and imminent launch of the IMS framework, the solution for the operators should consist in trying to converge these two worlds and make them coexist, taking the best characteristics of each in order to provide the customer with much better services. This is exactly the main objective of the WIMS 2.0 initiative, which pursues the analysis and achievement of the convergence between Web 2.0 and IMS, both from a strategic and technological point of view. As the

IMS: A Development and Deployment Perspective   Khalid Al-Begain, Chitra Balakrishna, Luis Angel Galindo and David Moro

main result, the initiative aims to design a new service architecture for the provision of innovative IMS and Web 2.0 converged services: the WIMS 2.0 service platform.

5.1 Impact of the Web 2.0 Disruption on the IMS and Telecom Evolution

5.1.1 Web 2.0 Phenomenon

After the 'dot-com crash' in 2001, the forecasts predicted a pessimistic future for the evolution of the web and its services. However, during the last few years, many new services (such as SalesForce, Twitter, etc.) and social network tools (such as LinkedIn, Facebook, etc.) have appeared, obtaining a great acceptance and success among final users: it is the Web 2.0 revolution (see Figure 5.1).

The key point behind the Web 2.0 revolution is the change in the philosophy for designing and developing services. Although currently there are a great variety of services, the main guidelines of this new philosophy can be summarized in two core concepts:

- The user is the centre; the main objective is to give users what they want. The user is now regarded as the main active driver of the service. The users create the service content, they can customize service features, they effectively affect the service evolution and they can even participate in service development, directly constructing modules and applications in order to fulfil their own needs.

Figure 5.1 The Web 2.0 map.

- Combination and flexibility. The worldwide adoption of Internet and IP connection capabilities, along with the appropriate use of remote procedure execution schemes (i.e. open web APIs), enables the mixture of service functionalities (mash-ups) and content (syndication). Thus, the Internet becomes the platform for developing and delivering new cost-effective services, virtually to any part of the world.

Due to the power of these two concepts, it can be assumed that the Web 2.0 philosophy is not just a trend, but the way for creating innovative and successful services, and it may also be applied to a variety of industries. For instance, in the enterprise arena, there is an increasing need for a change in the way to organize, innovate and create value within companies. Sharing and collaboration aspects of Web 2.0 are regarded as an opportunity to increase companies' revenue and reduce costs. A new type of organization, the Enterprise 2.0 (McAfee, 2006) is emerging, where Web 2.0 concepts and software are applied within the enterprise business model.

5.1.2   IMS and Telco 2.0

In contrast with the 2.0 revolution in the IT field, during the last few years telecom operators are having problems evolving and growing in mature markets. The main revenues still come from voice and the provision of broadband IP access, thus differentiating the offer is difficult and the only solution is to compete in price. In order to tackle these problems and provide new approaches, a new trend called Telco 2.0 (2006, 2009) was born, almost simultaneously to the Web 2.0 appearance. For the near future, this initiative assumes the existence of all-IP networks, where the service and the connectivity are separated. In this world, traditional basic services are offered without cost and users pay for innovative services satisfying their necessities. Telco 2.0 is not a technology; it is rather a new way of thinking and reformulating the telecom business models. After this 'exercise', operators should move from a network-centric to a user-centric approach, in which users 'take what they want' instead of 'what they are given' (open garden versus walled garden). It can be seen that these concepts are very similar to the Web 2.0 philosophy. In order to adapt to the new situation, operators first need to identify their strengths and weaknesses and select the role they can better perform. Telco 2.0 tends to think operators are not the best players for final service provisioning and proposes them to adopt the role of 'enabling platform providers for third party services'. The WIMS 2.0 initiative focuses on this role (which perfectly fits with the convergence with the Web 2.0 world) and champions IMS as the suitable platform upon which Telco 2.0 ideas can be realized. The sum of IMS plus the set of standardized service enablers, along with the use of appropriate SDP/SOA schemes for service composition, constitutes a platform that can be satisfactorily used to construct services horizontally according to the new philosophy, also by third parties. In fact, one of the most attractive usages of IMS is the possibility to integrate Telco capabilities within the back-end processes running in a corporation, creating SVA adapted to the client's necessities and emphasizing the efficiency in mobility.

5.1.3   Convergence Opportunity: WIMS 2.0

As previously mentioned, in the last few years there have been gradual changes, both happening at the technology side as well as the business side, that prompts a foreseen evolution in the mid-term of the telecoms market. In this scenario, a need to scout new ways of evolution for traditional operators arises. On the one side, the path towards all-IP networks and fixed mobile convergence allows services to be decoupled from the network and to adopt bearer-independent horizontal architectures. Together with the value loss of traditional telecom services like voice, this indicates the need to adopt a new business model rather balanced towards user-centricness, with users able to elect, configure and personalize services, and with a stronger position of services provided by specialized third parties. On the other side, the world of services offered in the web ecosystem has experienced a very relevant development with Web 2.0. Thus, a myriad of new services, appealing to the user since they (the users) are the catalyst factor of service offering and service evolution. In the 2.0 era, the user is the guest star. Additionally, the flexibility to develop and mash up new services in Web 2.0 and user participation in the content creation and population for those services are consolidating the idea that Web 2.0 represents the successful manner to develop and exploit services in the near future. Beyond that, considering that Web 2.0 holds an increasing penetration in mobile and telecom devices, it turns out that it might pose a menace for traditional telecom operators. Web 2.0 philosophy is in fact a new strategy that the Telecom 2.0 initiatives are adopting as a service creation guideline. In fact, the 2.0 trend clearly indicates to an operator that if aspirations are beyond being a mere connectivity provider, operators will enter the new ecosystem of service generation to look for convergence between the web and the telecom services. Thus, the operator can adequately hold a position in the new arena by leveraging its existing and unique assets: telecommunication services and infrastructures. Only by following these strategy might operators be able to retain a relevant role in the service delivery value chain as a service provider and/or a service capability provider. If not providing end-user services, at least the operator can be well positioned by evolving into the desired platform for service co-creation by integrating interesting telecom service features into final user applications. Within the scope of telecommunication networks, IMS is the suitable architecture to link the connectivity plane with the generation of new services and as a consequence to evolve towards convergence with the web. The rationale behind this is as follows:

- IMS is the central system that, irrelevantly of the access method, will control the delivery of the new wave of services.
- IMS is the multimedia in nature from its birth.
- IMS is completely based upon technology coming from the Internet (IP technology). Technology compatibility with Web 2.0 is guaranteed.
- Functional interworking with the Internet was considered from the initial design of the IMS.
- The IMS basic elements can be completed and enriched with a set of horizontal service capabilities – the service enablers – especially designed to ease service creation for in-house applications as well as by exposure of enablers to external agents, like for instance, Web 2.0 applications.

Taking into account all of these reasons, together with the current market momentum, the WIMS 2.0 initiative arises to mix the best of two worlds, the telecom and the web world, by utilizing the best technologies, services and business models available from both sides.

5.1.4   Mission and Vision of the WIMS 2.0 Initiative

WIMS 2.0 (Web 2.0 and IMS) (WIMS 2.0, 2008) is an initiative promoted by Telefonica and its R&D branch and founded by Luis Angel Galindo and David Moro. As an open initiative, other players in the ICT industry participate. WIMS 2.0 seeks convergence between Web 2.0 and the telecom new generation services based on IMS (IP multimedia subsystem), to create innovative, appealing and user-centric services and applications. These services will combine features from both Web and telecom worlds. On the one hand, relevant features like interactivity, ubiquity, social orientation, user participation and content generation will be adopted from Web 2.0 services. On the other hand, IMS will enhance WIMS 2.0 service features with a collection of capabilities like multimedia telephony, media sharing, push-to-talk, presence and context, online address book, etc., all of them being applicable to mobile, fixed or convergent telecom networks. In addition, in order to provide more versatile and enriched services and a short-term implementation, WIMS 2.0 also considers the utilization of pre-IMS telco capabilities like SMS/MMS messaging, circuit-switched video and voice calls, networked address books, etc. WIMS 2.0 comprises the strategies, technologies and service platforms that will allow telecom operators to achieve this convergence with innovative Web 2.0 services in all-IP networks. Following the Web 2.0 revolution and the Telco 2.0 initiative, a change in the telco traditional business model is recommended following the adoption of a user-centric and open-garden philosophy consisting of the enhancement of the openness, flexibility and freedom in the fashion services that are provided. IMS, along with its associated set of service enablers, has been identified as the base platform to support this convergence and business model change. WIMS 2.0 follows a two-sided approach; on the one side, it continues to offer IMS capabilities to the Web 2.0 community, so they can be included into web mash-ups. On the other side, it continues to exploit Web 2.0 services and technologies to enrich operator's services. These are the foundations of the WIMS 2.0 strategies geared to achieving convergence, which are explained in the following sections.

5.2   WIMS 2.0: The Service Focus

The two main goals of the WIMS 2.0 initiative (WIMS 2.0, 2008a) are to create innovative services mixing Web 2.0 and Telco World and to create a new fair ecosystem: the WIMS 2.0 ecosystem (WIMS 2.0, 2008b). The technical strategies for achieving the former and the principles that set the foundation of the latter are presented in this section.

5.2.1   The WIMS 2.0 Strategies

Given the aforementioned situation, Web 2.0 is now seen with growing expectations from different telecom players. It is the target field for the evolution of IP-based services (communication, content distribution, etc.) and telecom operators must be ready to adapt to the new market environment. Although changes always represent a risk, they also represent an opportunity if they are well managed, which is the main technical objective of the WIMS 2.0 (Web 2.0 and IMS) initiative: 'To establish the strategic and technical principles for an appropriate convergence between the Web 2.0 and the telecom networks by means of the IP multimedia subsystem. The outcome of this work will be a service architecture that, placed on top of IMS and its enablers, makes the desired convergence possible.'

The first step consisted of clearly defining the guidelines in order to achieve the convergence. The exposure of IMS capabilities to the Web 2.0 community seems to be a very useful approach; however, the WIMS 2.0 initiative also encourages operators to explore how to use Web 2.0 to enrich their own services. Based on these considerations, a two-sided approach, with different convergence guidelines, has been followed, as explained in the next subsections. It is also important to highlight that the WIMS 2.0 initiative actively seeks to demonstrate the presented ideas within a real market environment in the short term. Although some of the proposed convergence guidelines are identified as being fully applicable for the mid or long term, we assimilate the Web 2.0 idea of testing our concepts while they are in a beta phase. This provides an important feedback to elaborate further a better solution and represents a mechanism for an early introduction of the new services in the real market, as will be later explained.

5.2.1.1   Offering IMS Capabilities to the Web 2.0 Community

In alignment with Telco 2.0, the main objective of this convergence approach is to offer IMS capabilities from the telecom network to the Web 2.0 community. The final service is actually provided by a third party located in the Web 2.0 world, but the operator offers an added value. Two general guidelines are considered for this approach: incorporation of IMS capabilities into Web 2.0 services through open Web APIs, allowing the integration of IMS into Web 2.0 mash-ups. This potentially applies to any IMS capability, which would therefore be usable by/from any Web 2.0 service.

Two different, but interrelated, strategies are proposed here:

- WIMS 2.0 strategic line 1.1: mash-ups based on widgets or PSEs (portable service elements). IMS applications can be incorporated into Web 2.0 services in the form of web-widgets, which would be the technology-specific implementation of a PSE. The PSEs provided by the operator (or a third party) can be easily integrated by end-users into Web 2.0 sites and, once incorporated, they are able to handle the interaction with the operator's open web APIs, thus enabling the use of IMS communication capabilities from Web 2.0 sites. Obviously, this strategy is only valid for those web sites that allow the inclusion of web widgets, but there are already several relevant possibilities, such as iGoogle, Facebook, Blogger, etc. Specific application examples are: an instant messaging gadget for iGoogle, a call-me button for Blogger or a more sophisticated Facebook application providing voice and presence services for the members of a group. It can be seen that, in the short term, this strategy allows IMS to spread into and across the web, as if it were a viral infection.
- WIMS 2.0 strategic line 1.2: mash-ups based on APIs with direct incorporation into Web 2.0 mash-ups. Once IMS capabilities are exposed through open web APIs, they can be directly incorporated into any Web 2.0 service in order to provide complementary functionalities, just like any other regular mash-up. The final outcome is a complete integration of telecom capabilities in the resulting service. Due to the fact that the Web 2.0 service operation must be modified to consider the use of IMS APIs, the application of this strategy is not immediate, but for the mid and long term it enables a convergence of great impact and further applicability. A specific service

example could be 'real-time sharing of YouTube videos to mobile users' while maintaining an instant messaging session among the involved users. Further examples are left to the imagination of the reader.

Publication is of user-generated content enabled by IMS. Mobile handsets are expected to be one of the main sources of multimedia content in the future, which leads to the next strategic line:

- WIMS 2.0 strategic line 1.3: IMS-enabled user-generated content publication and generation. Starting from the short term, this convergence guideline aims to obtain new ways for content publication through the usage of IMS multimedia transmission capabilities. Operators can provide new solutions to receive the multimedia content from the user and automatically upload it to Web 2.0 sites. An example application could be 'YouTube real-time video generation', where users videocall YouTube to create a spontaneous clip. Other examples may explore the use of other types of user-generated data, such as IMS presence or other user's context information.

5.2.1.2   Exploiting Web 2.0 for Telecom Services

In this approach, the operator keeps the role of the final service provider. Accordingly, two different guidelines are considered for the enrichment of telecom services:

- WIMS 2.0 strategic line 2.1: inclusion of Web 2.0 contents and events in the operator services. Nowadays Web 2.0 services hold the successful content (the user-generated content). The WIMS 2.0 initiative champions new mechanisms and functionalities for obtaining generic multimedia content and events from Web 2.0. Among the content, we may find videos, advertisements, podcasts, news, etc. Among the events, we may find contextual information associated with social networks, new content publication alerts, etc. The introduction, integration and distribution of all this information within IMS services is of great interest, since it can enhance end-user's satisfaction and is applicable for the short term. As example services, we may consider special feed readers for a mobile handset, the inclusion of social networks events into IMS presence information or news, video and music distribution.
- WIMS 2.0 strategic line 2.2: IMS online applications (virtual terminal). Through the use of AJAX and other technologies, Web 2.0 applications have achieved major advances in the field of user interfaces. WIMS 2.0 favours the usage of these technologies in order to build IMS online applications. The benefits of such kinds of IMS applications are very important: ubiquity and a great simplification of service development and deployment, since service logic resides on the network and not on the terminal.

5.2.2   WIMS 2.0 Ecosystem: Creating the WIMS–WIMS Relationship

The history of mobile telecommunications is the history of a continuous competition. While in the fixed area, the situation is a little bit different due to the high costs of deploying end-to-end infrastructure, which caused a great entry barrier for new entrants; at the end of the day the situation in the mobile arena seems to result in the same situation.

For years, fixed operators have been competing in price. With a common product and service portfolio, and with not so many innovations, fixed operators are offering ADSL/VDSL/FTTH access to the Internet and, in the best of situations, they have an offer of IPTV services or they are trying to position in the home automation services area. With this situation, it is usual to see how fixed operators are increasing speed access in broadband offers or reducing the price day by day. This competitive situation has driven fixed operators to be considered by users as bit pipes, deleting from the client's mind any idea to be considered as service providers. The situation in the mobile arena is slightly different. From the beginning, the cost of deployment in a mobile network is much more reduced, in that way increasing the competence. Evolution in radio aspects is a constant in mobiles that allows increasing speeds, quality and user experience. These changes in technology at the network side have been ported to the user side in the form of changes in trends, turning mobile phones into fashion objects, which are renewed every one and a half years.

However, it is quite curious to see that, although mobile operators have a huge service portfolio compared with fixed, there is no differentiation between them. The service catalogue in any operator is more than 200 services, but only 5% of them are used by 90% of the clients, showing that client's demands are not aligned with operator's services. The user, who is not aware of the network technology, is not satisfied with the applications that are available. Perhaps this situation is due to the limitations of the technologies that were more focused in voice and related services based on IN, instead of data services, where it is easier to create innovative services more focused on the client's needs; perhaps it is due to the fact that operators are far away from client's needs and push services on to the market; or perhaps it is due to a high time-to-market, which causes lateness. However, in the end, the current situation is that mobile operators are, as with fixed, competing in price. It seems that nothing has been learnt from the old telecommunications.

The current telco ecosystem is failing. Only a brief look at the figures of most of the companies warns that something is wrong in this ecosystem. As the economical crisis is based on trust, here in the telco arena the word 'trust' is the key. Although an outer vision of the mobile landscape could conclude that relationships between operators are trust relationships, the reality is the opposite. Examples such as optimal routeing, mobile data roaming or IMS roaming demonstrate this situation. Standardization efforts to provide an optimal routeing solution were done, but in the end, more than solving the technical problems, what was needed to be solved, the trust between operators, remained under discussion in operator associations such as the GSMA and other lobby organizations. The distrust between operators had and has the consequence that, for example, a voice call originated by a local subscriber to a roamer camping in the same network as the originator will go to the home network of the roamer, wasting resources unnecessarily. A similar situation occurs when a roamer is accessing the Internet using only the SGSN of the visited network and the GGSN and the connection to the Internet of the home network; again in this situation, the 3GPP defined the mobile data roaming using the visited SGSN and GGSN, but because of distrust it is not being used. In the case of IMS, the 3GPP defined the situation of a roamer using the visited P-CSCF, but it is unlikely that access to the IMS services would be in that way. WIMS 2.0 believes in trust between all the actors as the main pillar to build win–win relationships: the WIMS–WIMS relationships. Only with trust between each other and with a fair ecosystem, where all the actors are looking for stable relationships considering not only its interests but also the interests of the rest of the participants, will we be able to create a lasting commercial environment that lays the foundations for a continuous ideas generation and innovation based on co-creation. In the WIMS 2.0 ecosystem, each of the actors provides value based on its expertise, trusting in the other participants, but with co-opetition, with the goal to create sustainable relationships, the WIMS–WIMS relationships between the different actors:

- Operators: offering telco services to the client, to the ISPs, importing Internet services and integrating in the telco services, offering them to the clients.
- ASPs/ISPs: offering Internet services to its user/clients, importing telco services and integrating in the Internet services, offering them to the clients.
- NEPs: supplying the required infrastructure to tackle these services and the underlying telecommunications networks.

5.2.3 WIMS 2.0 Services and Business Models

The general scenario of the services exposed here considers a short/medium-term time-frame. These services also consider a convergent operator offering to the Web 2.0 world a set of tools to create innovative services. These versatile tools are based on different strategies as exposure of Telco/IMS capabilities through APIs, the use of widget platforms offering the user the possibility to embed Telco/IMS functionalities in external webs, new ways for publishing and distribution of contents between Web 2.0 and the operator's network, etc. For each of the WIMS 2.0 strategies previously exposed, a possible service is proposed, showing the capabilities of the WIMS 2.0 architecture in order to create services.

5.2.3.1 Services for Strategic Line 1.1: Mashups Based on PSE

PSEs of Personal Communications for Blogs from Blogger: 'Call-Me Button' and 'IM Conversation'

The operator, in this case Telefonica, is a communication PSE provider, offering a catalogue of them for integrating into like blogs as blogs from Blogger (1999) (see Figure 5.2). Thus, a blog's owner configures this service easily using the operator web page, choosing the PSE and the blog where it will be integrated. The PSE logic lies in the operator's platform. The operator automates the process, which avoids any technical skills from the user. Visitors can contact the blog's owner automatically and anonymously using a voice call or IM. The ASP, in this case Google, provides the platform, Blogger, to include PSEs from third parties. For this service, a commercial agreement is not needed between actors.

How Does It Work for the User?

Telefonica and Blogger users can enhance their blog by including widgets in it. Browsing in the operator's web page containing the widgets, the blog's owner chooses the favourite PSE and follows the following procedure:

- Log on to the customer channel online client channel.
- Introduce the user credentials for Google.

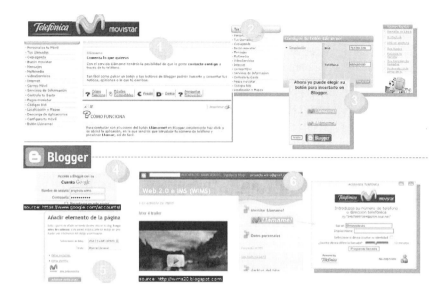

Figure 5.2 Telefonica PSEs for Blogger blogs.

- Choose the blog from Blogger, in case there is more than one.
- Personalize where and how it will be shown.

Telefonica offers the user the following PSEs:

- IM conversation. Visitors of the blog's owner can chat with the user from the blog web page on either mobile or PC.
- Call-me button. Visitors of the blog's owner can establish a call to a fixed/mobile number configured by the user when the widget is incrusted. The user's number is not disclosed to the visitors.

PSEs for FaceBook: Click-to-Multiconf and Localizanos

The operator, in this case Telefonica, is a communication PSE provider, offering a catalogue of them for integrating in Facebook (2004) user personal space. The PSE logic lies in the operator's platform, although access to Facebook info could be required in order to get data from the users. The ASP, in this case Facebook, integrates the external applications in the personal space of the users and also offers an applications directory where the available PSEs are listed.

For this service, a commercial agreement is not needed between actors, except for specific information from the user profile, such as the phone number.

How Does It Work for the User?

Users can personalize Facebook personal space with the Facebook applications of the operator in the following two ways:

Figure 5.3 Telefonica PSE for a multiconference in Facebook groups.

1. From Facebook, by searching in the applications directory of Telefonica in Facebook and following the instructions to include them in the personal space.
2. From the Movistar/Telefonica online client channel, Telefonica offers the user the following PSEs:
   - Click-to-Multiconf. This PSE enables users to establish a multiconference with all the users from a Facebook group using their fixed/mobile phones (see Figure 5.3). After installing the application in Facebook, the user can click on the multiconference application placed on the Facebook user main page. This application will show the groups which the user has joined and enables the user to click on any of them in order to establish a conference with all the members. The application calls to the phone number (fixed or mobile) stored in the Facebook user profile shows the conference just established and the users who joined. For those groups with a huge number of users, the application asks for a choice of members. Also, for those members who are not included by phone number in the Facebook profile, the application will ask the user to introduce the number manually, in case the user knows it.
   - Localizanos. This PSE enables the user to locate on a map from Google Maps all the members of a Facebook group (see Figure 5.4). In order to determine an accurate geographical position, mobile phone positioning is used, which requires Facebook group members to have introduced their numbers previously. In case these numbers were not in the profile, the application asks members who have not been located to introduce their numbers for introducing manually. Moreover, this application also shows on the map the presence status (available, absent, etc.) and the personal message configured by the member of the group in the mobile phone or in other devices. For just those members located on the map, the user can click on the individual icons representing the members to send a message (SMS, MMS, email) or to establish a call between the user's and member's phones.

Figure 5.4   Telefonica PSE for localizanos in Facebook.

5.2.3.2   Services for Strategic Line 1.2: Mashups Based on APIs

YouTube Real-Time Videosharing

The service provider, in this case YouTube (2005), offers this functionality as an added value to its video visualization services. The operator, in this case Telefonica, offers several capabilities that allows YouTube to invite IMS users to visualize videos under YouTube user demand (see Figure 5.5). This service requires a commercial agreement between actors. YouTube has to modify the current service to include this functionality. How does it work for the user? A user desires to visualize at the same time, in real-time, a video stored in YouTube with a friend. The user will see the video in YouTube while the friend of the user will see it, at the same time, on his/her mobile phone. This requires a new-generation multimedia mobile phone enable to receive invitations for real-time video. The steps to follow are:

1. Choose the video to visualize in YouTube.
2. Click on the 'Visualizacin compartida con un usuario mvil' icon.
3. Indicate the phone number of the person to be invited.
4. The user's friend will receive an invitation from YouTube with the user's identifica-tion to a shared visualization of the selected video.
5. Both user and friend could interact chatting using IM or rate and introduce comments.

5.2.3.3   Services for Strategic Line 1.3: User-Generated Content

Real-Time Video Generation

In this service, the operator, Telefonica, provides a new publishing functionality of user-generated contents in YouTube or other video websites accepting this kind of content upload-ing (see Figure 5.6). The service provider, in this case YouTube, offers this functionality as an added value to its video visualization services. A commercial agreement is not needed

Figure 5.5 YouTube real-time videosharing, powered by Telefonica.

Figure 5.6 Real-time video generation for YouTube, powered by Telefonica.

between actors as long as video providers offer interfaces or nonmanual mechanisms to publish the contents in the website.

How Does It Work for the User?

Thanks to this service, the user can capture a video in his/her mobile phone and publish it on YouTube without storing it or sending an MMS. The user has to activate the camera in the phone, dial a short number or a URL predefined by the operator (e.g. youtube@movistar.es) and call. All the video captured by the phone will be uploaded to a video in the YouTube's user account. When the call is finished, the service will provide simple instructions to rename the video, add tags, etc. The service requires mobile phones with videosharing or videocall.

Geo-referenced Publishing with Enriched Information

In this service, the operator, Telefonica, provides a new publishing functionality of user-generated contents in photo websites accepting this kind of content uploading. The service provider, only receives the user's photos in the usual way. A commercial agreement is not needed between actors, but if it exists user experience can be enhanced.

How Does It Work for the User?

This service enables the user to upload photos inmediately from the user's mobile phone to a photo website, adding automatically the geo-position of the user, obtained from the network, and other information provided by the user as comments or tags. This photo is included in an IM, where the user should indicate in the subject the album where the photo should be stored. The IM have to be sent a predefined email address. This service is available for all mobile handsets with IM, mobile email or MMS capabilities.

Automatic Publishing of Personal Icon and Message in Twitter

In this service, the operator, Telefonica, provides a new publishing functionality of user-generated contents in Twitter (2006) or other microblogging websites accepting this kind of content uploading (see Figure 5.7). The service provider, in this case Twitter, only receives the user's contents in the usual way. A commercial agreement is not needed between actors

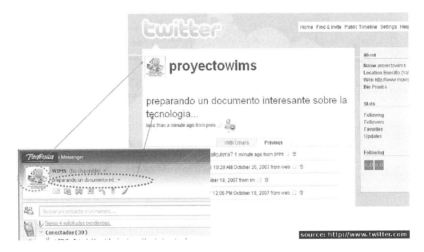

Figure 5.7 Automatic publishing of personal icon and message in Twitter, powered by Telefonica.

as long as providers offer interfaces or non-manual mechanisms to publish the contents on the website.

How Does It Work for the User?

This service enables the user to update the user's Twitter each time that the presence personal message is modified. Also, if the user desires, the presence status (available, not connected) during publishing can be shown. One of the advantages of this service is to write at one time and publish the information about what the user is doing on several websites. The service also enables the icon at the Telefonica's Communicator to be synchronized with the image shown in the user's Twitter.

5.2.3.4   Services for Strategic Line 2.1: Inclusion of Web 2.0 Contents and Events in the Operator Services

Photostream from Flickr as an Image to be Shown in Telefonica's Presence Service

In this service, the operator, Telefonica, offers the possibility to use a sequence of photos from Flickr chosen by the user to be shown automatically in the presence service offered by Telefonica (see Figure 5.8). This service is available for any terminal supporting presence capabilities. For this service, Flickr provides the contents to enrich the services. A commercial agreement is not needed between actors.

How Does It Work for the User?

The user has to follow the following steps:

1. Access the service configuration web page from the operator and introduce his/her credentials from CanalCliente or TelefonicaOnline.

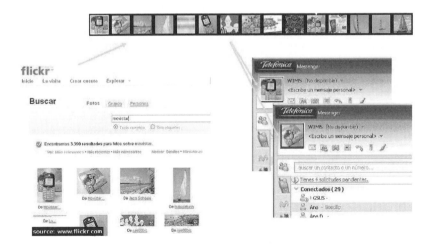

Figure 5.8 Photostream from Flickr as the image to be shown in Telefonica's presence service.

2. The service configuration web page accesses the Flickr search engine in order for the user to choose the sequence of images, using the search parameters defined in Flickr (tags, date, camera type).

3. After this, the user can configure the service defining the time when the image is changed and how the image is selected from the sequence (randomly, the most recent).

4. Once the service is configured, it works. New images introduced in Flickr will fulfil the criteria defined by the user and will be shown in the sequence.

Aggregator of Presence Information in the IMS Presence

In this service, the operator, Telefonica, aggregates the presence information, status info, about the user's contacts from different communities, services and social networks sites where they are connected, enabling the user to send messages to the user's contacts independently of whether they are connected (see Figure 5.9). The service adapts the message to the receptor website, using the formats accepted in it. A commercial agreement is not needed between actors.

How Does It Work for the User?

With this service, the user can know if his/her contacts are connected to the Telefonica network, to which Internet Services as como, in which Internet services are connected, and what they are doing at that moment, enabling the user to send messages to them. The information provided by the service is different for each of the websites:

- YouTube. The user can know which video is watching a contact at a precise moment and access to the personal lists, favourites, videos, etc., of the user's contact. Also, the user can send a message from the Telefonica's communicator to the contact's YouTube account.

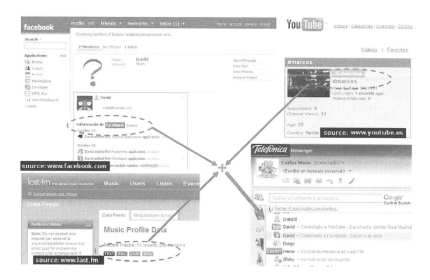

Figure 5.9 Aggregator of the presence information in the IMS presence.

- Facebook. A user can know when a user's contact to Facebook and to know about the activity of this contact. Also, the user can send a message from the Telefonica's communicator to the contact's Facebook mailbox.
- OpenSocial (2007). The user can know when a user's contact is connected at any of the social networks sites supporting the OpenSocial program (Orkut, Hi5, LinkedIn, etc.).
- Blogger and Twitter. The user can read the latest posts from a user's contact.
- LastFM (2007). The user can see what a contact is hearing.

My Music is Calling You

In this service, the operator, Telefonica, enriches the customized alerting tone/ring-back tone service adding the possibility of using the song that you are hearing at LastFM or one from your lists of the LastFM profile for a customized tone (see Figure 5.10). LastFM provides music from the musical taste of users. Also this info can be used to choose a song stored in the mobile device or from other music providers. A commercial agreement is not needed between actors, but for contents stored by LastFM user experience can be enhanced.

How Does It Work for the User?

With this service, the user can personalize his/her services in the operator's domain. The user has to access the web page from the operator domain and configure the service in order to be synchronized with LastFM as he/she desires; e.g. the service can configure a new ring-back tone for each call using the following list from LastFM:

- the most recent artist heard by the user this week;
- the most recent songs heard by the user this week;
- The 50 most heard artists or songs by the user.

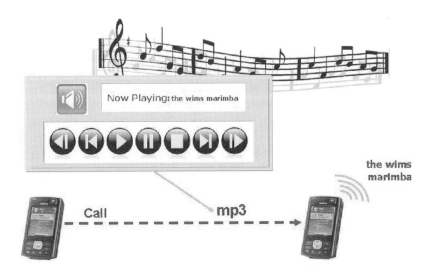

Figure 5.10 My music is calling you.

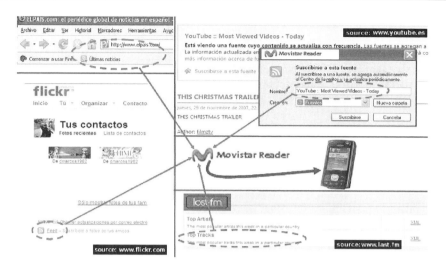

Figure 5.11 Movistar Reader.

Movistar Reader

This service provides a feed and multimedia content subscription and distribution to be delivered in mobile phones (see Figure 5.11). Web content providers supply RSS/ATOM or XML interfaces for feed readers. A commercial agreement is not needed between the operator and the rest of the actors; nevertheless, for those cases getting an audiovisual content, an agreement between the operator and the owner of the Web 2.0 site could enhance user experience.

How Does It Work for the User?

The user uses his/her mobile phone as a feed reader. The user has to subscribe to the feeds from the preferred websites with Movistar Reader and all the contents will be delivered towards his/her mobile phone. Thus, the user can receive updates from those websites ubiquitously. The user has to install the Movistar Reader application for PC and configure the browser to use Movistar Reader as the default feed reader. Then the user can browse normally and subscribe to the content updates chosen in the same way as with other readers. The difference is that the user can receive the feeds of the subscribed service directly in the mobile phone by SMS, MMS, IM, streaming, etc. The user chooses the way to deliver, when he/she subscribes to the service:

- For text and image channels (news, blogs) Movistar Reader allows the updates to be received using messaging, text-to-voice conversion or in the Telefonica communicator.
- For audiovisual channels (YouTube, LastFM) Movistar Reader allows an IM message to download or stream the content to be received. Also, it is possible to send using a multimedia session, but it requires an IMS-enable phone.

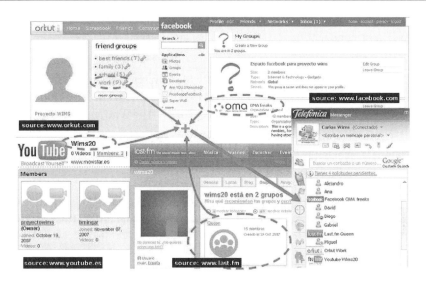

Figure 5.12 Aggregation of Web 2.0 groups in IMS groups.

Aggregation of Web 2.0 Groups in IMS Groups

This service enriches the user's contact address book. The operator, in this case Telefonica, provides an application for an address book integrated with different communication services in the handset (see Figure 5.12). Contact information about any contact in the address book can be used to any IMS communication services (IM, voice, etc.). Several SNSs and Internet websites provide information about the groups where the user is joined and the contact information of the members of those groups is included. A commercial agreement is not needed between the operator and the rest of the actors.

How Does It Work for the User?

The user can feed the address book with the contacts in the groups from social networks and Internet websites, where the user is joined. Thus, the user can communicate with all the members of a group easily, using communication services such as instant messaging, push-to-talk, etc. The user has no need to update any contact in the group. The user has to register in the service and select the social networks where the user is a member. For each of them, the different groups to be aggregated at the IMS groups also have to be selected by the user. The current service is available for some of the websites and communities more popular in the Internet:

- OpenSocial. Any group from the social network sites supporting the OpenSocial program (Orkut, Hi5, LinkedIn, etc.) will be a contact for the user's agenda automatically.
- Facebook. Any group from Facebook where the user has joined is included.

- YouTube. The user can notify his/her friends' group in YouTube about the new video uploaded or comment on the video that he/she is seeing from the YouTube group of the user's address book.

This service is available for any mobile phone with a presence address book.

5.2.3.5 Services for Strategic Line 2.2: IMS Online Applications (Virtual Terminal)

One of the barriers for deploying IMS services is the existence, for commercial purposes, of a huge number of mobile phones, in two lines: different models supporting IMS clients and a huge number of them for commercial deployments. To date, we are in a vicious circle, where mobile handset suppliers ask operators for the planned IMS services to be launched and operators ask suppliers for the roadmap of IMS enabled handsets; in the end, the situation is that there are no IMS services in the market, at least for the mobile market. From the WIMS 2.0 initiative, we think that new approaches can be open mixing both Telco and Internet worlds. The IMS online applications strategic line pursues a policy of running applications IMS without having a special device that is IMS enabled nor running an IMS client. The IMS applications in this line are running a server and the user connects to this applications browsing. These applications could suffer from two limitations:

- There are technical restrictions for accessing some functionalities available only using IMS features.
- A mobile phone with a good browsing experience is needed.

Nevertheless, both limitations can be assumed with not many problems. Most of the IMS features are available using IMS online applications, so that most of the service will not be affected. The second limitation has been removed thanks to the introduction of mobile broadband networks and, overall, the commercial availability of new mobile phones, like iPhone. iPhone breaks paradigms in the browsing experience and most competitors are working on creating devices with an enhanced browse, so it is foreseeable that in the short term there will be a huge number of mobile phones achieving this feature. In the WIMS 2.0 initiative, we have been working on that line with very satisfactory results using iPhone as a mobile phone. Studies done with users demonstrated that less than a 10% of the users perceived a difference in the performance, considering the applications shown as quite relevant for them and validating our strategic line.

5.2.3.6 Business Models

One of the goals of this book that you, the reader, who are reading these lines, will identify is to talk about the business model for IMS and WIMS 2.0. IMS was deployed commercially as another core technology and it still suffers from that consequence. As happened with GPRS, handset and network providers sold the benefits of the technology promising a huge amount of services that would be demanded by clients. Operators without previous expertise in selling services (operators were selling network capabilities as voice, forwards or other basics of the network) focused on selling the new technology: GPRS. The results are history and in the 'eyes' of all of the people who suffered from this wrong situation. It seems that all of the people involved in that situation should have realized the error made, but far from amending the situation, the telco community committed the same mistakes with UMTS, but, of course, enhanced: the huge amounts of money wasted produced the term 'unlimited money to spend'. Again there was the same situation: operators selling the technology without clear services defined caused the same results. IMS came with the lessons learned. Providers showed business cases needing to go ahead in the deployment, but the early days of IMS were a lack of services. Push-to-talk was the repeated example for all the providers and TTM was the justification to have this infrastructure deployed. The result is known for the reader.

In WIMS 2.0 we are not very innovative in business models, but we consider that business models and a fair ecosystem take the same line. Looking for fairness in the new convergent situation of the web and IMS/Telco do not imply a definition for new business models, but what is clearly implicit is a new way to share out incomes, costs and, at the end, revenues. Starting with this way of thinking, which must be house in mind during the entire development, the Web 2.0 way of thinking, it is obvious that a new member has to contribute to the community before getting something from it. Only with this behaviour, as mentioned by the visionaries from Win Win Consultores (2008), is it possible to build the new win–win relationships: the WIMS–WIMS relationships. All the business models considered by the WIMS 2.0 initiative are a combination of several models, but always having in mind the fairness flavour that emanates from its core. In order to tackle this chapter, an explanation of most of the existing digital business models has been given. Then we propose a selection of business models for each of the services previously detailed. We encourage the reader to develop his/her creativity by defining a new service under the scope of the initiative and, of course, the corresponding business models of each of them. WIMS 2.0 is an open initiative in pursuit of building this new ecosystem.

Business Models

Taxonomy Internet business models are one the most discussed issues, but also one of the least understood. Relevance, image, brand, community, etc., are some of the activities where money is involved, but there is no clear evidence of how it increases revenues compared with traditional business models. Business models have been defined and categorized in many different ways. Here we present our view, clearly influenced by frequent conversations with Win Win Consultores.

Advertising Model

This model is an extension of the traditional media broadcasters' model. In this case, the website provides services, like weblogs, email, etc., and content, normally, but not always, for free mixed with banner ads. There are several possibilities:

- Paid placement based on queries sells higher ranking link positioning like sponsored links or advertising, where the ranking mechanism is typically keyword driven to particular search terms in a user query. The best example is Google (1998a).
- Content-targeted advertising extends the paid placement to the rest of the web. Google AdWords (2009c) identifies the content of a web page and then automatically delivers relevant ads related to the content.

- Web portal is usually a search engine providing different services or contents. It is based on volumes of user traffic. A portal can have different objectives depending on the target market. An example is Terra (Terra Networks, 2000).
- A user registration site provides free access with registration in order to track users and their habits with the goal of selling targeted advertising campaigns. Any online newspaper requiring registration is an example.
- Contextual advertising systems scan the text of a website for keywords and returns advertisements to the web page based on what the user is viewing. Google AdSense (1998b) is an example.
- Behavioural marketing increases effectiveness of marketing campaigns using information collected on an individual's web-browsing behaviour, such as the pages visited or the searches made by one individual user, to select ads to be displayed for each user. An example is DoubleClick (2009).
- Classifieds are lists of items for sale or purchase as Craiglist (1995) does.
- Intromercials are animated full-screen ads placed at the entry of a site before a user reaches the intended content. The first site that launched this model was CBS Market Watch (1997).
- Ultramercials are click-through ads before reaching the intended content, like in Salon.com (Salon, 1995).

Merchant Model

This is the model of retailers and wholesalers of services and goods. There are several possibilities:

- virtual merchant, like Amazon Amaref (2009), that operates only on the web;
- catalogue merchant with a web catalogue and mail order;
- digital merchant, like Apple iTunes (2001), with both sales and distribution on the web;
- traditional retails with a web storefront.

5.2.3.7   Charging – Various Flavours

- Subscription. Users are charged a periodic fee to subscribe to a service irrespective of usage rate. It can be content services like Netflix (1997); trust services like Truste (1997), where a code of conduct is certified for the members; ISP like Telefonica (1924), offering network connectivity and related services on a monthly subscription.
- Freemium. Some variants of the services are free, but there are premium services, overall the user base is build. An example is Flickr.
- One-time fee like Mint (2005).
- Premium. 'A la carte' services like Apple iPhoto (2002) books for cross-selling services.
- Revenue share like ebay (1995), where the seller pays a listing fee and a commission scaled with the value of the transaction.
- Brokering model. In this model, brokers are the transaction facilitators bringing buyers and sellers together. It charges a fee or commission for each transaction enabled. There are several possibilities:

- Transaction broker provides a third party payment mechanism for buyers and sellers to settle a transaction. The most known example is PayPal (1998).
- Demand collection system. A buyer makes a binding bid for a product or service and the broker arranges the fulfilment. A pioneer example is Priceline (1997).
- Marketplace exchange provides a set of services covering the transaction process, from market assessment to negotiation and fulfilment, monitoring all the chain. BidPlace (2008) is a recent example of it.
- Virtual marketplace hosts online merchants that charges setup, monthly listing and/or transaction fees like in Amazon Web Services (2007).
- Buy or sell fulfilment. Users create orders for buying/selling products or services, where terms like price and delivery are included.

Infomediary Model

This model acts as a trusted agent, providing the opportunity and means for clients to monetize and profit from their own information profiles. There are different flavours:

- Advertising networks like DoubleClick enables advertisers to deploy large marketing campaigns by feeding ads to a network of member sites and analyse data from web users to increase marketing effectiveness.
- Online audience measurement services like the famous Nielsen (2008).
- Metamediary. Actors that ease the process of finding something demanded by connecting customers with the providers of goods and services they need to fill their needs, like in PCIQ (2007).
- Incentive marketing. This model offers incentives to clients like redeemable points or coupons for making purchases from associated retailers. The marketer sells data collected about users for targeted advertising. Toluna (2007) is an example of this model.
- Tryversity. In this model, products or services are provided to users in order to be tested by them and feedback is provided. An example is Latest in Beauty (2008).

Direct Model

This model allows manufacturers to use the web to reach buyers directly by reducing intermediaries. It produces more efficiency in operations, an improved customer care and direct information about client's preferences.

Community Model

This model is based on user loyalty:

- Open content like Wikipedia (2001), where people contribute with donations and voluntary work to the global community.
- Social network services, sites that provide individuals with the ability to connect to other individuals along a defined common interest. Social networking services provide opportunities for contextual advertising and subscriptions for premium services like in Facebook.

- Open source is software developed collaboratively by a global community of programmers who share code openly. In this model, companies like IBM get revenues from related services like system integration, product support, tutorials or user documentation.

On-Demand Model Based on Usage Rates

This model can be metered in usage or in portions.

Affiliate model

This model offers a percentage of revenues to affiliated partner sites. It is a pay-for-performance model: if an affiliate does not generate sales, it represents no cost to the merchant. Possibilities under this model are banner exchange, pay per click and revenue share based on click through of a purchased product.

Current WIMS 2.0 Business Models

This section covers the business models considered for the services explained above. It does not mean that any other business model can be applied to these services; in fact, the business models have been proposed based on the experience of the WIMS 2.0 founders.

1. 'Call-me button' and 'IM conversation' services. For these services, we propose to use a combination of advertising models (contextual or content-targeted models are the most suitable), binding with an on-demand model or a subscription model linked to the use of the PSE. Other alternatives could be to get incomes only from ads, based on the same models, and share a percentage of the revenues with the PSE provider. The target for these services would be a company, where it could be possible to have a model based on cross-selling instead of using an ads provider.
2. Click-to-Multiconf and Localizanos. These services could have the same business models as given in the previous point. An other possibility is to use behavioural marketing to get information about habits in the use of these services and then produce marketing campaigns increasing marketing effectiveness.
3. YouTube real-time videosharing. The model for this service could be an on-demand model or a combination of this with advertising models. It seems clear that induced traffic could be generated, producing more income. Suitable advertising models for this service would be intromercial or ultramercial when the content is displayed.
4. Real-time video generation. For this service, there are three main suitable business models: a subscription model, an on-demand model or an advertising model. For the last one, we think that it is possible to introduce ads while the video is uploading. Advertising models like user registration, needed for uploading the content in the personal space of the user, or contextual advertising based on user tags or in the content uploaded, in the case of existing tools to analyse video in real-time.
5. Geo-referenced publishing in Flickr with enriched information and automatic publishing of a personal icon and message in Twitter. For both, a subscription model or on-demand model is suitable. Also Flickr could offer this service as a premium service, being part of the freemium model.

6. Photostream from Flickr as an image to be shown in Telefonica's presence service. We think, only considering that Telefonica is selling the service, that this service can be monetized in two ways: subscription or advertising. For the advertising, the most suitable models are contextual and behavioural. Also, if the service provider is Flickr, with the role of a telco operator hidden, free or freemium models are possible. Flickr can get money from ads insertion or a premium service and share revenues with the telco.

7. Aggregator of presence information in the IMS presence. This service is open to be monetized using different business models. A subscription or on-demand business model is initially suitable, but ads can be included as a variant. Contextual advertising is interesting to include ads with words obtained from the information exchange in the user's communications. Behavioural marketing can be used for information obtained from the activity of the user in the social network sites or websites visited. Also, intromercial or ultramercial models are appropriate before establishing the communication or as a click-through step.

8. 'My music is calling you'. This cool service can be sold by a subscription or on-demand model, but there are other alternatives to be considered, such as incentive marketing if the tone used is music from an ad, digital merchant if the tone is paid to a merchant who then share revenues with the telco, intromercial where the user receives some ads previously to send the tone, and behavioural marketing or contextual advertising can be considered to introduce some ads removing the ring-back tone of the called (or the calling tone from the network in case the called has not predefined it).

9. Movistar Reader. A subscription model to this service is one of the first possibilities. Other models like contextual advertising, related with the info from the headlines received or behavioural marketing, from the info web searched are interesting, avoiding the initial barrier of the subscription price. An intromercial model based on the reception of a previous content to the MMS with the headlines and an ultramercial model in the case of the user click in some of the links of the MMS to access the complete news related with the headline are alternatives to the most traditional models.

10. IMS online applications. These services are based on the browsing capabilities of the mobile handsets. Traditional telco models like subscription are suitable, but noting that the user is surfing the web where the IMS application is residing, any kind of advertising model could be applied.

5.3 WIMS 2.0: The Technology Focus

The convergence between IMS and Web 2.0 results in an approach of the two worlds that, even if they share commonalties derived from the usage of the underlying IP technology, presents notorious differences:

- Business organization and conceptualization and its impact on technology. IMS was designed to provide control over an operated network, while the web is supported by a network with decentralized control.
- Service spectrum. IMS mostly provides interpersonal communication services, while services in the web sphere target a wider range of applications.

- Technology. The application-level protocol in IMS is SIP while HTTP is the protocol for the web.
- Identification of users and services is completely different in both worlds.

To achieve the WIMS 2.0 convergence strategies already described, it is required to complement the IMS with an adaptation layer capable of solving the differences between the two worlds and to exploit synergies to provide new capabilities. This layer will be the foundation to support the new convergent services. It is the WIMS 2.0 service platform.

5.3.1 Reference Model for the WIMS 2.0 Service Platform

This section introduces a reference model for the WIMS 2.0 service architecture. From a high level point of view, the WIMS 2.0 service architecture must be a service platform that, acting as an intermediary for the adaptation between both worlds, enables convergence in desired terms. The proposed reference model is reflected in Figure 5.13.

The above reference model defines a framework that identifies the required logical entities, as well as the relationships existing among them. Due to the different natures of the systems to be converged, this level of abstraction permits the main concepts and design principles to be settled prior to the definition of a concrete service architecture considering specific technologies. Two main types of entities are identified in the reference model: adaptation entities and entities for the provision of IMS online applications.

Within the adaptation entities, we also find two further subtypes: entities for the exposure of IMS capabilities and entities for the exchange of multimedia content and events. As a

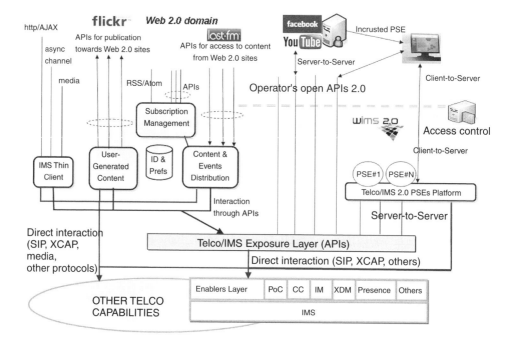

Figure 5.13 Architectural reference model for the WIMS 2.0 service platform.

relevant difference, for the interaction with the Web 2.0 world, the entities of the first group use APIs defined by the operator, while the entities of the second group use the APIs defined by the Web 2.0 services to be contacted. In fact, since the entities of the latter group are actually using functionalities of Web 2.0 services along with IMS capabilities, they can be considered as platforms for the development and execution of service mash-ups by the operator itself. The entities for the exposure and exploitation of IMS capabilities, and their functionalities, are the following:

- IMS exposure layer. This entity exposes IMS capabilities to the outer world through open Web APIs. On the outer side, it manages HTTP messages associated with the execution of a specific procedure, while, on the inner side, it interacts with IMS capabilities using the appropriate protocol (SIP or XCAP). Obviously, the interactions on each side must be properly interrelated. It can be clearly seen that it is actually a gateway between HTTP and IMS protocols. WIMS 2.0 holds the IMS exposure layer as a major axis for the architecture.
- IMS 2.0 PSEs platform. This entity holds and serves the operator's PSEs to be integrated into the Web 2.0 sites. PSEs implement the required logic to interact with a specific group of open IMS APIs (the ones considered above).
- Access control entity. To secure the use of the operator's APIs, this entity performs the required access control mechanisms.

The entities for the exchange of multimedia content and events, and their functionalities, are the following:

- Subscription management enabler. It subscribes to content and events of any kind generated in Web 2.0 services. Subscriptions may be made on behalf of the operator or directly on behalf of the final user. This entity talks with the content and events distribution enabler to inform about the existence of new content/events to be obtained.
- Content and events distribution enabler. It downloads, adapts and transmits content and events from Web 2.0 services to IMS users. To perform the transmission on the telecom network side, it uses a specific set of IMS capabilities.
- User-generated content enabler. It receives content from IMS terminals and, after adapting its format, it uploads this content to Web 2.0 services. To receive the content from the user, it uses a specific set of IMS capabilities.
- IDs and preferences. This database stores the relationships between IMS identities and the identities used by Web 2.0 services. The other entities of this group consult this database to perform the conversion of identities when using the Web 2.0 service APIs.

Finally, we find an individual entity for the provision of IMS online applications. On the one side, this entity interacts with IMS capabilities, acting as the final IMS client. On the other side, it shows a rich multimedia web interface to the user, according to the service being provided. This entity must also handle a continuous media interface towards the user, as well as an asynchronous interface to signal the need to refresh the web interface.

5.3.2 Web Technology Map for WIMS 2.0

The WIMS 2.0 reference model, as an abstracted description of a technological set of entities, aims at facilitating the dialogue between the telecom world and the Web 2.0, from the operator perspective. For achieving such a goal, the WIMS 2.0 platform eases the technology transition from the telco technology domain towards the pure Internet technology domain. As this is the goal, this section describes the relevant map of Web 2.0 technologies studied by the WIMS 2.0 initiative and selected for implementation of the key elements in the reference model for the WIMS 2.0 service platform.

Figure 5.14 depicts a summarized view of the entire web technologies considered, representing the domain of scope of each one of them. Before describing each of the different groups identified in Figure 5.14 and the relationship with implementation of the WIMS 2.0 service platform, it is important to stress the principles that organize this technology map:

1. Technologies have been located following the client–server paradigm that governs the web world. Within the server and the client sides, the different groups comprise technologies for implementation of similar functionalities that might be complementary, like, for example, the AJAX set of technologies, or represent an alternate choice to provide the same functionality, like, for example, JavaScript and ActionScript.
2. The HTTP protocol is the base for all interactions between remote elements, independent of the nature of those elements. These interactions are retrieval and manipulate web page contents and usage of web-ready APIs.
3. As a featured concept in Web 2.0, remote APIs are of paramount importance to support the mash-up of functionality within a given service. In the map, there are two variants of the usage of web-ready APIs:

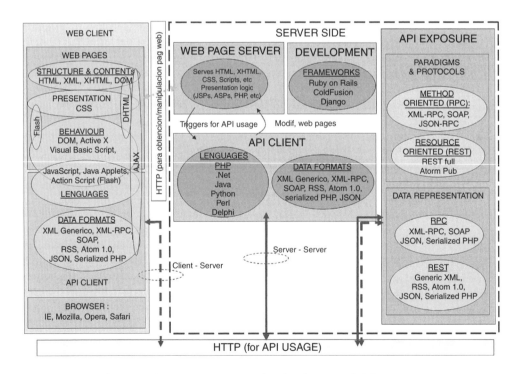

Figure 5.14 Map of Web 2.0 technologies for the WIMS 2.0 exposure strategy.

- Client–server scenario, where a browser directly invokes the API by sending an HTTP request to a remote server.
- Server–server scenario, where a server is the actor invoking the APIs provided by another server. The invoking server is acting as a web client as far as the usage of API is concerned. Normally this scenario is used to prevent the web browser realizing this kind of action, and thus all the service logic to handle the remote API is managed by the server.

Once the principles are explained, technologies are grouped as follows.

5.3.2.1 Client Side

The client side comprises all technologies that typically are used by web browsers. The following groups and subgroups are considered.

- Web page technologies related to presentation and processing of web pages. There are three technological levels:
  - Structure, comprising all technologies to support information, both contents and semantics, and the web page structure. HTML, XHTML, XML and DOM are technologies within this level.
  - Presentation, comprising technologies to control how the information is presented. CSS is the most representative one in this level.
  - Behaviour, which gathers all technologies used to implement the logic that control local actions and dynamic behaviour of the web page. This dynamic activity can result in modifications over the presentation and structure levels. VisualBasicScript, JavaScript, ActionScript and Active X are some of the ones considered.
- Technologies for web API clients in the web browser include the technologies in charge of process, form and send HTTP request towards remote APIs and receive and interpret responses according to specific data formats. Since these actions can be considered as dynamic behaviour of the web page, languages that rule this level are the same than those utilized in the web page behaviour level.
- Browsers are those alternative options as client platforms, like Internet Explorer, Mozilla Firefox, Opera, Safari and Google Chrome. Even if functionalities presented by these platforms are similar, some browses can present specific features that may affect the usage of certain technologies considered for the client side.

5.3.2.2 Server Side

This comprises technologies typically utilized in network infrastructures. Even though they are mostly associated with a web server, the role of a web API client shall also be considered. The following categories apply:

- Web page server technologies, which groups technologies to serve web pages such as software containers, servlets, JSPs, ASPs, etc. These are mature technologies that do not represent a key differential aspect of the web–telecom convergence.
- API technologies. The exposure of telecom capabilities in the form of web APIs enables distributed functionalities within the web world to be offered. In this group,

technologies for implementing the serving API logic are considered, and then they are valid for implementing both the client–server and server–server scenarios. There are different options when considering how the HTTP protocol is employed and how the functionality behind the API is modelled. In the WIMS 2.0 technology map, the RPC and REST approaches are considered. The map also considers options to encode the data exchanged between the API client and the API server, like XML, JSON, ATOM, etc.

- API client technologies for the server side. As previously mentioned, in the server–server scenario, web servers act are API clients, and in this particular case of the web loom, they act as web clients in order to invoke remote web APIs. Technologies are those languages at the server side to rule API invocation like PHP, Ruby, Java, Python, etc., as well as the data formats that the specific API is utilizing to encode the data exchange. As mentioned before, this is a feature of the API server and there are a variety of possibilities, most of them based on different variants of XML.
- Development frameworks, which include environments geared at easing the web service creation. These frameworks feature a high level view of the development process and a rapid deployment over differerent server platforms like Java or .NET. A popular framework for Web 2.0 application is Ruby on Rails.

5.3.3 Implementation Aspects

The technology map considered in the previous section addresses the technical needs for an implementation of the WIMS 2.0 service platform. Taking into account the reference model for the WIMS 2.0 service platform, the axial concept for convergence within that model is the exposure of capabilities through web-friendly APIs. Thus, this section describes guidelines on the technical details to implement the key entity in the WIMS 2.0 service platform.

5.3.3.1 IMS Exposure Layer

The IMS exposure layer and its implementation aspects are illustrated in Figure 5.15. It shows the technologies to be taken into account in this entity and the tools that will ease its usage from Web 2.0 applications. There are two aspects for the technical realization of this entity: on the one side, the web-friendly APIs that expose IMS service capabilities towards web applications and, on the other side, the internal logic to interwork and interact with the underlying IMS infrastructure in order to execute the functionalities exposed via the API.

The IMS exposure layer shall implement a gateway adapting the HTTP-based API primitives into SIP and XCAP protocol operations to enable communication between the web world and the IMS. This mission affects two levels of communication in the telecom side: the signalling plane and the media plane. Wherever possible, both planes need to be adapted from a pure technical perspective, although continuous real-time media is not supposed to be adapted by means of API approaches. This means that audiovisual media exchange between a telco endpoint and a web component is not processed through an API and thus not handled by the IMS exposure layer. The following IMS service capabilities constitute a nonexhaustive list of capabilities that can be exposed via the IMS exposure

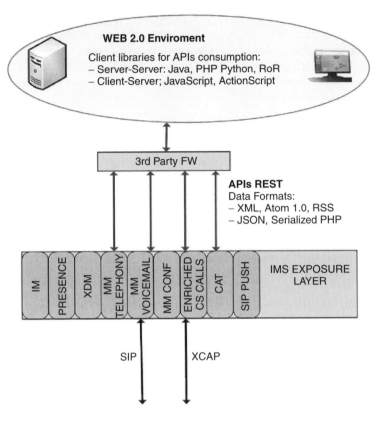

Figure 5.15 Implementation of the IMS exposure layer.

layer, and thus the underlying gateway shall be able to communicate with the enabling servers implementing these service capabilities:

- Presence. An API client might be able at least to subscribe, consult or modify presence information for a given user.
- XDM, to manage lists and contact items in the IMS domain.
- Instant messaging both in pager mode as well as session mode.
- Multimedia telephony, to provide a third party call ability to web applications and call control features of sessions established between two endpoints.
- Multimedia conferencing, to provide a web application with the possibility to schedule and create multimedia multiparticipant calls and manage them.
- Rich CS calls, to provide features related to media sharing combined with circuit-switched calls.
- SIP push, to provide web applications with the ability to push events and contents to IMS terminals.
- Customized alerting tone enables a web application the control over ring-back tones and personalized alerting tones (ring-forward tones).
- Multimedia inbox control.

To expose capabilities, the APIs of the IMS exposure layer will implement a resource-oriented model as a basis; i.e. the REST philosophy will be utilized to model the APIs for IMS capabilities. Since most IMS services inherit a session orientation, some RPC techniques might be added to the basic REST representation in order to obtain a similar result to that obtained when modelling telecom services such as WebServices. However, REST provides a simpler approach, which reduces the entry barrier for web developers to integrate and utilize telecom service features in web applications. Most Internet players in the Web 2.0 ecosystem provide APIs utilizing this resource-orientation philosophy. Besides, since REST is naturally stateless, scalability increases in comparison with pure RPC alternatives like SOAP and WebServices. Beyond that, the processing for a lightweight option like REST is not comparable to the SOAP computational cost. For technical realization of the REST APIs of the IMS exposure layer, there are several protocol alternatives. Among them, the atom publishing protocol, or AtomPub, is one of the most suitable, as well as being a standardized mechanism by the IETF. Figure 5.16 presents an example of the IMS exposure layer exposing presence by means of REST API. In this example, the presence APIs relies on some of the intrinsic features of REST; i.e. it works with resources, e.g. a presentity. The resources can be freely defined and can keep status (e.g. presence status). The client acts on the resources: get, modify (e.g. state), delete, etc., and these resources are always univocally identified by URIs. The actions on the resources are performed through

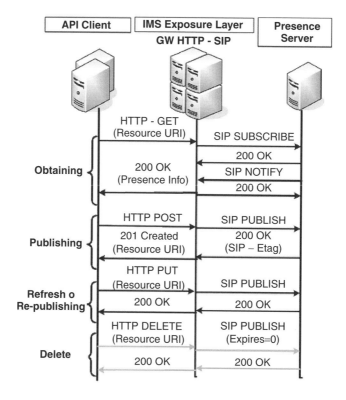

Figure 5.16 An example of the IMS exposure layer: a REST API for presence.

standard HTTP verbs: GET, POST, PUT and DELETE, which have a specific meaning in the context of each API. For instance, in this presence example HTTP GET is utilized to retrieve a presentity's status; HTTP POST can be used for publishing presence information; HTTP PUT for refreshing softstate presence information; and HTTP DELETE to erase this softstate information. The application of REST principles to the IMS presence is thus immediate.

Aside from the protocol aspects, the API exchange information is between the API client and the API server. To accommodate the expectations of a bigger number of developers, APIs shall provide different data formats so that the user can select the most suitable one according to the programming language and development tools the user utilizes. The IMS exposure layer at least shall provide the following data formats:

- client–server scenario: XML, Atom 1.0, RSS and JSON;
- server–server scenario: XML, Atom 1.0, RSS and serialized PHP.

This allows the use of the remote APIs and the exchanged data from most of the languages being utilized on the Internet. Furthermore, the telecom exposure layer might be simpler to utilize if the developers are provided with language libraries that handle the API client behaviour. This means that the developer does not need to deal with HTTP request and response construction and processing. Most Internet players providing functionalities via API are adopting this best practice, offering client libraries in a limited number of programming languages and also the bare REST interface with different data formats available for those programmers utilizing other programming languages besides those considered by the client libraries. For instance, the telecom exposure layer should offer client libraries JavaScript and ActionScript languages for client–server scenarios and Java, PHP, Python and Ruby for server–server scenarios. Developers utilizing other languages can still access the functionality by handling the REST APIs directly.

5.3.4   WIMS 2.0 Roll-Out

The WIMS 2.0 reference model described in the previous section in Figure 5.13 represents a logical representation of functional entities and the relationships those functional entities hold between them to achieve the target convergence. However, this reference model of a WIMS 2.0 architecture does not present nor represent the location of those functional entites in their context, the telecom IMS-enabled service delivery domain. Thus, Figure 5.17 depicts how the WIMS 2.0 reference model is deployed on to the service delivery domain of an operator and how the different entities of the WIMS 2.0 reference model can be either implemented by existing entities in the telecom service architecture or how new ones –specific for WIMS 2.0 –fit into the service delivery domain. The service delivery domain is modelled after the layered architecture presented in Chapter 1. In Figure 5.17 the functional entities of the WIMS 2.0 reference model that are new in the service delivery domain of an operator are marked in dashed lines, while existing ones are depicted utilizing dotted lines. For instance, a case for the later would be the 'access control entity' in the WIMS 2.0 reference model; here it is mapped on to a general gateway for securing the access of third parties to the service delivery domain, i.e. the integrator towards third parties of the SDP of the layered architecture. As for WIMS 2.0-specific entities, the IMS exposure layer is located at the service exposure layer of the layered architecture, while the rest of the WIMS 2.0

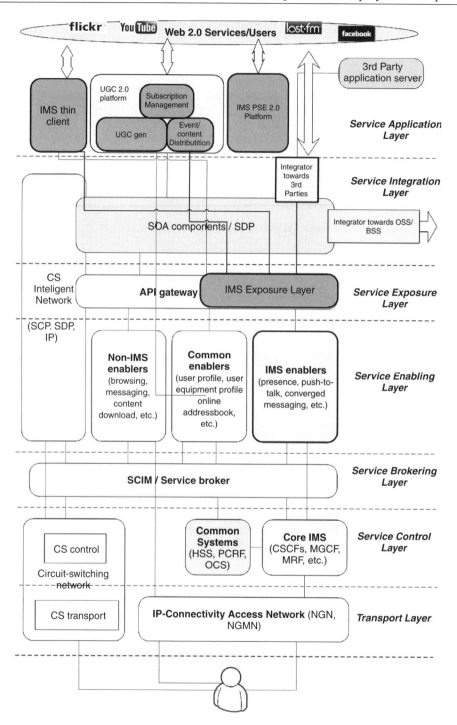

Figure 5.17 The WIMS 2.0 service platform within the telecom service delivery domain architecture.

functional entities are considered to be service platforms and thus they are represented in the service application layer. That would be the case for the IMS thin client, containing service logic for providing end-user application online representations of an IMS phone; the functional entities of WIMS 2.0 related with content and media delivery or generation are gathered under a service platform called UGC 2.0 at the service application layer, except for the database containing the user IDs and preferences, which is stored together with other user data at the user profile directory (a common enabler) at the service enabling layer; and finally, the IMS 2.0 PSE platform is regarded as a container for end-user applications, in fact the portable service elements.

6

The Way Forward – Paths to Follow

Chapter 5 has described how an IMS-enabled network can be exploited beyond its original conception to retain competitiveness and boost innovation in an operator business. That has been shown through the case of WIMS 2.0, which promotes the convergence of telecom services with the Web 2.0, based on a fully enabled IMS service architecture. This kind of strategy preserves the positioning of the telcos in the 2.0 era. However, the wide ICT landscape is a rapidly changing environment. This dynamic context requires flexible adaptation and an enthusiastic and firm pace to walk towards the future. New business models and new technologies modulate this evolution, without compromising cost optimization and exploitation of past investments, especially those in network infrastructure. This chapter describes and discusses potential paths to follow for operators looking to a future that is securing investments and profitability. In these paths of evolution, IMS-enabled service architecture plays a decisive role in supporting innovation and providing differentiation in service creation. After all, services and service construction are the essence in driving the digital evolution.

The present chapter presents service platforms and service creation trends in the mid term, as well as a set of initiatives in the R&D domain looking further into the future. The chapter takes the case of service platforms for operators first. Here, SDP, SOA, the marketplace of services and the open telco initiatives are discussed. Then the service creation is considered as this is an intricate aspect of the success of services and products. Service creation is examined from the perspective of all key contributors, including the user participating in the service generation, the professional developers and the challenges posed by devices in service development.

IMS: A Development and Deployment Perspective   Khalid Al-Begain, Chitra Balakrishna, Luis Angel Galindo and David Moro

6.1 Operators and Service Platforms

Innovation is an important asset for operators to exploit their service capabilities, which might not be explicitly exposed to service development and product building. There is a hidden potential in telecom networks for service creation that today is not fully exploited. In most industrialised countries, market saturation is leading to new technology solutions that target at leveraging already existing capabilities for new usages. In the technology sphere, this is materialized by a new service platform that unveils the potential of existing telecom architectures and underlying networks for service creation. As such, IMS hides a potential for service innovation that can be opened to the large ecosystem of service creators – both in-house and third parties – with the right tools and the right service platforms. If not, investments in IP multimedia will be relegated to pure technology trials.

This section reviews the trends for opening telco service capabilities to a wider space of third party developers in the so-called 'open telco' initiatives that are based on service platforms. It also discusses how these initiatives can be supported by service delivery platforms (SDP), which are geared beyond the open capabilities objective to also target reducing time-to-market for in-house application development. Finally, there is the creation or the telco contribution to a digital marketplace of open services and capabilities. Throughout the section, the role of IMS in opening service capabilities and flexible commercialization of open services in a marketplace has been sustained. It is argued that it provides new service capabilities and enablers to the catalogue for the sake of operator competitiveness and positioning in a fierce market of open services and service creation.

6.1.1 Open Capabilities

Operators are in a difficult situation in the digital service landscape. Their business is exposed to a continuous risk of saturation, and growth principles are not reflected on the seeds of seducing customers with appealing services. The majority of the income in the residential markets comes from basic services like traditional telephony and increasing bandwidth for pure Internet connectivity. Even though operators are in the loop of continuous developing and bringing a myriad of new value-added services, customers are still using the same basic services like messaging and basic voice calls, thus ignoring the remaining 90% of the service portfolio. Instead of these value-added services, customers seem to use the Internet services. Therefore, it can be concluded that operators are not engaging correctly with the user population in the value-added services and applications. Possibly due to that their business models are not attractive enough, and that they are disconnected from user needs and desires in terms of digital lifestyle. Services offered by telcos are not as innovative as those populating the wide Internet, where different patterns for building services, open innovation, cheaper development, proximity to users, blurry frontiers between users and developers are some of the bases of this effect.

There is an emerging ecosystem on application stores, software-as-a-service and a cloud full of service capabilities offered to Internet communities for development. This indeed channels and stimulates the creative energy of digital users. Some remarkable players of this trend in opening services and capabilities are the key Internet players, such as Amazon, Google, Yahoo and Salesforce. Each of them is targeting different segments: from platform and storage like Amazon, or CRM and productivity applications like Salesforce, to web Internet services like mail, maps, search which Google and Yahoo are offering.

On the other hand, telco companies have unique and valuable assets that potentially could be mashed up with the others in this cloud of service capabilities. This implies a change of thinking in the telecom business in the form up to the Internet trend. This means that the telcos would offer a portfolio of open capabilities in the form of reusable APIs, tools and SDKs to the wide Internet, to let developers and software companies of different sizes integrate telco services in new and innovative applications. An operator pursuing this new business model and providing the necessary platform for service creation and mash-up with features coming from other service providers can consequently qualify to become an 'open telco'.

By delivering this kind of platform, the 'open telco' obtains an important source of benefits for the local brands; customers and enhancements on time-to-market; and CAPEX utilization and income. A twofold principle for generating income governs this change of business thinking:

- An 'open telco' is able to create a new segment of customers, the developers, willing to pay for – or commit to – a set of high level assets that enable them to build new services.
- An 'open telco' can create new revenue streams based on the customer usage of new third party applications built on its open capabilities. Hence, customers will enjoy a new wave of services, increasing traffic and access business, which is shared with the third party that delivers and maintains each application.

The success of such 'open telco' initiatives depends on a set of factors that are indeed not technological. These include:

- Creating a lively community of developers to produce applications with a fair revenue-share model with the 'open telco'. Beta testers of the developed applications should also be engaged and encouraged. Thus, the community is the axial part of such an 'open telco' initiative, at least to foster the initial open platform utilization and to exploit marketing proximity between users and developers.
- Application store and customer access to innovation is an important part. Beta testers are the first front end for a new application trial, but the necessity to step into commercialization of a new application is the key for generating real revenue and traffic in the operator systems. Those applications with great impact in the market are expected to be included in the operator's portfolio and branding of the 'open telco' upon achieving a fair agreement with the owner. Thus, if an 'open telco' initiative achieves a critical mass, it can determine the agenda for future products of the associated telecom company, and probably others by propagation.

Nevertheless, there are drawbacks in the 'open telco' strategy, mainly derived from the clash of business models between the source, the telco company, which is naturally a closed and controlled business, and the Internet ambience, which is the initiator of the 'open' service provider trend. Since the Internet players are offering similar open platforms that usually do not look for direct revenue from users and developers and their services are not subject to a tariff, the success of an 'open' but 'priced' initiative such as those promoted by 'open telco' can reduce their acceptance possibilities to a minimum. Another remarkable factor for the potential failure of the 'open telco' initiative as a global trend is fragmentation. Telcos operate in geographically fragmented markets, while the Internet is regarded as virtually global. When examining the market situation, there is an increasing panorama of telco companies joining the 'open' trend. Within the established and de facto principles of an 'open telco' described here, there are variations from company to company when examining how they approach the community. The number and functionality of the offered open APIs, the technology and the business models behind are the substantial differences. In a comparative summary of a survey that has been conducted for a number of telcos, the most representative were: Telefonica's Open movilforum, O2 Litmus, Vodafone Betavine, Orange Partner, BT's Ribbit and Telenor Playground.

Table 6.1 summarizes the differences in the business models that were found in the above mentioned open initiatives. In particular, the table shows if the open community of a given telco recognizes the role of the developer and the role of beta tester, and if there is a channel to commercialize the developments via an application store, and how the parties are charged or revenue is shared.

From a pure technological viewpoint, there are commonalties and best practices to be recommended. The Web 2.0, as the first step towards a future Internet of services, has already shown that the Internet is 'The Platform' for constructing and delivering ubiquitous services through the dynamic mash-up of open services coming from different players, which exposes functionality of their services by means of open APIs. Functionality exposed by the 'open telco' is very dependent on its strategy and technical plans. However, there are commonalties between the different open telco initiatives in the sense that most of them are exploiting the base telco service assets like messaging, location, phone book and telephony features. Table 6.2 shows a comparison of the open capabilities offered by the studied open telco initiatives.

Table 6.1   Open business model comparison.

Initiative	Developer	Beta tester	Application store
Vodafone Betavine	Free of costs	Free	Yes, not linked to Betavine
Orange Partner	Price is dependent on the specific API; alpha and beta APIs are free	Role not defined	Orange Shop; revenue shared with developer
Telenor Playground	Free of costs	Not defined	Yes; revenue shared upon negotiation
Ribbit	Prices dependent on the planned usage	Not defined	Widget platform planned; revenue shared
Open movilforum	Charged per use to an MSISDN; free vouchers	Not defined	No
O2 Litmus	Free of cost	Yes	Yes; revenue share 70% for the developer

Data collected in January 2009.

Table 6.2   Open capabilities comparison.

	Vodafone Betavine	Orange Partner	Telenor Playground	BT Ribbit	Open movilforum	O2 Litmus
SMS send	Yes	Yes	Yes	Yes	Yes	Yes
SMS reception	No	Yes	Yes	Yes	Yes	No
MMS send	Planned	No	Yes	No	Yes	No
MMS reception	No	No	Yes	No	No	No
Call control	No	Yes	Yes	Yes	No	No
Multimedia conference	No	Yes	No	Yes	Yes	No
Device capabilities	Yes	Yes	No	No	No	No
Location	No	Yes	Yes	No	Yes	Yes
Wap push	No	No	No	No	Yes	No
Presence	No	No	No	No	Yes	Yes
Phone book	No	Yes	No	Yes	Yes	No
Voicemail	No	Yes	No	Yes	No	No
Authentication	No	Yes	No	Yes	No	Yes

Data collected in January 2009.

To retain relevance and leverage of its unique service assets, an 'open telco' operator offers its capabilities towards this new third ecosystem, thus enabling the easy introduction of telco functionalities within almost any third party application. The latter is not limited to just web applications but also to other non-web applications that simply use HTTP as a transport mechanism to join the Internet service fabric. The open APIs exposing the telecom capabilities should also be offered to third parties and to the Internet in a controlled and adapted manner. The open API concept can be used to meet both in-house and external service delivery needs. To design a future-proof solution, an open API model of telecom services has to have a web or HTTP-like view on them. To define open APIs of telecom capabilities, the combination of the representational state transfer (REST) (Fielding, 2000) philosophy (defined in Section 6.2.1) with blended operations of an RPC (remote procedure calls) is recommended. This will provide a simple-to-use API while still having a fine control of session-oriented services like telephony, which requires a richer set of operations to be exposed. Besides the remote API, a common practice in this open environment would be to provide a set of client libraries in different programming languages such as Java, PHP, Python, Ruby, etc. These libraries will handle the HTTP specifics of the invocation of API primitives from the application side (client side) towards the remote API server in the open platform of the operator. This can accelerate development and integration of the API into the application, becoming in fact a matter of attracting developers into the community of the given 'open telco'.

Table 6.2 describes the different approaches taken by the studied 'open telcos' for using remote API technology and the languages for the accompanying client libraries.

Besides the pure exposure of telecom capabilities by means of open APIs, the platform offered within an open telco initiative can be enriched with SDKs and other tools for easing mash-up generation for nonprofessionals users, i.e. user-generated service (UGS) tools. The open platform should also rely on mechanisms for registering and promoting

Table 6.3 API technologies and client libraries in open initiatives.

Initiative	API technology	Client library languages
Vodafone Betavine	REST	Python, Java, Ruby
Orange Partner	RPC (for some APIs); REST-like (for some APIs)	Java .NET
Telenor Playground	RPC: Parlay X (SOAP)	Java
Ribbit	REST	Flex (ActionScript 3) and Flash
Open movilforum	REST-like	PHP, Ruby, Python
O2 Litmus	RPC: SOAP	Java, .NET, Flex and PHP

Data collected in January 2009.

applications, i.e. the application store, and tools for managing the marketplace of services and applications behind the community. In this context, REST and other development techniques, UGS and digital marketplace tools are also discussed in this chapter The fact that IMS is an innovative technology and concept affects this kind of programme, providing a richer multimedia set of service features that can be exposed to augment the catalogue of open capabilities that operators are offering to the community of developers. Beyond that, the open communities are usually populated by pioneers in technology adoption, the so-called early adopters. New capabilities coming from the IMS world can be offered to a community eager to experience high end technology. This can be a twofold objective: on the one side, to maintain the sophisticated style of the community, as well as to gather a feedback outside the operator on service design patterns, usage concepts and other metrics about IMS services.

6.1.2 SDP and SOA

A service delivery platform (SDP) usually refers to a set of components within a telecom operator service architecture that is in charge of easing service creation, deployment and execution utilizing service capabilities and enablers. As the keystone objective of an SDP would be to reduce time-to-market of a service, the key characteristic for an SDP is thus the ability to ease the integration task of application logic with the enablers and capabilities in the telecom service architecture for the delivery of a service to users. The SDP thus enables the rapid transformation of a service concept into a full product including all the related service lifecycle management and commercial interactions. The SDP is positioned within the operator's service architecture, as described by the model in Figure 6.1, constituting the axial component of the service integration layer.

There is no global standard for SDP implementation, although there are some industrial de facto conventions. Upon the latter, a set of principles can be extracted for describing SDP following a compass-inspired interpretation. This is depicted in Figure 6.2.

- *North.* An SDP provides clear service interfaces for exposing telecom capabilities, both for in-house development as well as enabling external agents to develop services and applications using the operator as the platform, i.e. the service delivery platform. An example is the previously described 'open telcos' initiative, whose technical realization tends to be realized with the operator's SDP.

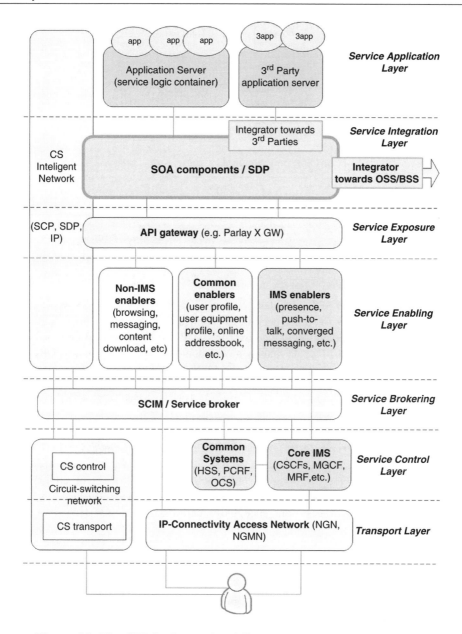

Figure 6.1 The SDP in the service delivery domain layered architecture.

- *South.* The SDP provides abstractions for the operator's enablers and service capabilities in the underlying service architecture, thus providing a simple means of invoking telephony, media and other service features from the applications by means of these abstractions. Therefore, the SDP simplifies the process of integrating an application container with the network infrastructure and enablers.

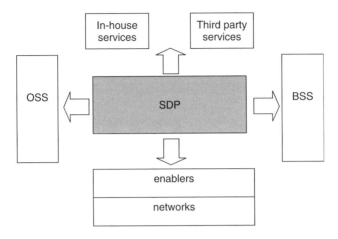

Figure 6.2 The compass-inspired approach for the SDP.

- *West and east.* The SDP includes connectors to the OSS and BSS systems of the operator, as well as for other back-office applications. By means of this connection to OSS/BSS, the services deployed over the SDP can be integrated into the company service and product lifecycle management, including operations that usually involve the service delivery architecture and the OSS and BSS systems, like the provision and activation of services; supervision and failure of management; administrative validations; billing and quality of experience (QoE) measurements. In this way, an SDP simplifies the integration of a new application towards the commercial and operation systems of the company.
- *Central.* The SDP typically provides a service creation environment for service design, and a service orchestration and execution environment, which represents the core of the SDP. Together with these core functions, there is a set of components in the SDP to enable policy control, security, user directories and other functions for business/operation process adherence and integration.

The adoption of the service-oriented architecture (SOA) (OASIS, 2008) in the SDP simplifies enabler deployment and facilitates other SDP requirements for composition, reuse, delegation, orchestration, policy enforcement and business/operation process adherence and integration. SOA has been traditionally associated with the IT domain, but now the trend is to reuse it in telco-oriented environments like the SDP. SDP and the election of SOA for its implementation have been deemed necessary in order to cope with a series of business drivers; in particular, the SDP focuses on the reduction of time-to-market for new services, as already mentioned. It is also expected that the SOA-based SDP allows the creation of a wider range of value-added services in a cost-efficient way, the implementation of an open service marketplace for third party interactions and the efficient integration with OSS/BSS.

The service-oriented architectures seek to save costs in investments and to provide flexibility to services offered by companies. SOA pursues a methodology to define and reuse software components and business processes that on many occasions are already developed and implemented. Thus, SOA is a paradigm of software architecture with a service orientation as the prime design principle. SOA can be defined as an architecture that provides uniform methods to offer, discover, interact and govern capabilities, which are modelled as SOA services. These services are loosely coupled and present clear interfaces to match business requirements and implement reutilization. Due to its neutrality, in an SOA environment services are accessible irrelevantly of the type of platform and technology of implementation. Thus, SOA is not linked to a specific technology for its realization and there are many to satisfy the implementation requirements, e.g. SOAP, RPC, DCOM, CORBA, WCF. Although SOA is technology independent, web services are the most extended technology together with their associated technologies like HTTP, XML, WSDL, SOAP and UDDI. However, a service-oriented SDP usually provides support for a higher range of technology in order to guarantee integration with a large variety of services and capabilities. Beyond that, there are ongoing trends to enhance SDP to support service orientation as well as resource orientation, a software philosophy embodied in some interface techniques like REST.

For constructing the infrastructure of the SOA-based SDP, a skeleton of the platform can be based on the a SOA suite. Typically, these elements, although not standardized, are part of the SOA product portfolio of the IT vendors with a clear leadership in this domain. The common key elements in a SOA suite across vendors' portfolios are the BPM, ESB, Registry and Compositor. These elements are depicted in Figure 6.3 and described next:

- BPM (business process management) is a set of technologies that host and execute business processes. BPM workflows can be long-lived sessions and can involve human interactions.
- ESB (enterprise service bus) is a logical service bus to ease service access and interoperability by means of decoupling between applications, service enablers and capabilities, on the one side, and the integration of different systems, on the other side. ESB basic functionalities are message exchange and routing, and translation between different technologies.

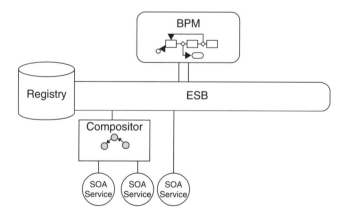

Figure 6.3 The elements of an SOA suite.

- Service registry provides the ability to register, discover and govern services in the service-oriented architectures. In run-time, the service registry is accessed by applications to locate and bind services. Thus, this registry contains a service catalogue and those parameters required for the execution of services.
- Compositor (or service orchestrator) utilizes and orchestrates different elements in the service architecture to produce combined services that are more complex and reusable by application. Sometimes, this is also referred as 'integrator', a piece of the SOA dedicated to the development of high transactional services based on reutilization of other services.

On top of these SOA elements, the functional components can be constructed, interlinked and connected to external pieces of the service delivery infrastructure and systems beyond the service delivery domain. Of all the elements, the ESB is regarded as the key element for providing the infrastructural support of the SDP, due to its integration ability.

A clear conclusion is that SDPs, which are usually implemented with IT technologies, will abstract the service enablers and service capabilities of a telecom network. Each of these technology domains is driven by different principles, where the telecom technological domain is usually based on asynchronicity and event-driven protocols, while the IT world, where SOA is positioned, is based on information exchange patterns and synchronicity. Usually the former is associated with optimized technologies like those associated with circuit switching, while the latter is rather oriented towards Internet-like processes.

The role of IMS in this clash of technology philosophies is precisely the ease of merging or abstracting capabilities from the telecom technology domain into the IT domain. This means that IMS capabilities should be easier to be abstracted and integrated into the SDP to become ready to be utilized by services developed on the SDP, since the technology foundations are the same or, at least, present a high degree of synergy. Besides, IMS provides a rich set of enablers and service capabilities that considerably augments the resulting portfolio of service features offered by the SDP to the service creators when building any new application over the SDP.

6.1.3 Digital Marketplace of Services

Nowadays, service providers recognize that access-line loss is progressing rapidly and needs the ability to introduce new services quickly in response to market needs generating new revenue streams. It is therefore clear that service providers can realize more revenue when SDP and business and operational systems (B/OSS) are fully integrated. With the back-office functions of provisioning, activation, service quality monitoring, customer intelligence and billing, service providers can deliver great customer experience and actively manage the service environment. Communications providers have an opportunity to satisfy consumer demands better through the new concept of a digital marketplace strategy, resulting in higher acquisition, retention and satisfaction rates. It begins by providing the digital environment through expertise and resources and delivering the digital lifestyle through applications, content, commerce and social networking. Creating a marketplace for next-generation services where Internet, media and telecommunications industries converge means that it requires the participation of service providers that have multiple specialities. These providers can make available distinct but complementary resources that can be combined and commercialized for a variety of customers using a variety of business models.

A digital marketplace is an advanced collaboration framework with three main pillars:

- The exposition of its capabilities from the network and OSS/BSS to third parties as reusable, manageable and chargeable services exposed through open APIs.
- The composition and aggregation of the exposed services with services offered by third parties.
- The agile marketing of these resulting products under several brand and business models beyond advertising, e.g. SLA-based business models. Taking these aspects into account, the digital marketplace of services can be defined as a business and management layer to be put on top of these 'service-oriented communities', where the participants expose, share and combine services to create new ones. The marketplace enables the end-to-end management of services from conception to cash across service provider's domains, the automated establishment of business agreements between the participants and the commercialization of these collaborative services and products under different brand and business models. Figure 6.4 shows the marketplace vision from Win Win Consultores (Consultores, 2009).

Since the collaboration and interworking between different industries is a key marketplace characteristic, standardization is essential: for instance, standardizing how a service provider can share its assets as services reusable by other service providers and, in the same way, can reuse the assets provided as services by other services providers without losing end-to-end control and management capabilities. In this manner, the marketplace concept is closely related to standards as the TM Forum (TMF, 2009b) service delivery framework (SDF) (TMF, 2009a). The aim of SDF is to enable service providers to be agile and competitive when managing the full service lifecycle in this new world, where services can either be developed by third parties or by themselves, and can also be customized, reprogrammed, composed, assembled with other services, etc. Therefore, the service delivery framework helps organizations maintain control over service lifecycle management across all execution environments (i.e. mobile, convergent telco, IPTV, broadband provider, etc.) and frees them to architect their SDPs as they choose. The operation of a marketplace can be explained by the following example:

- A digital marketplace exits where telcos and ISPs publish their services in order to allow other service providers to reuse these services. The service specifications include

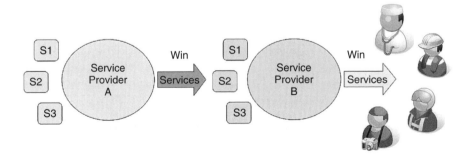

Figure 6.4 Marketplace by Win Win Consultores.

functional, management and business aspects like the price model, the payment model, QoS metrics, service levels, etc.

- An operator discovers in a digital marketplace the virtual meetings service from an ISP and decides to combine it with its high quality video on-demand service to create a new commercial service. As the first step, the operator and the ISP agree a service level agreement (SLA), including aspects like quality of service, support and troubleshooting model, brand model, business model and revenue sharing. Once the agreement is signed, the telco operator develops the new service and defines the related commercial offer for this service, which is published in the marketplace product catalogue.

- A customer discovers the product, contracts and uses it. The marketplace is in charge of monitoring the service level and the service usage (detecting SLA violations, sending events to the billing and service assurance systems, etc.), even when the services included in the product are distributed across the telco and ISP service provision domains. This new service is a win–win for both, the telco that offers a more attractive service and the ISP that reaches new markets and final clients who have access to an enhanced service.

As mentioned earlier, a marketplace enables the end-to-end lifecycle management of services (from conception to cash) across different service provider's domains, the automated establishment of business agreements between the participants and the commercialization of these collaborative services and products under different brand and business models. Therefore, platforms for opening the operator capabilities (like those behind Open movilforum, Vodafone Betavine, etc.) or tools for composing services (like Popfly, Pipes, etc.) are not a marketplace. There are several benefits provided by a digital marketplace of services:

- The marketplace is an enabler for collaborative innovation with other service providers and industries. The marketplace allows taking advantage of the collective intelligence in order to build more attractive services that use components provided by third parties.

- Time-to-market reduction. The marketplace owner can use its own services and services shared by others to create a new one.

- The digital marketplace is a new business with the possibility of reaching new market niches (the long tail).

- With the digital marketplace, the shop runner can have an enriched product portfolio for customers, enriched with services syndicated by other service providers. Moreover, customers do not have to worry about who is providing each individual service.

- Telco operators, who are expected to be the digital marketplace owners, will increase their revenues because other service providers will reuse their services to create new ones. New incomes are generated because of the increase in the network traffic and because of revenue-sharing agreements between service providers.

In addition to this, the marketplace adds value to all the participants in the environment, because it provides common functions like billing or service assurance, so the involved service providers do not have to take care to provide these functions to the final customers.

As service providers transition to an IP-based infrastructure, deploying an IP multimedia subsystem (IMS) network is a natural progression for telecom operators. IMS addresses the challenges of service interdependence, where simple service building blocks are combined to deliver more sophisticated end-user services that integrate multimedia, data and voice within a single user session. The openness and distributed nature of the marketplace will generate a new wave of innovative services by enabling the technology and business capabilities of service providers, still providing benefits from any actor in the value network, including operators, infrastructure vendors, content providers and others. Presence and location-based services are just a few of the exciting and potentially lucrative service enablers that are delivered with IMS. These service enablers are then combined with voice, value-added text and multimedia within a single user session to deliver simple, engaging and easy-to-use IMS end-user services.

Several researchers believe that IMS standards will stabilize in 2010 (Research, 2008), as by then it is anticipated that standards bodies, platform vendors and service providers will reach a critical-mass stage of consensus around architectures, security and broad service categories. By 2010, market growth will see an inflection point and the contribution of IMS gear revenues to the overall equipment market will ramp up as barriers to adoption fall. Therefore, most IMS services will be successfully integrated in the digital marketplace.

6.2 Developers and Service Creation

The convergence between the telecom world and the Internet is one of the engines of innovation nowadays. As discussed previously, many operators are launching initiatives to open their capabilities towards the web world, and there are an increasing number of developers expecting interesting features to build new services with features of both worlds. At this point, technology is crucial. Operators should offer their functionalities in a friendly way to developers and the latter should use the appropriate programming languages and frameworks to create applications that are mostly coming from the Internet web world. This section picks some of the key representative technologies of this new wave for developers and service creation. Examples on these are, for instance, REST as the paradigm for opening a service with distributed APIs that is reachable from any point on the Internet; Ruby on Rails as a representative framework for accelerating fast development of applications with minimum effort and maximum reusability of code; rich Internet application techniques and flavours as boosters of the creation of appealing applications with easy-to-use human interfaces; and AJAX as the interim step towards the power of RIAs, providing a set of combined mature technologies that together have consolidated the web as an online user interface with unprecedented interactivity and dynamicity.

All these technologies that are coming from the new web world are expected to be applied to telecom–web converged services and to the 'telecom-only' applications as well, as they have proven their convenience in easing the developers' task and for fostering creativity of service designers. The true multimedia experience that IMS services render to users is a perfect companion of these interactive-rich multimedia applications that can be built under the combination of RIA, AJAX and REST techniques. In fact, most of the technologies that have been presented in this section are perfect candidates for the realization of most of the service strategies pursued by WIMS 2.0 and presented in Chapter 5. For instance, RIAs and AJAX enable the technical realization of the IMS virtual client in the desktop, web and mobile environments and hence seamlessly extend the user's perception of the telecom services from the native client residing in telecom-specific devices, like mobile phones and IPTV set-top boxes. REST, as a key paradigm for a new-generation exposure layer of telecom services and capabilities has already been discussed in Chapter 5.

6.2.1   Opening Capabilities: REST

Representational state transfer (REST) is a term coined by Roy Fielding in his PhD dissertation (Fielding, 2000) to describe an architecture style of networked systems. The World Wide Web is the key example of RESTful design. REST is intended to evoke an image of how a well-designed web application behaves: a network of web pages (a virtual state machine), where the user progresses through an application by selecting links (state transitions), resulting in the next page (representing the next state of the application) being transferred to the user and rendered for their use. An important concept in REST is the existence of resources, each of which is referenced with a global identifier (e.g. a URI). In order to manipulate these resources, components of the network (clients and servers) communicate via a standardized interface, e.g. HTTP, and exchange representations of these resources (the actual documents conveying the information). An application can interact with a resource by knowing two things: the identifier of the resource and the action required. The application does, however, need to understand the format of the information (representation) returned, which is typically an HTML, XML or JSON document of some kind, although it may be an image, plain text or any other content.

6.2.2   Application Framework: Ruby on Rails

Ruby on Rails is an open source web application framework for the Ruby programming language that uses the model view controller (MVC) architecture pattern to organize application programming. It combines simplicity with the possibility of developing real-world applications writing less code than other programming languages. It is intended to be used with the agile development methodology, which is often utilized by web developers for its suitability for short, client-driven projects. Rails initially utilized lightweight SOAP for web services, which was later replaced by RESTful web services. Since version 2.0, Ruby on Rails by default offers both HTML and XML as output formats. The main principles of Ruby on Rails include 'Convention over Configuration' (CoC) and 'Don't Repeat Yourself' (DRY). 'Convention over Configuration' means a developer only needs to specify unconventional aspects of the application. Generally, this leads to less code and less repetition. 'Don't Repeat Yourself' means that information is located in a single, unambiguous place. For example, using the 'ActiveRecord' module of Rails, the developer does not need to specify database column names in class definitions. Instead, Ruby on Rails can retrieve this information from the database.

6.2.3   Development Techniques: AJAX

One of the main technologies in Web 2.0 is AJAX (asynchronous JavaScript and XML) (Gittleman, 2006). AJAX uses a last generation browser and allows more interaction in applications development with the advantage of not being needed to install anything in

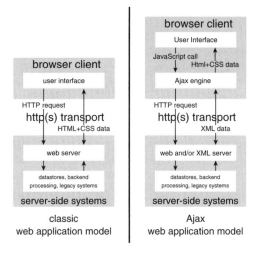

Figure 6.5 The AJAX model.

the client (see Figure 6.5). AJAX breaks the bad interactions from the request–response HTML forms, introducing an asynchronous request to the server with responses with partial reloads of the same web page, seeming a high interactive application and enhancing the user experience. AJAX has existed since the end of 1990s, but it is only now, with new browsers and more bandwidth, that it has been spread. Unlike other client technologies such as Java or Flash, the server keeps the application and interface status, not requiring an initial phase of application loading. It is based on standard web technologies (JavaScript, XML y HTTP), although not all browsers have the adequate support.

The main features are:

- AJAX allows building applications, rather than websites, enabling features to be developed that previously were out of the scope of the applications.
- Continuous interaction, similar to what the user is used to in common applications.
- Visual and interaction richness. Despite the name, the use of JavaScript and XML are not actually required and nor do the requests need to be asynchronous. AJAX is a valid technique used in multiple platforms and operative systems because of being based on open standards, namely:
- HTML and cascading style sheets (CSS) for presentation.
- Document object model (DOM) accessed with a scripting language by the user, specially ECMAScript implementations like JavaScript and JScript, for dynamic display of and interaction with data.
- XMLHttpRequest object for asynchronous communication.
- XML is the format used for the data transfer requested to the server, although any format can work.

With this technology, problems such as application distribution or the distribution of new versions or patches is solved, thanks to the use of a light client: the browser, which becomes a client-side platform container of service logic. The impact of these applications has faded the distinction between web and desktop applications, being a very interesting alternative for developers.

6.2.4 Rich Internet Applications

A rich Internet application (RIA) (Wikipedia, 2009) is a web application designed to deliver the same features and functions normally associated with desktop applications. Applications are rich or enriched by their ability to integrate multimedia capabilities such as sound, animation and video in addition to a greater degree of interactivity between the application and the client's machine. The user can thus instantaneously interact with the application without having to wait for roundtrips to and from the server because the entire application is loaded from the start. RIAs provide true two-way communication between clients and applications, moving away from the traditional page-to-page web navigation model. An RIA normally runs inside a web browser and usually does not require software installation on the client side to work. However, some RIAs may only work properly with one or more specific browsers. For security purposes, most RIAs run their client portions within a special isolated area of the client desktop called a sandbox. The sandbox limits visibility and access to the files and operating system on the client to the application server on the other side of the connection.

The development of enriched applications for the Internet responds to a necessity of elevating presentation levels (front end) so that they have the ease and richness that users expect. RIAs focus on increasing the optimization of the final user's experience and the reduction of the bandwidth and the server's load. There are some examples of RIA frameworks and technologies from different creators. In the remaining part of the section the Adobe, Microsoft and Google approaches on RIA are discussed.

6.2.4.1 Adobe AIR

Launched in early 2008 as the 1.0 version, the Adobe integrated run-time (AIR) is an interesting development framework that enables web designers and developers to use the technologies used to build websites to build applications that are used on the desktop. This implies technologies like Adobe Flex, Adobe Flash, HTML, JavaScript or AJAX. Adobe AIR runs on Windows, Mac OS X and recently on Linux. A version for mobile devices is under development.

A web browser enables a user to interact with the content and applications typically located on a website on a server. Adobe AIR builds upon capabilities and technologies used in the browser to enable deployment of applications on the desktop. Adobe AIR complements the browser by providing users and developers with a choice about how to deliver and use applications built with web technologies. Because they run on the desktop and not in a web browser they can offer more features and access and save files on the local hard drive or network. It can interact with online resources if connected or use local storage (via the included SQLite database) and synchronize data with an online source when next connected. Adobe AIR is also smart enough to know when it is and is not connected so you can create programmes that work with online data and/or use the local database as a fallback if the connection to the Internet is dropped. When compared to traditional desktop applications, Adobe AIR applications are simple to deploy, easy and cost-effective to build, have better web integration and will run on all three of the major operating systems.

6.2.4.2 Adobe Flash

This is a multimedia authoring and playback system from Adobe used to create RIAs. Adobe Flash in commonly used to create content such as web applications, animations, advertisements, games, movies and applications that accepts input from the user. Flash contents are also available for mobile phones and other embedded devices. ActionScript is the language used to write Flash programs. Several software products, systems and devices are able to create or display Flash contents, including Adobe Flash Player, a free client application which is available for most common web browsers, some mobile phones and other electronic devices (using Flash Lite). The files created using Flash are Shockwave Flash (SWF), played in Flash Player, and Flash Video (FLV), played in VLC, QuickTime and Windows Media Player with external codecs added. Flash content reaches over 98% of Internet viewers with over 30M SWF files online today. Moreover, around 70% of web video is in Flash.

On May 2008, Adobe announced the Open Screen Project, which aims to drive rich Internet experiences and create a consistent application interface across all devices like personal computers, mobile devices or consumer electronics. It was perceived that the main goal of the project is to bring Adobe Flash to mobile phones. A project advantage is that it abolishes licensing fees for Adobe Flash Player and Adobe AIR. It also removes restrictions on the use of the Shockwave Flash (SWF) and Flash Video (FLV) file formats and publishes application programming interfaces for porting Flash to new devices. Additionally, it publishes Flash Cast protocol and Action Message Format (AMF), which allows Flash applications to receive information from remote databases.

6.2.4.3 Adobe Flex

At the present time Adobe Flex version 3 is a collection of technologies released by Adobe for the development and deployment of cross-platform RIAs based on the proprietary Adobe Flash platform. Adobe Flex is a free, open-source framework for building highly interactive, expressive web applications that deploy consistently on all major browsers, desktops and operating systems. MXML, a declarative XML-based language, is used to describe graphic user interface layouts and behaviours, and ActionScript 3, a powerful object-oriented programming language, is used to create client logic. RIAs created with Flex can run in the browser using Adobe Flash Player software or on the desktop on Adobe AIR, the cross-operating system run-time. This enables Flex applications to run consistently across all major browsers and on the desktop applications, reaching more potential users. By using AIR, Flex applications can now access local data and system resources on the desktop. Since February 2008, Flex 3 is available as an open-source software through the Open-Source Flex SDK Project. Flex provides a modern, standards-based language and programming model supporting common design patterns.

6.2.4.4 Google Gears (Google, 2009b)

Created in 2007, Google Gears is free and open-source software offered by Google. Gears is a plug-in that extends the browser to create a richer platform for web applications. The broader goal is to close the gap between web applications and native applications by giving the browser new capabilities. For example, webmasters can use Gears on their websites to let users access information offline or provide contents based on customer geographical location.

Gears installs a database module, powered by SQLite, in the system client that stores data locally. Therefore a web application can synchronize the local data with the online service. Stored local information is mainly used instead of online information. Synchronization can wait until a connection is available or until the user wants to do it. Gears also has a desktop module that allows web applications to interact more naturally with the desktop and a geolocation module that lets web applications detect the geographical location of their users. As for now, Gears is compatible with Internet Explorer, Mozilla Firefox and Safari browsers. Gears was designed to be used on both Google and non-Google sites. Examples of web applications that currently make use of Gears are Google Reader and Google Docs.

6.2.4.5 Microsoft SilverLight   (Microsoft, 2009b)

This is a cross-browser, cross-platform implementation of the .NET framework for building media experiences and rich interactive applications for the web. Version 2.0 brings additional interactivity features and support for .NET languages and development tools. It is compatible with multiple web browser products used on Microsoft Windows and Mac OS X operating systems. Mobile devices, starting with Windows Mobile 6 and Symbian (Series 60) phones, will also be supported. A third party free software implementation named Moonlight is under development to bring compatible functionality to GNU/Linux. SilverLight supports the display of high definition video files so that Microsoft will do the heavy lifting of sending them over the net. This will enable smaller developers to deliver large and high definition files quickly and reliably without paying for content distribution network fees. Also, SilverLight applications are delivered to a browser in a text-based markup language called XAML. That is not difficult for web users once they land on a site. However, search engines, like Google, can scan XAML but they cannot dive into compiled Flash applications.

6.3   Users and Service Creation

One of the promises of the future Internet of services is to bring a huge amount of functionalities that can be combined easily to develop new applications or services that are more customized to the user premises. With an approximate rate of one new open API per day, the number combinations of those APIs to create new mash-ups is going up dramatically. Nowadays, it is easy to create mash-ups, but a knowledge of programming languages is still required to compose the new application, which limits the access of any user. Nevertheless, developers act in two main lines: creating applications based on common interests that can satisfy a group of potential users under demand or creating applications to satisfy some of the developer's necessities. Going to a more and more customized world, it still seems difficult to see a developer creating one application for only one user, targeted on the user's necessities or interests. However, what could be quite probable is to see one user developing applications to satisfy his or her own necessities if there are some tools to make this situation possible and easy. In fact, user-generated services (UGS) deal with this strategy. The goal is to create some tools that allow easy creation of a new service by only being familiar with the environment and the APIs available in it. Most of those tools are using click-and-drag functionalities for achieving such simplicity of use. Tools to create UGS are appearing on the market from different providers: Microsoft, Yahoo, Google, etc. In this section, Internet-oriented UGS tools like those by Yahoo and Microsoft are prenetted together with a telco-oriented solution coming from the European research space: OPUCE. Google also has a platform called Google Mashups Editor, but it still remains in close beta version limited and hence to a small number of developers.

6.3.1   UGS Based on Data Mash-ups: Yahoo Pipes

Yahoo Pipes (Yahoo, 2009) is the current UGS solution from Yahoo that provides a graphical user interface for building applications that aggregate web feeds, web pages and other services, creating web-based applications from various sources and publishing them (see Figure 6.6). The site works by letting users 'pipe' information from different sources and then set up rules that establish how the content is modified or retained (e.g. filtering). A typical example is a WIMS 2.0 case based on Flickr – a pipe that takes any RSS feed and adds a photo from Flickr based on the keywords of each item of the RSS stream. In this case, Yahoo Pipes redirects flows of data, a chained flow of actions, where the result of one serves as input for the next.

Pipes provides huge possibilities to handle and manipulate the content on feeds, translate them and filter for keywords, creating personalized filters. The pipes editor is a JavaScript

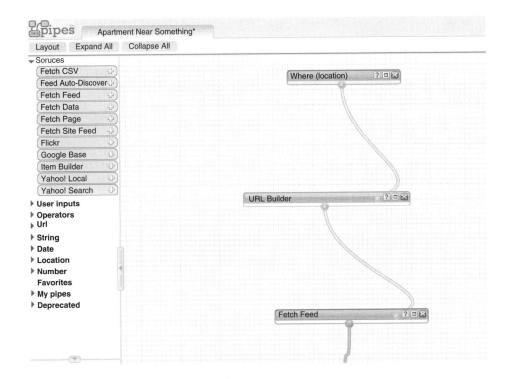

Figure 6.6 Yahoo Pipes.

tool that lets users create and edit pipes in an intuitive visual interface. The editor consists of the following pieces:

- Library, which lists available modules and saved pipes;
- Canvas, which is the main work area for assembling pipes;
- Debugger, which lets a user inspect pipe output at various stages.

A user can build and edit pipes by moving modules on to the Canvas from the Library and wiring them together.

6.3.1.1   Library

Library shows a list of all the available modules, which are grouped by functionality:

- sources: data sources that return an RSS feed;
- user inputs: input fields that a user of the pipe can fill in at run-time;
- operators: basic features like foreach, sort, count and filter;
- URL: modules for building and manipulating URLs;
- string: modules for handling strings;
- date: modules for manipulating dates.

The Library model also holds a list of user's favourite pipes belonging to other users and user saved pipes. Users can build new pipes by using those created previously by the same or other users, constituting these subpipes as new building blocks. This enables users to build useful complex components that can be reused across multiple projects. For the sake of simplicity, which is guaranteed in all aspects of the pipes–user interface, drag-and-drop functionality is enabled in order to add a module or subpipe to a new pipe and the user only has to drag it over to the canvas.

6.3.1.2   Canvas

The Canvas is the main work area for assembling and testing new pipes. The user can drag modules around and arrange them based on user decisions or using predefined layouts in the tool. The user has to wire modules together by clicking the output terminal of any module and then clicking on the input terminal of the module to be fed with data from the generator module. The editor will check the compatibility of the terminals (what kind of data that a terminal expects to emit or receive), showing them in orange to the user. Many modules have configurable parameters and input fields that can be filled by the user or supplied with information from another module.

6.3.1.3   Debugger

This shows the contents of the pipe immediately downstream from the currently selected module, allowing the inspection of each segment of the pipe to make sure it is behaving as expected. Once the pipe has finished, it can be published, making it available to the wide Internet. This case will allow people to clone a copy for their own use.

6.3.2   UGS Based on RIA: Microsoft Popfly

Microsoft Popfly (Microsoft, 2009a) is a tool that allows users to create web pages, program snippets and mash-ups using the Microsoft Silverlight (Microsoft, 2009b) RIA (rich Internet application) run-time platform and the set of online tools provided with it. It requires users to log on with their Windows Live ID that is the Microsoft commitment to provide a unique identifier. Popfly provides four tools based on SilverLight technology:

- Game Creator. A tool that allows users to create their own game or extend a game already built. It can be exported to the social network Facebook or be used as a Windows Live Gadget.
- Mash-up Creator. A tool that lets users fit together pre-built blocks in order to mash together different web services and visualization tools.

For example, a user could snap together a photo building block and a map block in order to get a geo-tagged map of pictures on a topic of their choice. Also Popfly presents an advanced view for blocks, which allows users to modify the code of the block in JavaScript, as well as providing users with flexibility in designing the mash-ups. Additional HTML code can also be added to the mash-ups. This tool also provides a function for auto-completion of HTML code and a preview function, with a live preview in the background as users link blocks.

6.3.2.1   Web Creator

This is a tool for creating web pages. Web pages are created without HTML coding and can be customized by choosing predefined themes, styles and colour schemes. Users can embed their shared mash-ups in the web page. Completed web pages will also be saved in each user's Popfly Space. The Popfly Space provides a quota of 100 MB per user where completed mash-ups and web pages are stored. Users also receive a customizable profile page and other social networking features. Public projects can be shared, rated or 'ripped' by other users. Popfly allows users to download mash-ups as gadgets for Windows Sidebar or embed them into Windows Live Spaces. Another feature of Popfly Space is the Popfly Explorer plug-in for Visual Studio Express to download the mash-ups and modify the coding, as well as perform actions such as uploading, sharing, ripping and rating the mash-ups. Popfly is more powerful than Yahoo Pipes, which is limited to RSS channels, as previously mentioned, enabling users to build mash-ups in minutes without any knowledge of either programming or writing a line of code. These mash-ups can be downloaded as gadget, iframe or added to the Popfly Space or some social network sites such as Facebook. One of the existing limitations, however, is in the source blocks, where a user finds Live resources and some third parties resources (Yahoo, Twitter, etc.). It is difficult to see some interesting resources from Google. Popfly allows users to create their own blocks, but it is not possible to choose sources not considered by Microsoft. Popfly interface is usable and powerful and it is focused on amateur developers and enthusiastic people creating new applications.

6.3.3   UGS Based on Telecom Convergence: OPUCE

OPUCE (open platform for user-centric service creation and execution) (OPUCE, 2009) aims to foster the evolution from the absolutely successful 'user-generated content' digital society to a 'user-generated services' one. OPUCE brings Web 2.0 benefits to telecommunications by providing tools that allow nonskilled users to build, share and control their own personalized convergent services, combining both web and communication capabilities. In brief, OPUCE leverages on WIMS 2.0 strategies for service convergence by adding UGS facilities and mechanisms to the WIMS 2.0 ecosystem. Some of the most remarkable advantages of OPUCE are:

- New convergent services. This covers web applications and telecommunication services (e.g. voice calls, SMS, MMS, etc.) interaction. Users will gain a real global experience, increasing satisfaction.
- End-users and customers involvement. End-users can participate fully in the creation and development process. This allows fast and early access to new services, generating a thriving space for service creation, adjusted to real market and user demands, and reducing services-associated costs.

Nowadays, there are several web mash-up solutions available, as mentioned earlier (Pipes or Popfly), but they have some strong drawbacks, such as the fact that they are only focused on combining Internet services and their fields of application are limited or unsuitable for professional usage. OPUCE is designed for users without programming skills, enabling them to mix Internet and telco capabilities in order to compose services to leverage traditional communication system constraints. Moreover, it offers smart solutions with a drag-and-drop graphical user interface, which allows a more engaging interaction for service creation and consumption, increasing user satisfaction. Figure 6.7 shows a comparison of the existing UGS platforms, with a detailed comparative description on fine grained features.

OPUCE considers three roles for interacting with the platform:

- Service Creator is the user who logs on the platform to create a new service combining base services, known as atomic enablers.
- Service Subscriber is the one who, after browsing a service catalogue or being notified when a new service of his interest has been created, can subscribe to it and can start using the service.
- Service Administrator has privileges to manage the OPUCE platform, being able to alter the lifecycle of a service or removing services that are no longer needed.

One of the distinctive features of OPUCE is the support of ubiquitous access to different applications from different end-user devices (PCs, classic telephones, mobile phones, PDA, etc.) via different wired and wireless networking technologies, at each stage of the service lifecycle including execution and creation. Figure 6.8 shows that OPUCE is positioned as a platform with two main features: simplicity and remixability between the telco world and the web.

OPUCE definitely enhances typical mash-up tools with telco-specific services support, including telco resources such as provisioning, delivery and management mechanisms. At the same time, it intends to be a pluggable solution that can be seamlessly integrated on top of existing telco service delivery infrastructures and open interfaces, by providing additional tools that greatly simplify the management and use of such environments, enhancing user satisfaction and quality of experience. In that sense, OPUCE complements the WIMS 2.0 so that the latter provides the foundation, i.e. the open service architecture and exposure layer of telecom service capabilities, to be integrated into the OPUCE editor.

OPUCE features	OPUCE	Web 2.0 Mashup tools						Telecom providers solutions					others	
		Popfly	pipes	Google mashup editor	Connected Services SANDBOX	Developer Portal	BEAHADSOFT Xtended	openkapow	Web2iCSD	betavine	Bubbletop	Olumaa	Acti5va	EzWeb
Telco services	✓	✗	✗	✗	✗	✓	✓	✗	✓	✓	✗	✗	✗	✓
IT services	✓	✓	✓	✓	✓	✓	✓	✓	✓	✓	✓	✓	✓	✓
Messaging	✓	✗	✗	✗	✗	✗	✓	✗	✓	✓	✗	✗	✗	✗
Presence	✓	✗	✗	✗	✗	✗	✓	✗	✓	✗	✗	✗	✗	✗
Voice calls	✓	✗	✗	✗	✗	✓	✓	✗	✓	✗	✗	✗	✗	✗
Audio/video conferences	✓	✗	✗	✗	✗	✓	✓	✗	✓	✗	✗	✗	✗	✗
Authentication	✓	✓	✗	✗	✓	✗	✓	✓	✓	✓	✓	✓	✗	✓
SMS	✓	✗	✗	✗	✗	✓	✓	✗	✓	✓	✗	✗	✗	✗
Call control	✓	✗	✗	✗	✗	✗	✓	✗	✓	✗	✗	✗	✗	✗
Easy services composition	✓	✓	✗	✗	✗	✗	✗	✓	✗	✗	✓	✓	✗	✓
Web Editor	✓	✓	✓	✗	✓	✗	✗	✓	✗	✗	✓	✗	✗	✓
Mobile Editor	✓	✗	✗	✗	✗	✗	✗	✗	✗	✗	✗	✗	✗	✗
content creation	✓	✓	✓	✓	✓	✗	✓	✓	✗	✗	✗	✗	✗	✓
Events oriented	✓	✗	✗	✗	✗	✗	✗	✗	✓	✗	✗	✗	✗	✓
clean service specification	✓	✓	✓	✓	✗	✗	✗	✓	✗	✗	✗	✗	✗	✓
service logic ("IF..." blocks)	✓	✗	✗	✗	✗	✗	✗	✗	✗	✗	✗	✗	✗	✗
User privileges / Rights management	✓	✗	✗	✗	✗	✗	✓	✗	✗	✓	✗	✗	✗	✓
Suscription to notifications	✓	✗	✗	✗	✗	✓	✗	✗	✗	✗	✗	✗	✗	✗
New services notifications	✓	✗	✗	✗	✗	✓	✗	✗	✗	✓	✓	✗	✗	✗
context adaptation	✓	✗	✗	✗	✗	✗	✗	✗	✗	✗	✗	✗	✗	✓
Full Service Lifecycle Mgmt.	✓	✗	✗	✗	✗	✗	✗	✗	✗	✗	✗	✗	✗	✓
Open to 3rd parties	✓	✓	✗	✓	✗	✓	✓	✓	✗	✓	✓	✓	✗	✓
User's community	✓	✗	✓	✓	✓	✓	✓	✓	✓	✓	✓	✓	✓	✓
Runtime configuration	✓	✗	✗	✓	✗	✗	✗	✗	✗	✗	✗	✗	✓	✗
Personalization based on user profile (Sme)	✓	✗	✗	✗	✗	✗	✗	✗	✗	✗	✗	✗	✗	✓
Tools for "building blocks" (Base services) creation	✓	✗	✗	✗	✗	✓	✓	✗	✗	✓	✗	✓	✗	✓
Mobile Portal	✓	✗	✗	✗	✗	✗	✗	✗	✗	✓	✓	✓	✗	✓
Gad gets widgets	✗	✗	✗	✓	✗	✗	✗	✗	✗	✗	✗	✓	✗	✓
View of the code generated	✗	✗	✗	✓	✗	✓	✓	✓	✓	✓	✓	✗	✗	✗
Suggest to components to connect to	✗	✓	✗	✗	✗	✗	✗	✗	✗	✗	✗	✗	✗	✗
View the results of the execution in the creation tool	✗	✓	✓	✗	✗	✗	✗	✓	✗	✗	✗	✗	✗	✓
Direct execution on the browser (no backend)	✗	✓	✓	✓	✗	✗	✗	✓	✗	✗	✗	✗	✗	✓
Designing tools	✓	✗	✗	✓	✗	✗	✗	✓	✗	✗	✗	✗	✗	✗

Figure 6.7 A comparison of the most relevant UGS platforms.

The OPUCE platform can be described in terms of the functionalities it provides to users, or the elements that conform to the platform and support those functionalities. Firstly, the OPUCE functionalities are described below:

- Service creation. An advanced service creation environment allows end-users to create easily and intuitively new services combining building blocks to control the workflow among them. This environment is available both for web and mobile users and enables them to cross-edit services (see Figure 6.9).
- Service description. Services are described in a multifaceted extensible approach where each facet describes a specific concern of the service: functional features, nonfunctional features and management features.
- Service lifecycle management. A feature to control and manage the whole service lifecycle from creation to withdrawal of the services through a set of components such as service deployment, service repository, service provisioning and service monitoring.
- Service execution. A functionality to provide a highly scalable and fault-tolerant hosting environment for running services.

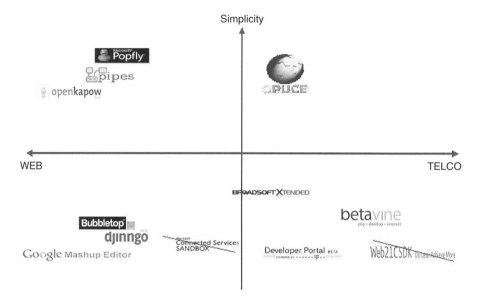

Figure 6.8 Positioning of the main UGS platforms.

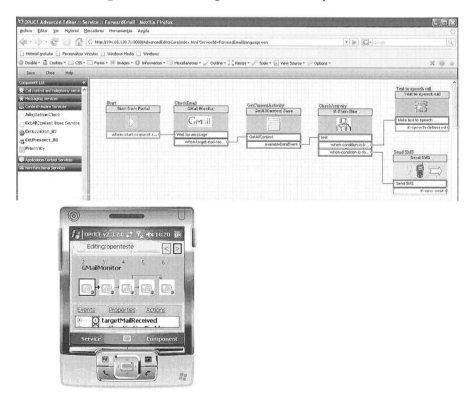

Figure 6.9 OPUCE web and mobile editor.

- Service advertising. An advanced service advertising system with learning engine and service recommendation will be built to facilitate the sharing of services between various levels of system participants – from operators to end-users.
- Service adaptability and personalization. The platform must be aware of user's profile parameter changes and must be capable of adapting services accordingly with those changes. The OPUCE platform is constituted of the following elements.
- Portal. This is used to allow service creators to compose new services from base services available and to contribute to the creation of service building blocks to be shared in a common repository. Through the portal, users can be managed by service administrators and service subscriptions directly by the users themselves.
- Service lifecycle manager. This provides interfaces to deploy and provision services, allowing them to be available automatically to service subscribers depending on its ambience, for instance. At run-time, the service lifecycle management is used to monitor execution characteristics of running services and tune their run-time parameters and deployment configuration.
- Service execution unit. This provides a distributed, highly scalable, fault-tolerant environment for running services.
- User information manager. This stores in a secured way critical and sensitive information about users, such as identity, profile, preferences and context.
- Service advertising unit. This allows users to share services and be notified about services using multiples channels. It also features an intelligent learning engine to match user characteristics and preferences to specific service features.
- Context awareness unit. This is used to personalize and adapt service executions based on certain user conditions (device features, media capabilities, ambience and user interface capabilities).

6.4   Devices and Service Creation

Applications development in mobile devices has suffered, from the beginning, from a huge fragmentation and an added complexity. While at PC platforms, implementation details were hidden by operating systems while in the mobile world the situation was not the same. Moreover, most of the initiatives were born with a not-so-open model as everyone was trying to maximize the return on investment by capturing a high market share. Innovation and development of applications for mobile devices has evolved slower than in the fixed world due to the necessity to modify each application for the different mobile device models and the limitations of those devices. There is no unified mobile device platform for easing development of applications. Instead, different aspects like screen resolution, APIs availability and configuration details have to be modified from one device to another. This causes an application developed for some of these mobile platforms not to work properly in the rest, hence increasing the price and the complexity of the application and building a barrier for developers. Likewise, any modification of the application requires the same arduous way, which results in disillusioning developers to evolve and enhance applications. Furthermore, given the targets for market penetration, an operator does not risk launching a new service unless it is supported by all relevant parts of the terminal park under its control. As network capabilities evolve, the intelligence required at the terminal side is usually higher for most innovative applications. This leads to an increasing need to deploy application logic at the terminal side in order to get a service in place. For instance, most of the innovation that evolved in telecom networks, such as those foreseen by an IMS-enabled telecom network, is based on the fact that new applications can be developed by distributing service logic at the network side and the terminal. An example is an IMS-enabled rich communicator with enhanced user interfaces or IMS-based games. This indeed requires developing client applications for mobile device platforms.

The current functionalities of mobile phones (computing speed, screen resolution, colour width in displays or Internet access speed) are similar to the functionalities of PCs ten years ago. The abstractions done by handsets manufacturers over the existing hardware are minimal in order to get the highest speed and the best real functionalities of each device. On the other hand, real efforts from manufacturers to merge with a common platform are short. The increase in smart phone devices allows similar development environments to be accessed to the PC/UNIX, but still there is great fragmentation and many limitations such as the high degree of specialization required to develop applications or the use of costly and proprietary development environments.

Before explaining future trends, it is necessary to mention the Java (J2ME) solution. A promising solution of the slogan 'develop one – run many' was hype. The reality is that to develop an application requires adaptation and tests for each of the handsets models where the application will run, not to solve existing problems.

6.4.1   Mobile Browser: Web-Based Applications

When examining terminal platform possibilities for creation and development of services overcoming multidevice challenges, the web browser appears to be a good candidate for developing cross-terminal applications by utilizing the web environment in the mobile device: the web browser. The Web 2.0 revolution has promoted web development techniques specifically targeted at adding functionality at the client side, like AJAX, described in the previous section.

The term AJAX has come to represent a broad group of web technologies that can be used to implement a web application that communicates with a server in the background, without interfering with the current state of the page. With the advantage of no requirement to install anything in the client, it uses the last generation browser. With this technology, problems such as application distribution or new versions or patches distribution are solved, thanks to the use of a light client: the browser. Mobile devices include browsing applications with varying features and complexity. A homogenization in this respect is a trend that has been happening since 2007, where expectations are that most of the 80% of feature phones in the market in 2010 will support AJAX and other Web 2.0 techniques.

6.4.2   Mobile Widgets/Gadgets

One of the tendencies inside the current web is to build applications that are developed and run in an application container. The use of preconfigured application frameworks and a remix of information allow this kind of application to be developed in a short time. These frameworks render the information and service presentation aspects at the mobile platform side, but the logic of the application runs in a remote container. The complexity lies in

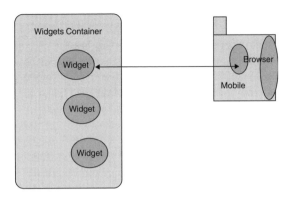

Figure 6.10 Widgets container.

choosing the right technology. Each of the manufacturers and ISPs has a version, which are seemingly the same but are usually incompatible. This is also the case with the so-called ODP (on-device portal), which is offered by most mobile handset manufacturers to the operators, where a small Java application containing the different widgets is installed in the mobile device (Figure 6.10). When the user runs this application a connection to a server is established in order to run the widget's functions. The problem here is the fragmentation due to the lack of interoperability between manufacturers. In order to offer common solution thinking to the user, different working groups in the World Wide Web Consortium (W3C) are working to develop a technology and a set of standards that allows interoperability between platforms.

If these applications are embedded in any web page and also run, without changes, in the mobile device, the problem of a unique user experience with applications and services designed easily at one time and run in many different environments would be solved. In parallel, one of the goals for these applications is to run 'offline' for those moments when it is not possible to have an adequate mobile coverage level, ensuring at least a minimum performance.

6.4.3 Mobile RIA Applications

Flash and Adobe technologies have evolved towards a simple solution, although proprietary, to develop interactive applications over a browser. The Adobe's challenge is to start a version of this software on ARM processors that would constitute a very adequate alternative to develop mobile applications due to the huge number of developers able to use this technology. Many current browsers have Flashlite, a reduced version of the Flash environment with fewer capabilities, but it still is not clear in which devices these will run. Finally, there is a way to run these Flash/FLEX applications as another application in the operating system, using AIR, although this platform is only available for PCs. Adobe's major competitor for mobile RIA platforms is Microsoft, with SilverLight, a more recent framework that will probably be catering for niches such as mobility. Microsoft plans for SilverLight to support mobile devices, starting with Windows Mobile and Symbian OS.

6.4.4 Android Environment

The Android environment proposed by Google (2008c) has the objective to increase advertisement and user profiling business by increasing the number of devices connected to the
Internet, with the inclusion of mobile devices in the mesh. Google proposes to offer the
same user experience already offered in PC environments. That is the 'look-and-feel' combined with the set of services and integration of all the pieces that Google provides. The
main problem in mobile devices is the high fragmentation in the existing operating system
offering that requires adaptation of applications on a case-by-case basis, resulting in different user experiences for a common application. To avoid this situation, Google has created
Android: a flexible, solid and free execution environment clearly focused on the user. The
main goal is to have the same user experience for the applications independent of the access.
Android works in two lines:

- To offer the best user experience, unified and attractive, independent of the devices.
- To reduce the production cost of a device with android up to 10%. In parallel with
  the development of Android, Google has built an ecosystem via the Open Handset
  Alliance (OHA:www.openhandsetalliance.com), which is based, at least initially, on
  creating win–win relationships between the members, each one pursuing a specific
  interest:
    - Device manufacturers: reducing costs in licenses and TTM due to the fact that
      Android is easily integrated into manufacturer's hardware.
    - Developers and content providers: making a reality from the paradigm 'write once,
      run many'.
    - Operators: an open environment to personalize and integrate services quickly and
      easily.

In order to ensure the success of Android, there is a need to have a critical mass
of handsets selling to feed the ecosystem. It seems that Nokia, focused on the S40 and
S60 series, or Sony-Ericsson and RIM, with proprietary platforms, are not viable candidates to play in this 'game'. HTC implemented the first Android mobile phone, the
G1, launched commercially by T-Mobile in 2008 (Google, 2008a). The rest of manufacturers, such as Samsung, Motorola or LG, could be candidates to implement Android
inside their devices. The high level architecture of the Android platform is represented in
Figure 6.11.

Based on the target requirements, Android presents the following features:

- Execution environment completely developed from J2ME.
- License model: kernel under GPLv2, user-space under Apache.
- 'Dalvik' virtual machine: executes code optimized for mobile devices (.dex) and
  manages memory more efficiently.
- HW requirements: processor ARM9 – 200 MHz, 128 MB RAM, mini/micro SD,
  screen QVGA TFT 16 bits. Optionally required is Qwerty keyboard, WiFi or GPS.
- SW requirements: Linux 2.6.
- Browser: based on KHTML, WebKit. Full-navigation, CSS, javascript, DOM, AJAX.
- Graphics: SGL for 2D and Open GL for 3D.
- Media framework: based on open code from PacketVideo.

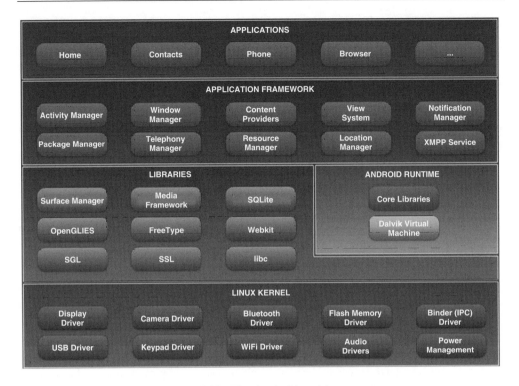

Figure 6.11 The Android architecture.

- Security model: applications have to declare the access to sensitive resources in order for the end-user to authorize the access in execution time.
- Specific APIs for Google services: GMaps, Media and XMPP.

The Android SDK allows the development of applications in Java and is perfectly integrated with Eclipse, although it has other additional tools. Android redefines the lifecycle of an application, delegating its own management process to Linux. Each application runs in a process, thus ensuring the system robustness. The design of the interfaces is based on XML with the objective to ensure compatibility between devices with different graphics capabilities. The use of XML enables the enhancement of the performance obtained using AJAX (Galindo, 2007).

In summary, Android provides a modular, robust and, what is more important, open framework based on Linux, which allows the interest of a much larger community to be captured. Android provides a tool for an environment with high fragmentation and difficult access for application developers. Android is open, offering opportunities for environment personalization never seen before in the mobile world. Nevertheless, there are open questions to answer in the future, such as the paradigm to develop an application that works in many heterogeneous handsets or to have the 'Google experience' homogeneously in any device, fixed or mobile. Considering the business aspects, Google is not only sure of its objective but also, for this adventure, knows that it is needed to create a right ecosystem.

6.4.5   The iPhone Toolkit

Apple designed the iPhone applications strategy based on AJAX applications, but instead of installing any software in the mobile device and after evaluating the low impact on the market, Apple decided to open it using a similar toolkit to the OSX.

The SDK is based on Objective C. The graphical and widgets toolkits are based on Cocoa and Carbon, allowing developers the quick creation of applications. The development platform that comes with many facilities already available to easy developments consists of four modules:

- Core OS is based on the OS X kernel and includes functions for power management, library system, security systems, sockets and network protocols implementation.
- Media for multimedia and graphics management. To highlight the support of 3D graphics using OpenGL ES and 3D positional audio, with open AL and Core Animation for animation creation.
- Core services provide access to files, threads management, network services management, preferences, address book, URLs, process and Core Location and geolocation management.
- Cocoa Touch, which is the framework to create tactile interfaces, enables a set of classes to handle events and controls (Multi-Touch, accelerometer, view hierarchy, localization, alerts, web view, people picker, image picker, camera).

Developers can have access to the same tools used internally by Apple for the software development for the iPhone. For instance, XCode is the integrated development environment with debugging support, which is the same as that used in Mac. Interface Builder is used to design application interfaces and to join the interface elements with the code based on a drag-and-drop tool. The iPhone simulator allows the running of the applications developed at Mac OS X without an iPhone handset. This tool also enables a debugger function of the applications running in the iPhone connected to the Mac. Instruments is a set of tools to check the application performance while they are running in the iPhone. For those developers programming in Mac, it is easy to create an application for the iPhone.

The installation model is based on iTunes with centralized and verified download during the connection. This model, says Apple, avoids malicious applications installation, but also is an income for Apple if the developer is planning to sell the application.

One important limitation, however, is that not all Apple applications are running in the background, which means that there is a need to save the status of each application and no asynchronicity is naturally possible. The iPhone core is the same as that of Mac. The iPhone is not a mobile phone with multimedia facilities, but a multimedia device that includes a telephone facility.

6.4.6   Maemo

Although it is not focused on mobile phones, but rather in tablets PC with WiFi connectivity, it is quite probable that Nokia is considering introducing this software in its top range of mobile phones, so-called 'smart phones'. It is based on a special distribution of Linux, called Maemo (Maemo, 2008), which will enable opening mobile devices to all existing Linux applications. It is a version of SuSe Linux (SuSe, 2008) adapted to the ARM processor, which is currently included in smart phones.

It is a very promising line, but it has to be developed for a set of mobile handsets that still neither exits nor have been announced. Maemo enables a quick development in this environment for Linux developers and portability from exiting Linux applications. Space disk limitations in mobile devices and instabilities are some of the current problems that should be solved to create a solid line of evolution.

6.4.7 Linux-Based Devices

Linux-based mobile phones represent a promising line. Although there are some examples of devices available from 2007 it was in 2008, with more than 20 devices launched on the market, that this trend started to get market share. The earlier-mentioned Android environment is an important point for Linux mobile phones. Three handset manufacturers have implemented phones running Android software during 2008: HTC, with the HTC G1 model (Google, 2008a) sold by T-Mobile; Qigi, with the Qigi i6 model (Google, 2008b), which will be sold in the Chinese market; and Kogan Technologies, with the Kogan Agora model (Arora, 2006). Qigi will be commercialized in 2009, although most of the market parameters are defined.

Before closing this chapter, it is necessary to mention the OpenMoko manufacturer (OpenMoko, 2008a) and the 'crusade' that they are following. In 2007, this company decided to change its strategy to create an open-source project (OpenMoko, 2008b). They opened not only the software of its devices but also released the CAD files. In 2008, they announced the release of the schematics for its products. They are selling the Neo FreeRunner phone for advanced users and they have planned to sell for the mass market when the software becomes stable. The Openmoko stack, which includes a full X server, allows users and developers to transform mobile hardware platforms into unique customized products. The license itself gives this possibility. OpenMoko represents a completely different way of creating mobile handsets in a very fragmented world, where the user and his/her necessities are still out of the scope of the manufacturers.

6.5 Research and Development

Beyond the current established trends, the R&D community is working in differently termed initiatives to push forward the future of the ICT panorama of services and technologies. Some of them need to be considered when observing paths to follow in the evolution of the IMS itself or the telco infrastructure in a broader perspective. These could be specific items, like the immediate strategy to leverage on IMS-enabled terminals by exploiting the potential of mobile browsers as applications containers or the future protocols considered for a multimedia telecom infrastructure that may supersede the SIP or even the IMS as a concept, or broader topics, like the considerable next-steps that may notoriously affect the ICT business, the so-called Web 3.0 and Telco 3.0. Acting as an umbrella, the Future Internet is the embracing initiative with clear political influence to sustain the digital economy, and is also considered.

In all these initiatives, the decisive role of R&D investment will shape the landscape of ICT and, more specifically, the telco business of tomorrow where, as already mentioned in this book several times, the IMS component might prove to be a booster from the technological side.

6.5.1 IMS Exposure to Browser

Developers who have ever tried coding up a mobile client application noticed that the huge variety of mobile operating systems makes it difficult to build rich applications that work on every device. This also applies to the case of developing applications that benefit from the IMS service capabilities. A trend in simplifying this development is to create web applications that run in every mobile device independently of its operator or manufacturer. Developers could use the same coding skills to create mobile applications with simple technologies and developing frameworks coming from the web. This approach is followed by Adobe in the AIR framework, which enables the creation of PC desktop applications in a similar manner as if they were web applications. Besides this, developing mobile applications utilizing the mobile web browser as container for the client-side application logic has benefits in terms of upgradeability of the application. Web applications are inherently easier to upgrade and modify their look-and-feel without forcing the user to download new software. Beyond this, if these mobile web applications could work offline, users would be able to use them when they are disconnected from the network.

Nevertheless, for applications that require interpersonal communications or other enabling capabilities like those provided by IMS – or more generically, the telecom network via the terminal device – there is a growing need for mechanisms to expose terminal capabilities to the web application running in the mobile web browser. Following this principle, the browser acts as a service container with the ability to provide applications with a local environment. This comes with controlled access to the terminal features, including IMS services and other terminal specifics, like the accelerometer, GPS, vibrator and camera. In addition, the cloud computing principles are applied to telecoms: local contact lists, call logs, multimedia local store, messaging local store, etc.

In this sense, nowadays there are some initiatives that might evolve towards the concept of IMS and terminal capability exposure to the browser. Among those OMTP Bondi and Google Gears Mobile are discussed in this section.

6.5.1.1 OMTP Bondi

The open mobile terminal platform (OMTP, 2008b) is a forum founded in 2004 by eight mobile network operators. The OMTP was set up to discuss standards with manufacturers of cell phones and other mobile devices with the aim of simplifying the customer experience of mobile data services and improving mobile device security. Bondi is a new initiative launched in 2008 by the OMTP. The name takes reference to an Australian beach following the corporative slogan 'It's safe to surf'.

The OMTP Bondi initiative tries to solve the problem that an application written for one phone must be rewritten again and again if it is to work on all phones. This is the mobile platform fragmentation problem. The cost to the mobile industry of this inefficiency is huge. This situation creates barriers that slows down time-to-market, limits market size and increases customer confusion.

OMTP seeks to remedy this problem by helping to address the way in which the existing Web 2.0 environments are moved on to mobile devices. Mobile devices offer new capabilities to web service developers that make them very desirable, but also present new security issues. OMTP is defining the key interfaces that enable the mobile web platform to access sensitive functions on the mobile device. OMTP is defining appropriate security mechanisms that will enable new services and create user trust.

By delivering high level requirements, draft specifications and a reference implementation (OMTP, 2008a), which will cement those requirements in a solid foundation, OMTP is seeking to provide a consistent, highly desirable and secure interface from web applications to billions of devices in the future. The Bondi solution lies on the following functionality:

- Application packaging. Defines the precise mechanisms by which an application can both declare its identify and define any dependencies it may have on either device capability or other software components.
- Extensible APIs. Where the functional interfaces define static mandatory APIs this work stream analyses the mechanism by which new APIs can be dynamically installed on the phone.
- Policy management. Defines the mechanisms by which policies can be changed/ managed by remote entities.
- Security policy definition. A precise, flexible and interoperable definition of when each application can access a particular resource, where an applications identity is determined by its security credentials.

The standards that Bondi is seeking will allow a web page or a widget access to specific functions of a phone, like start calls, send SMS and access to the contacts within the device address book or geo-location functions. To make it possible, Bondi is also putting a lot of focus on solving the security and privacy issues. We have to avoid situations where a widget can take a photo, send uncontrolled SMS messages or start a call without customer authorization. Things like this will damage the consumer trust in these innovative functionalities.

OMTP have joined the World Wide Web Consortium (W3C) as a member. This membership of W3C will allow OMTP to make submissions directly into the key W3C activities affecting Bondi. These will, for example, include work within the W3C Web Applications Working Group helping to standardize widgets. The ability to protect customers from malicious services originating on the web is paramount, as is the consistency of the interfaces for the developer community. By working with W3C, OMTP will be able to help deliver consistency and security and assist in making web services ubiquitous across different handsets.

6.5.1.2 Google Gears Mobile

Gears, created in 2007, is free and open source software offered by Google (2009b). Gears is a plug-in that extends the browser to create a richer platform for web applications. The broader goal is to close the gap between web apps and native apps by giving the browser new capabilities. For example, webmasters can use Gears on their websites to let users access information offline or provide contents based on customer geographical location.

After the introduction of Gears for the wide web environment, Google launched the mobile version. Gears for mobile (Google, 2009a) is an open-source mobile web browser plug-in launched in 2008. It lets developers create web applications with added speed or functionality, such as offline capabilities that allow web applications to work without an active Internet connection. If users are using a mobile web application and suddenly lose your cell connection, users can access Gears-enabled mobile web applications offline. Initially, Gears for mobiles is only available for Internet Explorer Mobile on Windows Mobile 5 and 6 devices. However, Google has said that it plans to expand support to other browsers and cell phone platforms. Presumably, that includes Opera, Safari on the iPhone and its own Android software.

In practice, Gears for mobiles would allow you to write a mail by Gmail, create a document using the text editor of Google Docs or use the Google spreadsheet while the network connection is down. Thus, users could send messages or documents using a short time connection. This could be very useful to customers who are scared by the mobile Internet high prices to use these services. Nowadays, in European countries, most customers do not use their mobile phone to access Internet services because of this. Gears is useful in case of lost Internet connection, accidentally or not.

6.5.2 Future Session and Service Protocols for IMS

Is SIP really in danger of its usage in IMS? If this is so, what are the menaces? Does any protocol really exist that might make SIP obsolete? As a matter of fact, it is not possible to answer 'Yes' to the first and third questions. At least the Internet does not reveal any exact substitute for SIP in its application to telecom networks, but this does not ensure the emptiness of the list of menaces. In fact, there are a few protocols that, according to certain studies (or professional opinions), perform better than SIP at tasks generally performed by SIP or where SIP is present somehow. SIP seems to be basic for core signalling, and that is a big difficulty itself to get SIP made obsolete, but, with regards to the media exchange, there are a few competitors that may enable the shift. In this section, comparisons with other protocols are offered (proprietary protocols are considered out of the scope of this book), and pros and cons are analysed in order to reason whether those protocols could become a preferred option to substitute SIP in a certain areas.

6.5.2.1 H.323

This is an ITU-T recommendation that used to be the only protocol for the VoIP calls setup until SIP (IETF standard) appeared. SIP offered more simplicity, ease of development, processing time reduction and overhead reduction. All these contributed to why SIP has become more popular than H.323 (Levin, 2003).

However, both protocols have evolved along the years and have learnt from each other, and even more work is currently being done to extend their specifications. SIP has become more complicated and H.323 has been simplified, so their performance is located now more or less at the same level. The literature considers that H.323 may be, again, a SIP competitor.

In terms of functionality and services that can be supported, H.323 and SIP are very similar. However, supplementary services in H.323 are more rigorously defined and therefore fewer interoperability issues are expected to arise. Furthermore, H.323 has better compatibility among its different versions and better interoperability with the PSTN. The two protocols are comparable in their QoS support. In general, SIP primary advantages are its flexibility to add new features and its relative ease of implementation and debugging (Papageorgiou, 2001).

However, one key concept that might make H.323 impose upon SIP is the fact that H.323 specifies services while SIP just specifies the signalling needed for the services to be delivered. For service-oriented platforms, H.323 implementations might be more interesting.

6.5.2.2 H.325

Since 2005, ITU-T study group 16 has been working on a project that aims to widen the multimedia communications vision depicted by H.323. The project, called the 'Advanced Multimedia System (AMS) Project', relies on the new protocol H.325 to drive the development of a third generation multimedia terminal and system architectures able to support emerging, media-rich applications that fall outside the bounds of traditional call-based communication platforms. These applications include highly converged media applications involving multiple personal and public devices, enterprise systems and network services in support of communications, collaboration and entertainment (ITU-T, 2007). The main difference between H.325, on the one hand, and H.323 and SIP, on the other hand, is that H.325 is focused on applications and device enablement rather than call setup. It has been decided that the H.325 system will utilize XML for signalling.

AMS is based on the following architectural components:

- Container. This is the device that represents the user to the network (e.g. a desk phone, mobile phone or softphone application).
- Application protocol entities (APEs). These are the applications that register with the container to provide the user with voice, video and data collaboration capabilities.
- Service nodes (SNs). These are the network entities that enable the container to establish communication with a remote entity, facilitate NAT/FW traversal and provide their network-based services.
- Application servers (AS). These are elements in the network that provide various services.

The AMS architectural development process has several important functional goals:

- AMS will have a diverse application support model, such that many different types of applications can utilize the architecture. Separation of base signalling architecture and applications will ensure that application developers do not have to be concerned with the underlying signalling details.
- Security and privacy will be built into AMS and will be both policy aware so that they can be effectively managed and interdomain aware so that they can function in a diverse, global environment.
- Network boundary awareness will be built into AMS, so that AMS systems function well between networks using different addressing universes, such as NAT, IPv4/v6 and the PSTN.

- Strong versioning, tight vocabularies and efficient signal coding methods will be employed to ensure that interoperability problems and coding errors are minimized.
- Location awareness will be included as a core application functionality of AMS. This will support emergency management requirements, but will also enable presence applications and new applications to take advantage of location information.

H.325 is currently under a standardization process. An initial release is expected in 2010. Therefore, it is too early to say that H.325 will make SIP obsolete, since it is necessary for both protocols to be implemented for a proper comparison.

6.5.2.3 XMPP

The extensible messaging and presence protocol (XMPP) is a protocol for streaming extensible markup language (XML) elements in order to exchange structured information in close to real-time between any two network endpoints. While XMPP provides a generalized, extensible framework for exchanging XML data, it is used mainly for the purpose of building instant messaging and presence applications (Saint-Andre, 2004). Although SIP can carry XML bodies, it is not optimized for messaging and presence applications (its main purpose is the session establishment). For these applications, there is an SIP extension: SIP for instant messaging and presence leveraging extensions (SIMPLE).

SIMPLE may interoperate with XMPP but, unlike XMPP, it is not optimized for the advanced exchange of structured data and provides only a subset of messaging and presence functionality. In this sense, it might be stated that XMPP is an SIP substitute for messaging and presence applications.

Further on, if we try to see whether XMPP can stand in for SIP at more scenarios, we find 'Jingle', which is an XMPP extension for the transport of RTP traffic (another typical use of SIP). Jingle also supports the SIP-like capabilities of early media and session (re-)negotiation, but does not support as many codecs as SIP. Besides the applications variety, XMPP does not imply big implementation costs (thanks to the Jabber open-source community) and there already exist several IETF standard specifications (RFC 3920, RFC 3921, RFC 3922, RFC 3923). The question is: Is all this enough to say that XMPP might make SIP obsolete? A deeper study is needed, but it really looks like a well-positioned candidate.

6.5.2.4 IAX

The inter-Asterisk exchange protocol is an application-layer control and media protocol for creating, modifying and terminating multimedia sessions over Internet protocol (IP) networks. IAX was developed by the open-source community for the Asterisk PBX and is targeted primarily at voice over Internet protocol (VoIP) call control, but it can be used with streaming video or any other type of multimedia (Capouch, 2007). IAX is an 'all in one' protocol for handling multimedia in IP networks. It combines both control and media services in the same protocol. In addition, IAX uses a single UDP data stream on a static port, greatly simplifying the network address translation (NAT) gateway traversal, which is the main advantage of IAX over SIP. IAX employs a compact encoding which decreases bandwidth usage and is well suited for the Internet telephony service. In addition, its open nature permits the additions of new payload types needed to support additional services. Since IAX is dedicated to Asterisk devices, it is not a real competitor against SIP.

6.5.3 Web 3.0: from the Social Revolution to a More Usable Internet

As mentioned in earlier chapters of the book, the main change produced by Web 2.0 is the social revolution that happens in the Internet. The user has become an active actor, 'the leading man/woman', controlling and deciding over the contents and applications placed on the Internet and actively placing contents. Web 2.0 technologies were born at the end of the 1990s and the beginning of 2000s, but were successful some years later. When attempting to talk about a potential Web 3.0, the main feature would be how to include more usability in the Internet and, at the end, make it an easier world for the end-user. Figure 6.12 shows how the Internet will evolve to 2020 (Spivack, 2008).

There is no clear definition of the Web 3.0 term yet, but most researchers talk about emerging technologies such as Semantic Web, enhanced broadband technologies such as FTTH (fiber to the home) or LTE (long-term evolution), modular web applications and enhanced computer graphics to massive 3D deployments on the Internet. Thus, Web 3.0 represents a new era in the web, where the web structure is enriched thanks to an enhancement of collaboration between people, search quality, advertisements, etc. The new web will represent more usability in the search, from the keyword search where users and service providers need to have common patterns to produce good search results to natural language or semantic search enhancing the performance of the search.

A common term under Web 3.0 is Semantic Web. What is expected under this term is to create a vision of information understandable by computers. Today's computers need

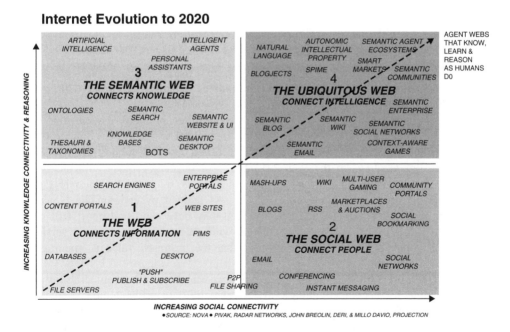

Figure 6.12 Internet evolution.

human guidance because web pages are designed to be read by people, not computers. With Semantic Web, computers could perform most of the work around finding, sharing and combining information on the web. As Tim Berners-Lee coined (Berners-Lee, 1999):

I have a dream for the Web [in which computers] become capable of analyzing all the data on the Web – the content, links, and transactions between people and computers. A 'Semantic Web', which should make this possible, has yet to emerge, but when it does, the day-to-day mechanisms of trade, bureaucracy and our daily lives will be handled by machines talking to machines. The 'intelligent agents' people have touted for ages will finally materialize.

A semantic approach to the web has some advantages, such as more accurate searches, less computing resources required, working using both structured and nonstructured data and allowing more intelligent applications to be created that share data between them. However, today there are not many tools using this approach; scalability is a current limitation and there is a challenge concerning who is creating the required metadata. Semantic Web is considered as an open database over the current web. Some key issues are:

- Intelligence is inside data and out of the applications.
- The meaning of data and the way to use it is part of the data (auto-descriptions) enabling applications to be more intelligent thanks to the data analysed.
- Data can be shared and linked more easily. The Semantic Web is composed of open standards to create this new web.
- RDF (resource description framework). This stores data like triples (a subject–predicate–object expression).
- OWL (ontology web language). This defines systems of concepts called 'ontologies'.
- SPARQL (SPARQL protocol and query language), for queries in RDF.
- SWRL (Semantic Web rule language), for rules definition.
- GRDDL (gleaning resource descriptions from dialects of languages). This transforms data to RDF.

There are more possibilities than RDF/OWL to express semantics in the web, like micro-formats, chain of tags, controlled taxonomies and vocabularies, etc.

After seeing many changes in the Internet and related technologies, it seems that users, except those digital natives, spend more time that expected to accept new functionalities. In fact, one common characteristic of companies is that most of them still have not yet adapted their organizations to the Web 2.0 trends and business models with the goal to increase sales or to maintain the current position in sales, as stated by Antonio Fumero from Win Win Consultores (Consultores, 2008). As defined by experts like Nova Spivack (2008) in trends about the Internet, Web 3.0 may have success during 2010–2020. Therefore, it is perhaps still too soon to define with accuracy when the new version of the web will come, but what is quite clear is that the promised value is important for the end-user.

On this front, users are demanding more and more usability on the web. Usability needs to join the simplicity, accuracy, intelligibility and the ability to reduce tedious activities by the user. Whether Web 3.0 will bring those capabilities to the user is still to be demonstrated, but what has happened with Web 2.0 is an irreversible social change. More and more users are joining the Web 2.0 effect, of course without knowing about the underlying technologies.

If technologists have learned the lesson of communicating clearly the value to the users, the new web will be adopted quickly. If not, then Web 3.0 will take some more years to be accepted by the user, or, what is worse, could be deprecated by users. An underlying question is: If people are happy about more and more activities being controlled by computers, are we building the Matrix?

6.5.4   Telco 3.0

Telco 2.0 is known as the paradigm or industry model where telecom companies try and get benefits from sources that are not their own. The business model is impacted by the transformation in relationships with both users and competitors (Telco 2.0, 2006):

- Users evolve from simple consumers to 'prosumers' (producers + consumers). Telco service providers expose a series of services and enablers that bring together numerous applications and types of content from a variety of sources to enable people to produce and consume composite services.
- Competitors change into 'coopetitors': partnership and cooperation are invesiged in order to achieve service synergies. Telco 2.0 embraces the principles of Web 2.0, which shows that open interfaces are to be used. This means that anyone (user or company) can create and upload applications and services. With this, telco operators start to behave as Internet content providers. Keeping all this in mind, Telco 3.0 can be regarded as the addition of Telco 2.0 plus a series of features that can be grouped into two big fields (Canal, 2007):
  - Enabling infrastructure. More flexible, more effective, less expensive Infrastructures tend to be self-sustained.
  - Innovative technologies. These involve P2P, autonomics, semantics (ontologies/folksonomies), etc. Like Telco 2.0, Telco 3.0 embraces the principles of Web 3.0, which is regarded as an evolution of Web 2.0, where machines can gather, elaborate and combine data and information semantically, much as humans browse the web today, as described in the previous section.

A WebTelco 3.0 vision is elaborating the model of an open, distributed and flexible service ecosystem built as an ecology of components/agents learning and self-adapting to prosumers' behaviours. In particular, data, services and telco-ICT enablers are, in the same way, abstracted by lightweight components/agents capable of self-organizing semantically to serve (in an adaptive and goal-oriented way) prosumers' needs (Cascadas, 2009). From the business point of view, together with the 'coopetition', the concept of 'economy of abundance' arises; by offering a lot of services for free, telco service providers may gain and develop new sources of income.

Telco 3.0 does not count with a closed set of defining characteristics so far. Telco 3.0 is a wide, vague enough concept, so that it becomes extended with any single new research project that is started in the area. The main lines of investigation that support Telco 3.0 at the moment are the following:

- Future Internet services: Web 3.0 based, user-centric ecosystem based, a search for maximizing user experience and satisfaction. Open APIs are available for 'prosumers'.

- Distributed service eco-systems based on autonomic agents or components and peer-to-peer technologies and solutions – the situated and autonomic communications (SAC) paradigm.
- Situated communications (Sestini, 2004): reacting locally on context changes, ranging from sensor networks to virtual networks of humans and considering technological, social or economic strategic needs.
- Autonomic communications (Sestini, 2004): network elements autonomously interrelated and controlled, learning the desired behaviour, with self-organizing and radically distributed features.
- Service platform virtualization.
- Security, resilience.
- Self-organization, evolution, management, healing, synchronization, etc.
- Situation awareness: self-ware, knowledge-ware, sensor-ware, service-ware.
- Ambient intelligence: ubiquitous computing, ubiquitous communication and intelligent user interfaces.
- Pervasive computing and communications.
- Creation of distributed and dynamic ontologies and folksonomies.
- Technologies enabling collaborative services with direct interactions of prosumers.
- Interaction of new paradigms with society: social networks, user-generated contents and services.

Telco 3.0 seeks to find the following industrial benefits:

- Optimization of Telco-ICT service frameworks (Capex savings). Since all services will be developed within a common framework, capital investment is needed only for building this framework.
- Simplification of management (Opex savings) of Telco-ICT service frameworks. Since all services will be developed within a common framework, common management for all services/enablers may be implemented, yielding operational cost reductions.
- Introduction of innovative SAC services. The SAC paradigm (described above) contributes to the optimization of the service framework and the simplification of management. The 'situated' feature intends to improve the user experience; thus SAC services are more attractive to clients who aim for service personalization and environment customization. The 'autonomic' feature enhances and simplifies fault handling and resilience, as well as making possible cooperation with partners and open 'prosumption', which are new business opportunities after all.
- Research in business processes modelling for workflow enhancement. Improving processes efficiency avoids waste, reduces time-to-market and, in general, increases performance. Research in business processes modelling is a long-term activity since, due to technological innovations, an optimal process one year may be obsolete and poor performing the following year. Business processes modelling is a general worry in companies nowadays, and a number of organizations are leading research activities in this sense. For instance, the TeleManagement Forum (TMF, n.d.) is developing standards for telco business processes.
- Partnership synergy. Since Telco 3.0 is 'coopetition' oriented, sharing the service framework with partners will yield enhanced, composite services that will be ready to use within shorter terms.

- New sources of income yielded from 'economy of abundance'. The typical example that illustrates the concept of 'economy of abundance' is its appliance to music. Illegal downloads are provoking the condition that selling records is less and less a good deal. Applying 'economy of abundance', all music should be downloaded for free and revenues will come in the form of concert tickets (since the music is delivered for free, it can reach more people, including those who would not have bought an album in the first place; as a result, more people become fans), merchandizing, clicks on the adverts on the webpage where you can download the music, etc.
- Revenues from services not made by the Telco company. Users and other companies will develop a number of services. The Telco company may charge a small fee for development toolkits and make business from developed services in a direct way (selling the service) or indirect way (like advertisement add-ins, for instance).
- Benefits from the 'Long Tail' model. The theory that the Internet enables businesses to sell a large selection of goods in small quantities to a large population of customers at minimum cost was popularized as the 'Long Tail' by Chris Anderson in a wired magazine article in 2004 (Anderson, 2009).
- Future Internet development. Web 3.0 is the next step: what will come afterwards? What will be Web 4.0? What will be the business opportunities? The only way to find out is by taking the next step forward, joining the WebTelco 3.0 paradigm.
- Churn reduction. This is a logical consequence of clients' satisfaction, motivated by the availability of better services at a lower cost.

6.5.5 Future Internet

With over a billion users worldwide, the current Internet is a great success – a global integrated communications infrastructure and service platform underpinning the fabric of the world economy and society in general. However, today's Internet was designed in the 1970s for purposes that bear little resemblance to current and foreseen usage scenarios. The effect of that is summarized in the following points (*et al.*, 2008b):

- The Internet has grown from an experimental research network to an infrastructure supporting the economy as well as the provision of societal services.
- The Internet is becoming pervasive and ubiquitous, with already 25% of the world population having access to it.
- It has enabled user and consumer empowerment, through the emergence of e-Commerce and social networks.
- It has helped the modernization of public administration through the emergence of e-Government, e-Health, e-Education. Internet use is also expected to contribute significantly to solve emerging challenges such as climate change and energy efficiency.

The Organization for Economic Co-operation and Development (OECD, 2008) has politically acknowledged the importance of the Internet in the present days and in the future for the economic, cultural and social development of modern societies. OECD's Seoul declaration (OECD, 2008) for the future of the Internet economy stresses the achieved commitment to promote policies to facilitate the expansion of Internet access and use worldwide; the promotion of an Internet-based innovation, competition and user choice; to secure critical information infrastructures and ensure the protection of personal information intellectual property rights; and to create a market-friendly environment for convergence that encourages infrastructure investment, higher levels of connectivity and innovative services and applications.

Mismatches between original design goals and current utilization are now beginning to hamper the Internet's potential. A large number of challenges in the realms of technology, business, society and governance have to be overcome if the future development of the Internet is to sustain the networked society of tomorrow. Thus, multiple initiatives in the scope of the future Internet have been launched in different regions worldwide with the objective of sustaining the competitiveness of the regions concerned. In the US, the most well-known initiatives are GENI and NetSe initiatives of the US National Science Foundation (www.geni.net). In Japan, the AKARI initiative supports a Japanese new generation network (NWGN) project. For these two initiatives, clean slate approaches are considered with a time-to-market horizon in the 2015 to 2020 range.

Perspectives in Europe forged the emergence of the Future Internet Assembly (FIA, 2009), kicked off on the occasion of the Bled Conference of March 2008. As the Assembly narrates, the FIA gathered sector actors in the research space to define a European approach towards the future Internet, also leveraging the achievements of many national research initiatives. The FIA also addresses nonresearch issues, in view of easing the future deployment of the technologies and systems, such as standardization and international cooperation.

The main goal of the FIA is to confront the various technological and architectural approaches deriving from the multiplicity of usage scenarios and requirements applying to a future Internet. It is expected that this will help the emergence of cross-sector consensus and pave the way towards the later emergence of common architectural approaches and standards, since there is no agreed definition of what a future Internet should cover, both in terms of research and nonresearch issues. Early research work on the future Internet has primarily concentrated on networking issues and on the need to support a range of innovative applications with enhanced management capabilities and a particular focus on security issues. On the other hand, it is more and more widely accepted that future Internet issues are driven by a very large set of new application requirements. It is hence nowadays commonly admitted that the future Internet needs to be tackled from a holistic perspective, beyond the 'network' perspective but with a broader 'technology focus', which is mostly related to industrial (ICT) competitiveness. In the case of Europe competences and leadership in ICT, this includes notably technical areas such as mobility, very high rate end-to-end broadband, security and even 3D media. It is, however, also clear that a 'technological only' focus would not be the right approach, and therefore a need emerges to take into account the applications dimension very early in the process, eventually leading to future Internet technologies. With respect to applications, a key feature of many. Future Internet initiatives relates to the provision of testbeds and experimental facilities, which foster opportunities to make novel classes of innovative applications available and can have huge economic and social impacts outside the ICT sector. A typical example would be sensor-based applications for remote health care and environmental or energy efficiency control. Experimental research facilities offer a good opportunity not only to evaluate the viability of the technical solutions but also to consider more effectively and dynamically the application requirements as part of the research process. User-driven innovation is supported, by enabling user access to the experimental research facilities. It is more and more recognized that the availability and mastering of complex experimental research facilities and testbeds provide a competitive advantage in terms of effectiveness of R&D and innovation processes.

The need to make the future network and service infrastructure trustworthy is another key aspect to be addressed in designing the future Internet. Trust, security and privacy protection deserve specific attention as they have to be addressed horizontally, across all possible usage and technological scenarios. With these considerations in the landscape of the European vision, the following high level objectives are considered to drive research pursuing a future Internet:

1. To accommodate unanticipated user expectations together with continuous user empowerment.
2. To become the common and global information exchange environment of human knowledge.
3. To lever and evolve information and communication technologies, capabilities and services to fulfil increased quantity and quality of Internet use.
4. To support open cultural, scientific and technological exchanges across all regions and cultures, as well as within single communities.
5. To be ubiquitously accessible and open (at physical, connectivity and information levels).
6. To be secure, accountable and reliable without impeding user privacy, dignity and self-arbitration.
7. To support mobility, have widespread ubiquitous availability and be capable of assisting society in emergency situations.
8. To support effective and efficient performance management features based on context, content, etc.
9. To support innovative business models that allow for all entities equal access to the services and service provision markets.
10. To be carbon neutral, energy efficient and environmentally sustainable.

In order to cope with these high level objectives, finer challenges are depicted in the future landscape for the Internet to address a multiplicity of new usage patterns and the plethora of service requirements not foreseen when the Internet was designed. Even if there is no agreed definition or standard describing the nature of the future Internet, consensus on most of these challenges is driving the research. These challenges, described below, can serve as a set of technical features that may describe what the future Internet would resemble:

- Mobility, to allow a native support of moving scenarios and an open Internet fully available on the move.
- End-to-end very high throughput, to support data-intensive usage scenarios, such as medical imaging transfer and processing.
- Security and trust, involving processing of sensitive personal data or financial transactions.
- Internet of things to support device connectivity (RFID, sensors) and coupling of virtual world data with physical world information in a scalable and efficient way.

- Internet of services: service architecture enabling dynamic, secure and trusted service compositions and mash-ups while allowing the user a proactive role (user-generated services).
- 3D and media intensive support: 3D virtual environment, possibly coupled with physical world information, beyond games.
- Negotiated management and control of resources: dynamic and predictive network management, infrastructure observability and controllability to support variability of the business model, from a best-effort low level of control towards full real-time management of quality of service, security level, etc.
- User-controlled infrastructure: ad hoc network and service composition, user-driven deployment scenario and control of connectivity.

Beyond research, it is considered that important nontechnological issues have also to be covered. International cooperation with those regions having initiated large activities is seen positively, especially if it provides opportunities to run global experiments through interconnected research facilities. International cooperation is also deemed necessary for what concerns the standardization effort as well as interoperability testing that will have to be undertaken once novel technological approaches are identified at the EU level.

Finally, the future Internet should take due account of emerging societal challenges, in addition to more classical issues such as user acceptability and the need to increase further the role of the Internet as an enabler of innovation ecosystems. Such additional issues include green computing and support for the energy efficiency of multiple user sectors.

6.6   Concluding Remarks

The chapter has attempted to provide a possible snapshot of the future. The authors understand, however, that this is a very fragile and unpredictable domain with numerous established players and also new players who may influence the future. With that in mind, we believe that the book, in general, sheds new light on the prospects of IMS in a way that has never been done before. It also gives the most comprehensive evaluation of current standing and future possible positioning of IMS in the service-oriented landscape. In its final part, the book also provides a unique vision of the road ahead. We hope that it will help researchers as well as industry players and the strategy makers in their efforts to move forward in the currently turbulent sector.

References

3GAmericas (2008) Transitioning to IPv6; http://www.3gamericas.org.

3GPP (1999a) Bearer services (BS) supported by a GSM public land mobile network (PLMN), TS 02.02, Third Generation Partnership Project (3GPP).

3GPP (1999b) General packet radio service (GPRS); Serving GPRS support node (SGSN) – Visitors Location Register (VLR); Gs interface network service specification, TS 09.16, Third Generation Partnership Project (3GPP).

3GPP (1999c) High speed circuit switched data (HSCSD): Stage 1, TS 02.34, Third Generation Partnership Project (3GPP).

3GPP (1999d) The Third Generation Partnership Project; http://www.3gpp.org.

3GPP (2001a) General packet radio service (GPRS); Mobile station (MS) – Serving GPRS support node (SGSN); Subnetwork dependent convergence protocol (SNDCP), TS 04.65, Third Generation Partnership Project (3GPP).

3GPP (2001b) UMTS Phase 1, TS 22.100, Third Generation Partnership Project (3GPP).

3GPP (2001c) UTRAN TDD low chiprate, TR 25.937, Third Generation Partnership Project (3GPP).

3GPP (2002) 3GPP, technical specifications and technical reports for a GERAN-based 3GPP system, Technical Report, ESTI.

3GPP (2005a) Customized applications for mobile network enhanced logic (CAMEL); CAMEL application part (CAP) specification for IP multimedia subsystems (IMS), TS 29.278, Third Generation Partnership Project (3GPP).

3GPP (2005b) UTRAN high speed downlink packet access (HSDPA), TR 25.950, Third Generation Partnership Project (3GPP).

3GPP (2005c) Telecommunication management; charging management; charging principles, TS 32.200, Third Generation Partnership Project (3GPP).

3GPP (2006a) General requirements on interworking between the public land mobile network (PLMN) and the integrated services digital network (ISDN) or public switched telephone network (PSTN), TS 29.007, Third Generation Partnership Project (3GPP).

3GPP (2006b) IP multimedia subsystem (IMS) messaging; Stage 1, TR 22.940, Third Generation Partnership Project (3GPP).

3GPP (2006c) Multimedia broadcast/multicast service (MBMS) in the GERAN; Stage 2, TS 43.246, 3rd Generation Partnership Project (3GPP).

3GPP (2006d) Requirements for evolved UTRA (E-UTRA) and evolved UTRAN (E-UTRAN), TR 25.913, Third Generation Partnership Project (3GPP).

3GPP (2006e) Telecommunication management; charging management; charging data description for the IP multimedia subsystem (IMS) TS 32.225, Third Generation Partnership Project (3GPP).

3GPP (2007a) Generic access network (GAN); Stage 2, TS 43.318, Third Generation Partnership Project (3GPP).

IMS: A Development and Deployment Perspective   Khalid Al-Begain, Chitra Balakrishna, Luis Angel Galindo and David Moro
© 2009 John Wiley & Sons, Ltd

3GPP (2007b) Internet protocol (IP) multimedia call control protocol based on session initiation protocol (SIP) and session description protocol (SDP); Stage 3, TS 24.229, Third Generation Partnership
Project (3GPP).

3GPP (2007c) IP multimedia core network subsystem (IMS) multimedia telephony service and supplementary services; Stage 1, TS 22.173, Third Generation Partnership Project (3GPP).

3GPP (2007d) IP multimedia (IM) subsystem Cx and Dx interfaces; Signalling flows and message
contents, TS 29.228, Third Generation Partnership Project (3GPP).

3GPP (2007e) IP multimedia subsystem (IMS) emergency sessions, TS 23.167, Third Generation
Partnership Project (3GPP).

3GPP (2007f) IP multimedia subsystem (IMS) messaging; Stage 1, TS 22.340, Third Generation
Partnership Project (3GPP).

3GPP (2007g) IP multimedia subsystem (IMS); Stage 2, TS 23.228, Third Generation Partnership
Project (3GPP).

3GPP (2007h) Messaging service using the IP multimedia (IM) core network (CN) subsystem; Stage
3, TS 24.247, Third Generation Partnership Project (3GPP).

3GPP (2007i) Multimedia broadcast/multicast service (MBMS); Stage 1, TS 22.146, Third Generation
Partnership Project (3GPP).

3GPP (2007j) Multimedia messaging service (MMS); Stage 1, TS 22.140, Third Generation Partnership
Project (3GPP).

3GPP (2007k) Network architecture, TS 23.002, Third Generation Partnership Project (3GPP).

3GPP (2007l) Numbering, addressing and identification, TS 23.003, Third Generation Partnership
Project (3GPP).

3GPP (2007m) Presence service; Architecture and functional description; Stage 2, TS 23.141, Third
Generation Partnership Project (3GPP).

3GPP (2007n) Voice call continuity (VCC) between circuit switched (CS) and IP multimedia subsystem
(IMS); Stage 2, TS 23.206, Third Generation Partnership Project (3GPP).

3GPP (2007o) Evolved universal terrestrial radio access (E-UTRA); Long term evolution (LTE) physical layer; general description, TR 36.201, Third Generation Partnership Project (3GPP).

3GPP (2007p) Improved network controlled mobility between LTE and 3GPP2/mobile WiMAX radio
technologies, TR 36.938, Third Generation Partnership Project (3GPP).

3GPP (2007q) Organization of subscriber data, TS 23.008, Third Generation Partnership Project
(3GPP).

3GPP (2007r) IP multimedia system (IMS) Sh interface; Signalling flows and message contents, TS
29.328, Third Generation Partnership Project (3GPP).

3GPP (2007s) Location services (LCS); Service description; Stage 1, TS 22.071, Third Generation
Partnership Project (3GPP).

3GPP (2007t) Multiple input multiple output (MIMO) antennae in UTRA, TR 25.876, Third Generation
Partnership Project (3GPP).

3GPP (2008a) Conferencing using the IP multimedia (IM) core network (CN) subsystem; Stage 3, TS
24.147, Third Generation Partnership Project (3GPP).

3GPP (2008b) Evolution of policy control and charging (Release 7), TS 23.803, Third Generation
Partnership Project (3GPP).

3GPP (2008c) Introduction of the multimedia broadcast/multicast service (MBMS) in the radio access
network (RAN); Stage 2, TS 25.346, Third Generation Partnership Project (3GPP).

3GPP (2008d) Policy and charging control over Rx reference point, TS 29.214, Third Generation
Partnership Project (3GPP).

3GPP (2008e) Rationale and track of security decisions in long term evolution (LTE) RAN/3GPP
system architecture evolution (SAE), TR 33.821, Third Generation Partnership Project (3GPP).

3GPP2 (2000) The Third Generation Partnership Project 2; http://www.3gpp2.org.

6MAN (2008) IPv6 maintenance; http://www.ietf.org/html.charters/6man-charter.html.

Aboba, B., Beadles, M., Arkko, J. and Eronen, P. (2005) The network access identifier, RFC 4282, Internet Engineering Task Force.

Amazon (2007) Web Services, URL http://amazon.com/.

Amazon (2009) URL http://www.amazon.com/.

Anderson, C. (2009) The Long Tail blog; http://longtail.typepad.com/the-long-tail/.

Apple (2001) iTunes, URL http://www.apple.com/itunes/.

Apple (2002) iPhoto, URL http://www.apple.com/ilife/iphoto/.

Arkko, J., Lindholm, F., Naslund, M., Norrman, K. and Carrara, E. (2006) Key management extensions for session description protocol (SDP) and real time streaming protocol (RTSP), RFC 4567, Internet Engineering Task Force.

Arora, K. (2006) Android phone pops up Down Under; http://www.linuxdevices.com/news/ NS2005328761.html.

Asterisk (2008) Asterisk Project.

Audet, F. and Jennings, C. (2007) Network address translation (NAT) behavioral requirements for Unicast UDP, RFC 4787, Internet Engineering Task Force.

AVT (2008) Audio-video transport; http://www.ietf.org/html.charters/avt-charter.html.

Baker, M. and Stark (2002) The application/xhtmlxml media type Technical Report 3236, Internet Engineering Task Force.

Bang, N. D. Jorstad, I. and Jonvik, T. (2007) Analysis of IMS business requirements; http://www.icin. biz/files/programmes/Session6A-2.pdf.

Beck, A. and Ensor, B. (2007) IMS and IPTV service blending lessons and opportunities, *Journal of the Institute of Telecommunications Professionals*.

Berners-Lee, T. (1999) Weaving the web, ISBN:0062515861.

BidPlace (2008) http://www.bidplacesb.com

Blogger (1999) URL http://www.blogger.com.

Bormann, C., Cline, L., Deisher, G., Gardos, T., Maciocco, C., Newell, D., Ott, J., Sullivan, G., Wenger, S. and Zhu, C. (1998) RTP payload format for the 1998 version of ITV-T Recommendation H.263 video (H.263+), Technical Report 2429, Internet Engineering Task Force.

Burger, E. (2006) A mechanism for content indirection in session initiation protocol (SIP) messages, RFC 4483, Internet Engineering Task Force.

Caida (2003) Top problems with the Internet; http://www.caida.org/publications/presentations/2003/.

Calhoun, P., Loughney, J., Guttman, E., Zorn, G. and Arkko, J. (2003) DIAMETER base protocol, RFC 3588, Internet Engineering Task Force.

Canal, G. (2007) Telecom Italia Labs, the services network: mashing up Telco, SOA and Web 2.0, HP communications world; http://h41112.www4.hp.com/events/cw2007/pdfs/presentations/ 281107-LV-1530.pdf.

Capouch, B. (2007) IAX2: inter-Asterisk eXchange, Version 2, Internet-Draft draft-guy-iax-03, Internet Engineering Task Force (work in progress).

Carpenter, B. and Moore, K. (2001) Connection of IPv6 domains via IPv4 Clouds, RFC 3056, Internet Engineering Task Force.

Cascadas (2009) Cascadas Project, Workshop Web 3.0 and Telco 3.0; http://www.cascadas-project. org/docs/Workshop@TI.pdf.

CBS (1997) Market watch, URL http://www.marketwatch.com.

Cerf, V. (2002) The Internet is for everyone, RFC 3271, Internet Engineering Task Force.

Chen, J.C. and Zhang, T. (2004) *IP-Based Next-Generation Wireless Networks*, John Wiley & Sons, Inc.

Choi, D. and Fischer, C. (2006) Transition to IPv6 and support for IPv4/IPv6 interoperability in IMS, *Bell Labs Technical Journal*, **10**, 261–170.

Consultores (2008) Win Win Consultores: creating a more trust ecosystem; URL http://www. winwinconsultores.com.

Consultores (2009) Vision of the marketplace http://www.winwinconsultores.com.

Craiglist (1995).

CTIA (2005) International associations for wireless telecommunications industry; http://www.ctia.org/.

Cuevas, A., Moreno, J., Vidales, P. and Einsiedler, H. (2006) The IMS service platform: a solution for next-generation network operators to be more than bit pipes, *IEEE Communications Magazine*, **44**(8), 75–81.

DIME (2008) Diameter maintenance and extensions; http://www.ietf.org/html.charters/dime-charter. html.

DoubleClick (2009) URL http://www.doubleclick.com/.

Durand, A., Fasano, P., Guardini, I. and Lento, D. (2001) IPv6 Tunnel Broker, RFC 3053, Internet Engineering Task Force.

ebay (1995) URL http://www.ebay.com/.

Ericsson (2008) Rich communication suite – a business initiative, http://www.ericsson.com/ technology/whitepapers/.

Ericsson (2009) Service development studio (SDS); http://www.ericsson.com/mobilityworld/sub/open/ technologies/ims_poc/tools/sds_40.

ETSI (1993) GSM Technical Report 03.47, Version 4.1.0, ETSI.

ETSI (1996) High speed circuit-switched data (HSCSD), Stage 2 (Release '96), GSM 03.34, Technical Report Version 5.2.0, ETSI.

ETSI (1998) European digitial cellular telecommunications system (Phase 2+); GPRS base station subsystem (BSS) – serving GPRS support node (SGSN) interface network service, Technical Report Recommendation GSM 08.16, Version 7.1.0, ETSI.

ETSI (2002) Electromagnetic compatibility and radio spectrum matters (ERM); Base stations (BS) and user equipment (UE) for IMT-2000 third-generation cellular networks; Part 2: Harmonized EN for IMT-2000; CDMA direct spread (UTRA FDD) (UE) covering essential requirements of Article 3.2 of the R and TTE Directive, Technical Report Version 1.1.1, ETSI.

Facebook (2004) URL http://www.facebook.com.

FFMPEG (2008) FFMPEG Project, as seen on 08.08.2008.

FIA (2009) Future of the Internet assembly (FIA), portal of the European future of the Internet, http:// www.future-internet.eu/.

Fielding, R.T. (2000) Architectural styles and the design of network-based software architectures; http://www.ics.uci.edu/fielding/pubs/dissertation/top.htm.

Flickr (2009) URL http://www.flickr.com.

FMC (2004) Fixed mobile convergence; http://www.thefmca.com/.

FOKUS (2006) Open IMS core; http;//www.openimscore.org/.

Furht, B. (1995) A survey of multimedia compression techniques and standards. Part i: JPEG standard, *Real-Time Imaging*, **1**(1), 49–67.

Galindo, L.A. (2007) Web 2.0 movil: camino del Teleco 2.0 mundo Internet.

Garcia-Martin, M. (2006) A session initiation protocol (SIP) event package and data format for various settings in support for the push-to-talk over cellular (PoC) service, RFC 4354, Internet Engineering Task Force.

Garcia-Martin, M., Henrikson, E. and Mills, D. (2003) Private Header (P-Header) extensions to the session initiation protocol (SIP) for the 3rd-Generation Partnership Project (3GPP), RFC 3455, Internet Engineering Task Force.

Gittleman, A. (2006) *Review of 'Foundations of Ajax by Ryan Asleson'*, APress, *ACM Queue*, **4**(1), 62.

Google (1998a) URL http://www.google.com/.

Google (1998b) AdSense, URL http://www.google.es/intl/es/ads.

Google (2008a) Android debuts; http://www.linuxdevices.com/news/NS6676946579.html.

Google (2008b) Android phone launches in China; http://www.linuxdevices.com/news/NS6676946579. html.

Google (2008c) What is Android?; http://code.google.com/android/what-is-android.html.

Google (2009a) Google Inc., Google Gears Mobile; http://mobile.google.com/support/bin/topic.py?topic=14078/.

Google (2009b) Google Inc., Google Gears; http://gears.google.com/.

Google (2009c) AdWords URL http://www.google.es/intl/es/ads/.

Gouraud, C. (2006) IMS lantern; http://wwww.imslantern.blogspot.com.

Gourraud, C. (2007) Using IMS as a service framework, *Vehicular Technology Magazine, IEEE*, **2**(1), 4–11.

GSMA (2007) A guide to the GSMA's SIP trials focusing on practical interworking of IMS using SIP over the evolved GRX; http://gsmworld.com/our-work/programmes-and-initiatives/ip-networking/.

GSMA (2008a) GSMA rich communication suite (RCS); http://gsmworld.com/our-work/programmes-and-initiatives/rcs/index.htm.

GSMA (2008b) GSMA SIP trial outcomes; http://www.gsmworld.com/index.htm.

Greene, N., Ramalho, M. and Rosen, B. (2000) Media gateway control protocol architecture and requirements, Technical Report 2805, Internet Engineering Task Force.

Groves, C. and Lin, Y. (2007) H.248/MEGACO Package Registration Procedures (draft-groves-megaco-pkgereg-00), Technical Report, Internet Engineering Task Force (work in progress).

Handley, M., Jacobson, V. and Perkins, C. (2006) SDP: session description protocol, RFC 4566, Internet Engineering Task Force.

Huitema, C. (2003) Real time control protocol (RTCP) attribute in session description protocol (SDP), RFC 3605, Internet Engineering Task Force.

Hwang, J., Kim, N., Kang, S. and Koh, J. (2008) *A Framework for IMS Interworking Networks with Quali of Service guarantee* IEEE Computer Socie Washington, DC.

ICT (2008) Conference Session Report on the Future of the Internet European Commission Information Society and Media.

IETF (1986) Internet Engineering Task Force; http://www.ietf.org.

IMS-Communicator (2004) IMS Communicator Project; http://imscommunicator.berlios.de/.

IMSForum (2008) IMS Forum; http://www.msforum.org/.

IMSForum and NGNForum (2008) IMS and NGN report card; http://www.imsforum.org/content/public-documents.

ITU (1989) International Telecommunications Union; http://www.itu.int/.

ITU (2006) International Telecommunications Union; http://www.itu.int/ITU-D/ict/statistics/.

ITU-T (2007a) H.324: termina for low bit rate multimedia communication.

ITU-T (2007b) ITU-T, Study Group 16, draft initial project description for the advanced multimedia system.

JCP (2003) JSR 116: SIP Serolet V1.0.

JCP (2005a) JSR 164: SIMPLE Presence.

JCP (2005b) JSR 165: SIMPLE Instant Messaging.

JCP (2008a) JSR 287: Scalable 2D Vector Graphics API 2.0 for Java ME.

JCP (2008b) JSR 289: SIP Servlet V.1.1.

JCP (2009a) JSR 180: SIP API for J2ME.

JCP (2009b) JSR 281: IMS Services API.

Jennings, C., Peterson J and Watson M 2002 Private Extensions to the Session Initiation Protocol (SIP) for Asserted Identity within Trusted Networks. RFC 3325, Internet Engineering Task Force.

Khlifi, H. (2008) IMS application servers: roles, requirements and implementation technologies, *IEEE Internet Computing*.

Khlifi, H. and Grégoire, J.C. (2008) IMS application servers: roles, requirements, and implementation technologies, *IEEE Internet Computing*, **12**(3), 40–51.

Koodli, R. (2005) Fast handovers for mobile IPv6, RFC 4068, Internet Engineering Task Force.

Levin, O. (2003) H.323 uniform resource locator (URL) scheme registration, Technical Report 3508, Internet Engineering Task Force.

LastFM (2007) UR http://www.lastfm.es/

Latest in Beauty (2008) URL http:www.toluna-group.com.

Lightreading (2008) What's up with IMS? – Lightreading Technical Report; http://www.lightreading.com.

Loyola, I. (2008a) Report from the National ICT Research Directors Working Group on future Internet European Commission Information Society and Media.

Loyola, I. (2008b) Report from the National ICT Research Directors Working Group on future Internet European Commission Information Society and Media.

McAfee, A.P. (2006) Enterprise 2.0, the dawn of emergent collaboration, *MIT Sloan Management Review*, **47**(3), Spring.

Maemo (2008) http://maemo.org.

Mani, M. and Crespi, N. (2008) How IMS enables converged services for cable and 3G technologies: a survey, *EURASIP Journal of Wireless Communication Networks*, **8**(3), 1–14.

Marjou, X., Rajagopal, P., Said, M. and Ganesan, S. (2008) Session description protocol (SDP) offer/answer model for media control protocol, Internet-Draft draft-marjou-mmusic-sdp-rtsp-01, Internet Engineering Task Force (work in progress).

MEDIACTRL (2008) Media control framework; http://www.ietf.org/html.charters/mediactrl-charter.html.

MEXT (2008) Mobility extensions for IPV6; http://www.ietf.org/html.charters/mext-charter.html.

Microsoft (2009a) Microsoft Popfly; http://www.popfly.com/.

Microsoft (2009b) Microsoft SilverLight; http://silverlight.net/.

Mint (2005) The best way to manage your money, URL http://www.mint.com/.

MIPSHOP (2008) Mobility for IP: performance, signaling and handoff optimization (MIPSHOP); http://www.ietf.org/html.charters/mipshop-charter.html.

Montpetit, M. and Ganesan, S. (2007) IPTV – a device, any time, anywhere, in *Proceedings of SCTE 2007 Emerging Technologies*.

MSF (2008) Multi-service Switching Forum; http://www.msforum.org/.

Netflix (1997) URL http://www.netflix.com.

NGN (2006) Next generation networking; http://www.ngnuk.org.uk/.

NGN (2007) TISPAN NGN functional architecture, Release 1, ES 282 001.

Nielsen (2008) Online, URL http://www.nielsenonline.com/solutions.jsp?section=sol_1.

Niemi, A. (2004) Session initiation protocol (SIP) extension for event state publication, RFC 3903, Internet Engineering Task Force.

OASIS (2008) Reference architecture for service oriented architecture version 1.0 http://docs.oasis-open.org/soa-rm/soa-ra/v1.0/soa-ra-pr-01.pdf.

O'Connell, J. (2007) Service delivery within an IMS environment, *IEEE Vehicular Technology Magazine*, **2**(1), 12–19.

OECD (2008) The Seoul Declaration for the future of the Internet economy, Organisation for Economic Co-operation and Development.

OMA (2008) Open Mobile Alliance http://www.openmobilealliance.org/.

OMTP (2008a) Open mobile terminal platform (omtp), bondi architecture & security public working draft v0.43 http://www.omtp.org/.

OMTP (2008b) Open Mobile Terminal Platforms http://www.bitstream.com/wireless/.

OpenMoko (2008a) http://ww.openmoko.com.

OpenMoko (2008b) Openmoko,Wiki; http://wiki.openmoko.org/wiki/MainPage.

OpenSocial (2007) The web is better when it's social, URL http://code.google.com/intl/es/apes/opensocial/.

OPUCE (2009) Open platform for user-centric service creation and execution, the OPUCE Consortium; http://www.opuce.eu.

O'Reilly, T. (2005) what is Web 2.0: Design patterns and business models in the next generator of software, August 2005; http://www.oreilly.com/pub/a/oreilly/tim/news/2005/09/30/what-is-Web-20.html

OSS (2008) Open source OSS Project; http://openoss.sourceforge.net/.

PacketCable (2004) PacketCable CMS to CMS signaling, Specification PKT-SPCMSS-I03-040402.

PacketCable (2005a) PacketCable 1.5 architecture framework, Technical Report PKT-TR-ARCH1.5-V01-050128.

PacketCable (2005b) PacketCable multimedia, Specification PKT-SP-MM-I03-051221.

PacketCable (2007a) PacketCable 2.0; http://www.packetcable.com/specifications/specifications20.html.

PacketCable (2007b) PacketCable architecture framework, Technical Report PKTTR-ARCH-C01-071129.

Papageorgiou, P. (2001) A comparison of H.323 vs. SIP, University of Maryland at College Park.

Parlay (2008) Parlay/OSA Specifications, as seen on 08.08.2008, http://www.parlay.org/en/specifications/.

PayPal (1998) URL http://www.paypal.com/.

PCIQ (2007) Computer support services, URL http://www.pciq.com/.

Platform-A (2008) Launching next generation advertiser marketplace exchange, URL http://corp.aol.com/press-releases/2008/09/platform-a-launching-next-generation-advertiser-marketplace-exchange.

Priceline (1997) URL http://www.priceline.com/.

Research (2008) Next generation perspectives; http://www.tmforum.org/TechnicalPrograms/ServiceDeliveryFramework.

Richardson, I.E.G. (2003) *H.264 and MPEG-4 Video Compression: Video Coding for Next-generation Multimedia*. John Wiley & Sons, Inc.

Roach, A.B. (2002) Session initiation protocol (SIP) – specific event notification, RFC 3265, Internet Engineering Task Force.

Rosenberg, J. (2007) Indicating support for interactive connectivity establishment (ICE) in the session initiation protocol (SIP), Internet-Draft draft-ietf-sip-ice-option-tag-02, Internet Engineering Task Force (work in progress).

Rosenberg, J. (2008) TCP candidates with interactive connectivity establishment (ICE), Internet-Draft draft-ietf-mmusic-ice-tcp-06, Internet Engineering Task Force (work in progress).

Rosenberg, J. and Schulzrinne, H. (2003) An extension to the session initiation protocol (SIP) for symmetric response routing, RFC 3581, Internet Engineering Task Force.

Rosenberg, J., Schulzrinne, H. and Kyzivat, P. (2004a) Caller Preferences for the Session Initiation Protocol (SIP). RFC 3841, Internet Engineering Task Force.

Rosenberg, J., Schulzrinne, H. and Kyzivat, P. (2004b) Indicating user agent capabilities in the session initiation protocol (SIP), RFC 3840, Internet Engineering Task Force.

Rosenberg, J., Schulzrinne, H., Camarillo, G., Johnston, A., Peterson, J., Sparks, R., Handley, M. and Schooler, E. (2002) SIP: session initiation protocol, RFC 3261, Internet Engineering Task Force.

Rosenberg, J., Weinberger, J., Huitema, C. and Mahy, R. (2003) STUN – simple traversal of user datagram protocol (UDP) through network address translators (NATs), RFC 3489, Internet Engineering Task Force.

Rosenberg, J., Peterson, J., Schulzrinne, H. and Camarillo, G. (2004a) Best current practices for third party call control (3PCC) in the session initiation protocol (SIP), RFC 3725, Internet Engineering Task Force.

Rosenbrock, K., Sanmugam, R., Bradner, S. and Klensin, J. (2001) 3GPP-IETF standardization collaboration, RFC 3113, Internet Engineering Task Force.

SailFin (2008) Sailfin Project; https://sailfin.dev.java.net/.

Saint-Andre, P. (2004) Extensible messaging and presence protocol (XMPP): core, RFC 3920, Internet Engineering Task Force.

Salon (1995) URL http://www.salon.com/.

Schulzrinne, H. (2004) The tel URI for telephone numbers, RFC 3966, Internet Engineering Task Force.

Schulzrinne, H., Rao, A. and Lanphier, R. (1998) Real time streaming protocol (RTSP), RFC 2326, Internet Engineering Task Force.

Schulzrinne, H., Casner, S., Frederick, R. and Jacobson, V. (2003) RTP: a transport protocol for real-time applications, RFC 3550, Internet Engineering Task Force.

SEMS (2008) SIP express media server; http://www.iptel.org/sems/.

Sestini, F. (2004) Situated and automatic communications in IST, IST 2004 Networking Workshop; ftp://ftp.cordis.europa.eu/pub/ist/docs/fet/comms-64.pdf.

SIP (2008) Session initiation protocol; http://www.ietf.org/html.charters/sip-charter.html.

SIPForum (2008) SIP Forum; http://www.sipforum.org/.

SIPP (n.d.) Sipp Project; http://sipp.sourceforge.net/.

SIPPING (2008) Session initiation proposal investigation; http://www.ietf.org/html.charters/sipping-charter.html.

SIPS (2007) Open SIP Server Project; http://www.opensips.org/.

Sparks, R. (2004) The session initiation protocol (SIP) referred-by mechanism, RFC 3892, Internet Engineering Task Force.

Sparks, R. (2007) SIP: basics and beyond, *Queue*, **5**(2), 22–33.

Specifications g (1989) 3GPP Specifications; http://www.3gpp.org/.

Spivack, N. (2008) Internet evolution to 2020; http://colab.cim3.net/file/work/SICoP/2007-04-25/InternetTo2020.pdf.

SuSe (2008) http://en.opensuse.org/Welcome-to-openSUSE.org.

Swale, R.P., Mart, P.A., Sijben, P., Brim, S. and Shore, M. (2002) Middlebox communications (midcom) protocol requirements, RFC 3304, Internet Engineering Task Force.

Telco 2.0 (2006) Telco 2.0 blog; http://www.telco2.net/blog/2006/11/; http://www.telco2.net, 2009.

Telefonica (1924) URL http://www.telefonica.es.

Terra Networks (2000) URL http://www.terra.com.

ThunderHawk (2008) ThunderHawk, web browser for mobile platforms, as seen on 08.08.2008, http://www.bitstream.com/wireless/.

TISPAN (2000) Telecoms and Internet converged services and protocols for advanced networks; http://www.etsi.org/tispan/.

TMF (2009a) Service delivery framework (SDF), Telemanagement Forum; http://www.tmforum.org/TechnicalPrograms/ServiceDeliveryFramework.

TMF (2009b) Telemanagement Forum; http://www.tmforum.org/.

Toluna (2007) On-line market research service URL http://www.toluna-group.com.

Truste (1997) URL http://www.truste.org/.

Turletti, T. and Huiterna, C. (1996) RTP payload format for A.261 video streams, Technical Report 2032, Internet Engineering Task Force.

Twitter (2006) URL http://www.twitter.com.

UCT (2007) UCT IMS Client Project; http://uctimsclient.berlios.de/.

University (n.d.) Clean slate design for the Internet; http://cleanslate.stanford.edu.

v6ops (2008) IPv6 operations; http://www.ietf.org/html.charters/v6ops-charter.html.

Waiting, D., Good, R., Spiers, R. and Ventura, N. (2008) Open source development tools for IMS research, in *TridentCom '08: Proceedings of the 4th International Conference on Testbeds and Research Infrastructures for the Development of Networks and Communities*, pp. 1–10, Institute for Computer Sciences, Social-Informatics and Telecommunications Engineering (ICST), Brussels, Belgium.

Wikipedia (2001) The free encyclopedia, URL http://en.wikipedia.org.

Wikipedia (2009) http://en.wikipedia.org/wiki/RichInternetapplication.

WIMS 2.0 (2008a) WIMS 2.0 initiative, by Luis Angel Galindo and David Moro; URL http://www.wims20.org.

WIMS 2.0 (2008b) WIMS 2.0 ecosystem, by Luis Angel Galindo and David Moro; URL http://www.irr-events.com/IIR-conf/Telecoms/DocumentView.aspx?EventID=1800sDocumentID=2691.

Yahoo (2009) Yahoo Pipes; http://pipes.yahoo.com/pipes/.

Yavatkar, R., Pendarakis, D. and Guerin, R. (2000) A framework for policy-based admission control, RFC 2753, Internet Engineering Task Force.

YouTube (2005) Broadcast yourself, URL http://www.youtube.com.

Index

IMS: A Development and Deployment Perspective   Khalid Al-Begain, Chitra Balakrishna, Luis Angel Galindo and
David Moro
© 2009 John Wiley & Sons, Ltd